COSMIC PINWHEELS

Spiral Galaxies and the Universe

COSMIC PINWHEELS
Spiral Galaxies and the Universe

Ronald James Buta
University of Alabama, USA

World Scientific

NEW JERSEY · LONDON · SINGAPORE · BEIJING · SHANGHAI · HONG KONG · TAIPEI · CHENNAI · TOKYO

Published by

World Scientific Publishing Co. Pte. Ltd.

5 Toh Tuck Link, Singapore 596224

USA office: 27 Warren Street, Suite 401-402, Hackensack, NJ 07601

UK office: 57 Shelton Street, Covent Garden, London WC2H 9HE

British Library Cataloguing-in-Publication Data
A catalogue record for this book is available from the British Library.

COSMIC PINWHEELS
Spiral Galaxies and the Universe

ISBN 978-981-121-668-8 (hardcover)
ISBN 978-981-121-747-0 (paperback)
ISBN 978-981-121-669-5 (ebook for institutions)
ISBN 978-981-121-670-1 (ebook for individuals)

For any available supplementary material, please visit
https://www.worldscientific.com/worldscibooks/10.1142/11725#t=suppl

Desk Editor: Nur Syarfeena Binte Mohd Fauzi

Typeset by Stallion Press
Email: enquiries@stallionpress.com

To my parents
Barbara Pogachnik Buta and Daniel Adrian Buta, Sr.
and to my wife and daughters
Deborah Ann Crocker, Sarah P. Crocker-Buta,
and Katherine J. Crocker-Buta

Preface

Galaxies are probes of the structure and evolution of the universe. Like cosmic signposts, these massive star cities, each containing from billions to hundreds of billions of stars, populate our universe as far as our instruments can see. The way galaxies look, the manner in which they are distributed in space, and the ways they move, carry information on how the universe formed and evolved. Most galaxies are so far away that their light takes hundreds of millions to billions of years to reach us.

Many galaxies present a *spiral form*. This is an interesting shape to find among such large and massive objects, but it is not unique because nature provides many examples of spiral-shaped phenomena, such as ammonite fossils, chambered nautiloids (both fossil and living), the bark impressions of ancient scale trees, hurricane cloud patterns, sunflowers, DNA molecules, water flowing down a drain, etc. A galaxy is a spiral-shaped object on a much larger scale than any of these other objects, being typically 100,000 light years or so across. Gravity is believed to be the dominant force that molds the spirals. This is the same force that has sculpted the Earth into a sphere and keeps our feet on the ground.

Few stories in the history of astronomy compare with the magnitude of the discovery of the spirals and the concept of galaxies. On a night in April 1845, William Parsons, the Third Earl of Rosse, peered through the eyepiece of the 72-inch "Leviathan of Parsonstown," then the world's largest telescope, and saw, for the first time, the spiral structure of the M51 nebula. The swirling arms of the grand nebula, later christened the Whirlpool, were a newly found characteristic of the mysterious nebulae that peppered the sky away from the Milky Way, the term astronomers use to describe the appearance of our Galaxy as a milky band of light crossing the night sky. Parsons had built the telescope himself, on the grounds of his home in Birr, Ireland, for the expressed purpose of testing the most popular hypothesis of the time, that the nebulae were distant star systems like the Milky Way but were too far away for existing telescopes to resolve into individual stars. From the late 18th century until about the mid-19th century, the chief discoverers of nebulae had been the German English astronomer William Herschel and his son John. Together they catalogued thousands of nebulae and star clusters found using telescopes with an 18.7-inch speculum metal mirror.[1] At the time, an 18.7-inch telescope was very powerful compared to what was generally available, but nevertheless most nebulae defied resolution with such a telescope. Parsons had the right idea: the key to resolving the nebulae and establishing their nature was higher light-gathering power, which could only be achieved by building a larger telescope. The 72-inch mirror of the Leviathan of Parsonstown gathered 15 times as much light as the Herschels' telescopes. With such power, Parsons was bound to see something that no one had seen before.

It was nearly 80 years after Parsons' discovery that astronomers successfully achieved what he had tried to do. But there was more to the spirals than merely being distant star systems like the Milky Way (or "island universes," as they became popularly known). The study of spirals in the nearly 175 years since Parsons' great night has told us much of what we know about the Universe, including how stars evolve, the scale of the universe, how galaxies evolve, the existence of supermassive black holes, the expanding universe, the existence of dark matter, the Big Bang Theory, the origin and properties of interstellar dust, the discovery of supernovae, the origin and evolution

---

[1]Speculum metal is an alloy of tin and copper that, when polished, is highly reflective.

of chemical elements, and finally the possible origin and evolution of life.

Spiral galaxies are relevant to us because we live in a spiral galaxy. The evidence for the spiral nature of the Milky Way is fairly well-established. We can't see the spiral arms directly because we live inside the Milky Way, and our perspective is an edge-on view of its structure. Many other galaxies are similarly viewed, but many are not, and it is from the more face-on cases that we see the variety of shapes and possibilities of spiral structure in galaxies. Spiral galaxies are places where stars are actively forming and dying. We can say spiral galaxies are "where the action is" in the universe.

Superlatives are often used to describe spiral galaxies, like elegant, artistic, graceful, beautiful, wonderful, awe-inspiring, and magnificent. The best term is "majestic." The majesty of the spirals is what drew me into their study nearly 50 years ago. I was fascinated by the way they look, by how they were discovered, by the knowledge of what they are and how far away they are, and by what humanity has learned about the universe from the study of spirals.

With this book, I would like to take the reader into the fascinating world of spiral galaxies. It is interesting to examine spirals more closely than just as swirling masses seen in photographs. We can ask: what led to Parsons' discovery of spirals? What role did photography play in the study of spirals? What kinds of stars make up a spiral galaxy, and in particular, what kinds of stars make up spiral arms? What role does interstellar gas and dust play on the appearance of spirals, and how varied are their patterns? How have spirals contributed to our understanding of galaxy structure and dynamics? How far away are spirals, and how luminous and massive are they compared to other kinds of galaxies? Why are some galaxies not spirals? How do spirals rotate, and how is the mass in spirals distributed with distance from the center? How common are spirals, and how are they distributed in the universe? What is it like to observe spirals? How are galaxies catalogued and measured? And finally, what causes spiral structure in such massive objects?

While most of the book is about the properties of spirals, it is also partly a memoir of the time when I was a student and coworker of Gérard Henri de Vaucouleurs (1918–1995), the late Professor of Astronomy at the University of Texas at Austin who was one of the 20th century's most influential astronomers. Characterized as being from the "age of heroes," Gérard's legacy lives on in much of modern extragalactic astronomy.

On the title of the book, spiral galaxies are often called "pinwheels" because some look like the popular children's toy. One can imagine the spirals spinning like pinwheels in the "cosmic breeze" of intergalactic space. Pinwheels are usually colorful toys, and it is interesting that we live in an age where astrophotography has become so sophisticated that it can be used to show what many spirals look like in color.

Acknowledgments

I am grateful to University of Texas Professor Gérard de Vaucouleurs and his wife, Antoinette, for bringing me into the world of galaxies more than 40 years ago, for teaching me how to observe and measure galaxies, and for the inspiration they provided for the study of galaxies. I am also grateful to Michigan State University Professor Susan M. Simkin for taking me on as a graduate research assistant in the mid-1970s and assigning to me the analysis of a photographic plate of the spiral M94, a galaxy with a bright star-forming ring. The study of such rings became an important part of my career as an astronomer.

I am also grateful to Lia Athanassoula, David L. Block, Albert Bosma, Gene G. Byrd, Francoise Combes, Sébastien Comerón, Harold G. Corwin, Jr., George Contopoulos, Bruce and Debra Elmegreen, Kenneth C. Freeman, Tarsh Freeman, Gerry Gilmore, Agris J. Kalnajs, William C. Keel, Johan Knapen, John Kormendy, Eija Laurikainen, Marshall L. McCall, Preethi Nair, Stephen C. Odewahn, Pertti Rautiainen, Vera C. Rubin, Allan Sandage, Heikki Salo, Mark Philip Schwarz, Kartik Sheth, Beatrice Tinsley, Sidney van den Bergh, Piet van der Kruit, Lourdes Verdes-Montenegro, Raymond E. White III, James D. Wray, and Xiaolei Zhang, among many others, for further inspiring my interest in the subject of this book. In addition, I thank G. Anselmi, Adam Block, John Blakeslee, Serge Brunier, Noel Carboni, Kem Cook, Ken Crawford, J.-C. Cuillandre, Giuseppe Donatiello, Davide De Martin, Alan Dyer, Holland Ford, Don Goldman, Andrew Lockwood, Robert O'Connell,

Rolf Wahl Olsen, Federico Pelliccia, and Leo Shatz, for permission to use some of their excellent astrophotographs in this book. Thanks also to NASA, the ESA, and the Hubble Heritage team for the many excellent and related images they have made publicly available. I am also grateful to Carl Clements, Harold G. Corwin, Jr., Deborah Ann Crocker, and Lia Athanassoula for reading and providing feedback on parts of the manuscript, and to Lia Athanassoula, Andrea Ghez, Robert L. Hurt, John Kormendy, and Mark Reid for use of several illustrations. Thanks also to Steve Janowiecki for information on the frequency of clear nights at McDonald Observatory.

I would like to thank William Clere Leonard Brendan Parsons, the 7th Earl of Rosse, and Alison, Countess of Rosse, for their hospitality and graciousness during my visit to Birr Castle in April, 2010. It was a pleasure and an honor to visit the site where the great "Leviathan of Parsonstown" made its fabulous discovery of spiral structure in galaxies, and to see the excellent replica that now stands in its place.

The book has made use of the NASA/IPAC Extragalactic Database (NED) that is operated by the Jet Propulsion Laboratory, California Institute of Technology, under contract with the National Aeronautics and Space Administration. Funding for the creation and distribution of the Sloan Digital Sky Survey (SDSS) Archive has been provided by the Alfred P. Sloan Foundation, the Participating Institutions, NASA, NSF, the US Department of Energy, the Japanese Monbukagakusho and the Max Planck Society. All of the black and white galaxy pictures in this book were prepared using the Image Reduction and Analysis Facility (IRAF), which is written and supported by the National Optical Astronomy Observatories (NOAO) in Tucson, Arizona. NOAO is operated by the Association of Universities for Research in Astronomy (AURA), Inc., under cooperative agreement with the National Science Foundation.

Cover: *Hubble Space Telescope* image of M101 (NGC 5457), a spiral galaxy 22 million light years away in the constellation Ursa Major. Image credit: NASA, ESA, STScI/AURA.

Other books coauthored or coedited by Ronald J. Buta:

Third Reference Catalogue of Bright Galaxies, by Gérard and Antoinette de Vaucouleurs, Harold G. Corwin, Jr., Ronald J. Buta, Georges Paturel, and Pascal Fouqué, Springer, 1991.

Pennsylvanian Footprints in the Black Warrior Basin of Alabama, edited by Ronald J. Buta, Andrew K. Rindsberg, and David C. Kopaska-Merkel, Alabama Paleontological Society, 2005.

The de Vaucouleurs Atlas of Galaxies, by Ronald J. Buta, Harold G. Corwin, Jr., and Stephen C. Odewahn, Cambridge University Press, 2007.

Footprints in Stone: Fossil Traces of Coal-Age Tetrapods, by Ronald J. Buta and David C. Kopaska-Merkel, University of Alabama Press, 2016.

Contents

Chapter 1

Spiral Encounters

The night sky teems with celestial objects that look like shining clouds in space. Originally known as "nebulae" (from the latin *nebula*, or cloud), these objects have played a major role in our understanding of the universe. Nebulae represent some of nature's most important astrophysical phenomena. Today, the term "nebula" applies mainly to real clouds of interstellar gas and dust, like the Orion Nebula, but prior to 1920, galaxies were also referred to as nebulae. Galaxies are not shining clouds, but are massive gravitationally-bound systems that typically contain hundreds of billions of stars. Their light is mainly due to the unresolved glow of these stars. The fact that galaxies look nebulous is due to their enormous distances. Galaxies are the most distant objects we can see with telescopes.

Galaxies started being discovered as nebulae mostly in the second half of the 18th century. One particularly important case was discovered in October 1773 by the French astronomer Charles Messier (pronounced Mess-ee-ay, 1730–1817). The famous observer, best-known for his compilation of a catalogue (the "Messier Catalogue") of bright nebulae and clusters, still in use today, was observing a comet with a small telescope when he came across a "very faint nebula, without stars, near the eye of the Northern Greyhound" [constellation Canes Venatici, the Hunting Dogs], "below the star Eta of ... the tail of Ursa Major." He recorded that the new object "is double, each has a bright center" and that "the two atmospheres touch each other" with "one even fainter than the other." He reported its position "on the chart of the comet observed in 1773 and 1774"

1

[a comet that he himself is credited with discovering] and that the object was "reobserved several times."[1]

Little did Messier know that this object, which became number 51 in his famous catalogue, would one day lead to the discovery of a fascinating aspect of nebulae. Messier's telescope (a four-inch refractor)[2] was not powerful enough to reveal any details in the two components of the system. To Messier, neither member of the close pair of nebulae was especially bright, but when observed nearly 70 years later with a much larger telescope, Lord Rosse's 72-inch "Leviathan of Parsonstown," the brighter component of the pair revealed an incredible spiral pattern, a swirling mass of faint light surrounding the brighter component and filling the space between the two objects. The discovery was made at a time when there was no real understanding of the nature of nebulae — only philosophical speculations. Thus began humanity's journey that led from the cataloguing of nebulae in the sky to the modern understanding of galaxies.

This journey is described in this book, but before getting into that I want to define some terms relevant to the narrative, and also to present a personal memoir on how I entered the world of the spirals and what it was like to observe spirals not only from a historical perspective, but also from a modern research perspective.

First, nebulae (whether galaxies or clouds of interstellar gas) are accessible to anyone. A large telescope is not needed to see them — for example, many can be seen with only a 4-inch telescope. Larger telescopes naturally reveal more faint nebulae and more details in brighter nebulae. The main requirement for seeing nebulae and for studying them effectively is a dark sky, dark enough at least to see the Milky Way. Second, nebulae are what astronomers often describe as "deep-sky objects," i.e., they are objects much larger than stars and much farther away than typical naked eye stars. Star clusters are also considered to be deep-sky objects. Distances to deep-sky objects

---

[1]See www.messier.seds.org/xtra/history/m-cat81.html.
[2]A type of telescope that uses a double convex lens as the main light-collecting element.

in the Milky Way are typically hundreds to tens of thousands of light years, where 1 light year is the distance a beam of light travels in 1 year. Distances to galaxies are much greater, from millions to billions of light years. In astronomy, space and time are intertwined. For example, M51 lies at a distance of 24.6 million light years, meaning we are seeing it as it appeared 24.6 million years ago. In general, when we look out into the night sky, we look back in time.

The majesty of the spirals is highlighted in Figure 1, which shows modern digital images of M51 (left frame) and NGC 1365 (right frame).[3] These are the two most easily visible spirals in the sky. The beautiful M51 system is indeed remarkable, as it shows not only the bright spiral pattern of the main component, but also peculiar signs of gravitational interaction around the smaller component. Neither component of the M51 system has an especially bright center, which may explain why Messier described the pair as "very faint."

While M51 was well-positioned for viewing with Lord Rosse's Leviathan, the same was not true for NGC 1365. This is because the two galaxies lie in widely differing parts of the sky. M51 is in the far northern sky while NGC 1365 is in the southern sky. Lord Rosse's telescope was built on the grounds of Birr Castle in central Ireland, a location whose latitude is 53.1° north. This high northern latitude effectively placed NGC 1365 out of the reach of the Leviathan. NGC 1365 is a barred spiral, meaning it is a spiral galaxy where the arms break from near the ends of a bright linear feature called a bar. It lies 68 million light years away in the constellation Fornax, the Furnace, and is a more massive, more luminous galaxy than M51.

NGC 1365 was the first galaxy whose spiral arms I saw with a telescope. At the time, I was a graduate student at the University of Texas at Austin. Over a period of 7 years, I used the McDonald Observatory 30, 36, and 82-inch telescopes for a variety of observing

---

[3]These and all other black and white images in this book are in units of magnitudes per square arcsecond, which are the standard units of what astronomers call surface brightness, or brightness per unit area. The light displayed in these images usually is restricted to a given part of the electromagnetic spectrum by a piece of glass known as a filter. For example, a *B*-band image is restricted to blue light, while an *I*-band image is restricted to infrared light.

projects, mostly involving measuring the apparent brightnesses of galaxies. McDonald Observatory is located in the Davis Mountains of west Texas. The altitude of the observatory is 2,000 m or 6,800 ft above sea level. The latitude is 30.7° north, which gives better access to the southern sky than did the latitude of Birr, Ireland.

I saw the spiral arms of NGC 1365 on a partly cloudy night in 1977 with the McDonald 30-inch Boller and Chivens reflector. Not only were the arms plainly visible, but also dust lanes in the bar could be seen cutting into the bright central region. Prior to this observation, I had never seen a spiral galaxy with such a large telescope, and had not realized that a modern 30-inch telescope could take an observer into the domain of the spirals almost as well as Lord Rosse and his assistants experienced it in the 1840s–1870s.

The instrument I was using on the telescope, a photometer, was the primary tool for measuring apparent brightnesses of celestial objects at the time. Apparent brightness is a numerical measure of how bright a star, galaxy, or other kind of object appears, basically a number that can be placed in a catalogue. Knowledge of apparent brightness can be used along with knowledge of distance to derive an object's *luminosity*, or amount of radiant energy it emits into space every second. Luminosity is one of the best ways of comparing the physical properties of different types of objects, like stars and galaxies.

Astronomers use the *magnitude* system to quantify apparent brightness. The origin of magnitudes is simple: historically, the brightest stars in the sky were said to be of the "first magnitude," while the faintest stars visible to the naked eye on a dark, moonless night were said to be of the "sixth magnitude." In the 1850s, Norman Pogson (1829–1891) determined that first magnitude stars are about 100 times brighter than sixth magnitude stars. The modern magnitude system was born by defining five magnitudes to be exactly a factor of 100 in relative brightness, with one magnitude corresponding to a factor of 2.512, two magnitudes to a factor of 2.512^2, etc.

The photometer I used on any given observing run with the 30, 36, or 82-inch telescope was equipped with a wide-field finding eyepiece having a magnification ranging from about 300 times on the 30-inch to about 800 times on the 82-inch. I recall being amazed at how well-shown the spiral arms of NGC 1365 were in the 30-inch, a telescope with less than half the diameter of the Leviathan's mirror. This underscores the differences between a mid-19th century reflector and a mid-20th century reflector. Apart from technological improvements in mounting and tracking, the former generally used speculum metal mirrors to collect light, while the latter use silver- or aluminum-coated glass mirrors having considerably higher reflectivity than does speculum metal. When newly polished, a speculum metal mirror, made typically of two-thirds copper and one-third tin, reflects about two-thirds of the light that shines upon it. In contrast, a silver-coated glass mirror reflects 90% of the light incident upon it. When a secondary mirror is added to the mix, a telescope of a given aperture with a silver-coated glass mirror would collect nearly twice as much light as a telescope of the same aperture, but having speculum metal mirrors. Speculum metal had a further disadvantage in that it was less resistant to tarnishing over time than silver-coated glass mirrors.[4]

It is indeed unfortunate that Lord Rosse could not observe NGC 1365 with the Leviathan. I wonder how the course of nebular astronomy might have been affected if NGC 1365 and M51 switched places in the sky, so that NGC 1365, and not M51, had been the first spiral discovered. Figure 1 shows how distinctly different the two objects are in the inner regions: M51 is a normal-looking spiral while NGC 1365, with its strong apparent bar, is not. How would astronomers have explained the way the spiral arms of NGC 1365 bend sharply far from the center, rather than winding almost directly from the center? This is a relevant question because one popular view during the late 18th and 19th centuries was that nebulae were solar systems in formation. This "nebular hypothesis" was the main thesis of Pierre-Simon Laplace (1749–1827), who mathematically developed the idea in 1796. Although this was well before Lord Rosse made his

---

[4]See https://en.wikipedia.org/wiki/Speculum_metal; wikipedia is the online encyclopedia.

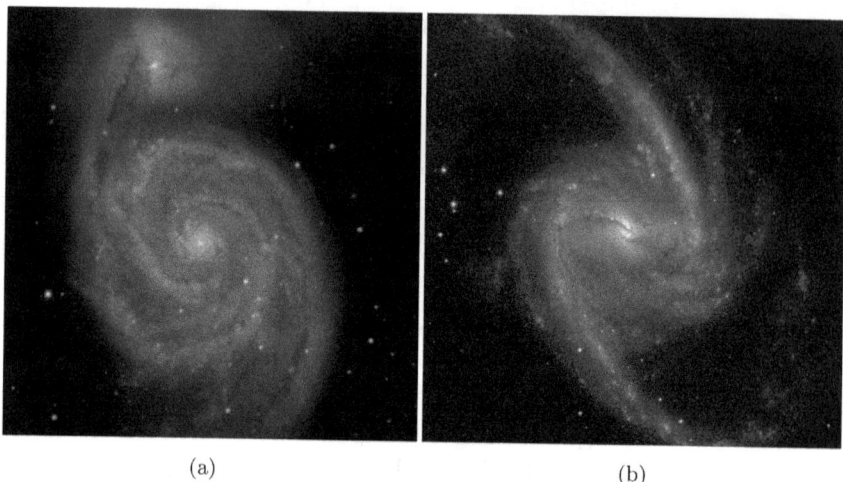

(a)   (b)

Fig. 1.   The two most conspicuous spirals in the sky: NGC 5194-5 (M51) in Canes Venatici (a) and NGC 1365 in Fornax (b). These blue-light images are from the de Vaucouleurs Atlas of Galaxies published by Cambridge University Press, 2007; reprinted with permission.

discoveries of spiral structure, the theory envisioned the formation of a star with accompanying rings of matter that could form planets. Spirals were part of a related theory proposed in 1900 and 1901 by Thomas Chamberlin (1843–1928)[5] and Forest Moulton (1872–1952),[6] who envisioned the formation of a solar system by a star passing close to another star and, due to tidal forces, drawing out gases in a spiral arrangement, the material of which could collect into larger, orbiting bodies. Although a fairly complex idea, it likely would have been easier to fit a regular spiral into this theory than a barred spiral.

Most of the photometric work I did at McDonald Observatory involved measuring the apparent brightnesses of dwarf galaxies, which were generally so faint that they could only be observed around new moon. This project, which was assigned to me by my eventual PhD thesis supervisor Gérard de Vaucouleurs, virtually guaranteed that I was at the observatory only when the sky was as dark as it could get. The totality of my work on the dwarf galaxies was carried

---

[5]T. C. Chamberlin, *Astrophysical Journal*, **14**, 17 (1901).
[6]F. Moulton, *Astrophysical Journal*, **11**, 103 (1900).

out during 19 observing runs covering about 80 nights from 1977 to 1983, and is described in Chapter 7.

1.1. Chilean Interlude

For students of astronomy, entry into research in a given field often begins with a major observing run on a large telescope. My entry into the world of spiral galaxies came in the summer of 1981. It was summer in Texas where I lived at the time, but it was winter where I went in early July. I was attending my first observing run with the 4-m telescope of Cerro Tololo Inter-American Observatory (CTIO), a collection of professional telescopes located on a 2,200 m (7,200 feet) high mountain-top in northern Chile. I was still a graduate student at the University of Texas at Austin at the time, and although UT Austin was connected to a major astronomical observing facility, I needed a larger telescope and more sophisticated instrumentation than were available at McDonald Observatory. The data were needed for my PhD dissertation project, which was eventually titled "The Structure and Dynamics of Ringed Galaxies." A ringed galaxy is a galaxy showing a prominent ring-shaped feature as part of its structure. My goals were to study the structural and star-forming properties of galactic[7] rings using photography in different colors, in order to better understand the nature of the rings. Rings in galaxies are often associated with spiral structure and my dissertation, being focussed on such a special aspect of some spiral galaxies, could shed light on the nature of spiral galaxies as a whole.

At the time I started my dissertation project, I was very much a northern hemisphere observer. My dissertation observing proposal was mainly intended for the 4-m telescope of Kitt Peak National Observatory (KPNO), and in it I had to explain what I wanted to do, why I wanted to do it, why it was important, what instruments were to be used, and how much telescope time I needed. For the latter,

---

[7]In this book, the term "galactic" with a lower case g will refer to other galaxies or galaxies in general; however, when a capital G is used, as in "Galactic," it is specifically referring to the Milky Way, our home Galaxy.

I had the audacity to ask for 10 dark-of-the-moon nights mostly spread across one semester. KPNO is located in southern Arizona, and is a popular observing site for American students of astronomy. I really wanted to observe there and felt I had a great sample of galaxies to work with in the north. Nevertheless, CTIO has a 4-m telescope virtually identical to the one at KPNO (Figure 2). I took my KPNO proposal, shortened it, changed the names of the galaxies, asked for only five nights, and submitted a parallel proposal to CTIO. I received the CTIO time I asked for, but not the KPNO time. As a result, I became a southern hemisphere observer, which took me to the alien skies of Chile in 1981. It was my very first trip south of the equator. This being so, I was excited not only about observing at CTIO, but also of seeing the night sky and the famous bright stars and other objects that lie relatively close to the south celestial pole and which therefore cannot be seen from much of the US. In particular, I wanted to see the Southern Cross and Alpha and Beta Centauri, Alpha being the nearest star system to the Sun.

Getting to CTIO was a fairly slow process in 1981. I first flew from the US to Santiago, the capital of Chile. Then I took a 7-h bus ride into the Andes Mountains to La Serena, a resort town on the Pacific Ocean located 400 km (250 miles) north of Santiago. CTIO was approximately an hour's drive from La Serena. As my plane was landing in Santiago, I noticed it was rather cloudy in that area and became concerned about the weather. No one wants to travel a long distance to an observatory only to be clouded out. But as the bus approached La Serena, we went higher and higher into the mountains, and eventually the sky was almost completely clear. There were only low-lying clouds over the ocean. I recall as the sky darkened, I began to see bright stars through one of the large windows on the bus. However, the only stars I saw initially were bright northern ones I was already familiar with, such as Arcturus. Because travel in Chile is mostly north or south, the bus to La Serena rarely deviated and so I kept on seeing only familiar northern stars. Then at one point the bus either took a sharp turn to the left or may have stopped briefly, and I saw, standing fairly high, the Southern Cross and Alpha and Beta Centauri. Then I knew I was in the southern hemisphere!

Fig. 2. The CTIO 4-m Blanco Telescope. The large black cylinder at the front end is the prime focus can that, prior to the CCD revolution, was where an observer would sit to take photographs using 8 × 10-inch plates. National Optical Astronomy Observatories photo.

The sky over Chile is indeed alien to most Americans. There are many constellations visible from Chile that never rise above the horizon from most locations in the continental US. Constellations like Tucana the toucan bird, Dorado the swordfish, Octans the octant, Crux, the southern cross, Volans the fish, Triangulum Australe, the southern triangle, and much of the huge ship constellation called Argo Navis, are not visible from Texas. In addition, the two nearby galaxies called the Magellanic Clouds are not visible from any continental US location but are easily seen from all locations south of the equator. From Texas, southern constellations like Sagittarius and Scorpius are easily seen, but do not rise very far above the horizon. In July, these constellations are well placed for night-time viewing, and are popular targets for amateur astronomers because of the prominence of the Milky Way in those directions and the large number of star clusters and galactic nebulae seen. But from Chile in July, these same constellations stand *overhead* at midnight, and this provides a perspective on the Milky Way that northern hemisphere dwellers just do not see. What is significant about the Milky Way in Sagittarius is that the very center of our Galaxy lies in the direction of this constellation. Harlow Shapley (1885–1972), a prominent early 20th century astronomer, first suggested in 1918 that the high concentration of globular clusters in Sagittarius proved that the center of the Galaxy was far from the Sun at a point in that constellation.[8]

We can't actually see the center of our Galaxy directly because of the large amount of interstellar dust between us and the center. Nevertheless, the Milky Way is bright in Sagittarius, and is prominent even from northern locations. In July 1981, I was at CTIO for five nights, all of which were clear with little or no moonlight. The sky was darker than I had ever seen before. I had one night before my run started, and I recall walking around the mountain top gazing up at the stars. I was stunned by the Milky Way. When Sagittarius was overhead, the Milky Way was extremely bright. It looked like a giant bird with enormous "wings" (Figure 3). The body of the bird was the central bulge[9] of our Galaxy, and the wings were the extended disk that defines its shape. The Milky Way in Sagittarius and Scorpius is riddled with star clouds, dust lanes, and scattered bright patches.

---

[8]H. Shapley, Publications of the *Astronomical Society of the Pacific*, **30**, 42 (1918).
[9]The bulge refers to the central, more spherical component of our Galaxy and other galaxies.

Fig. 3.   The Milky Way towards the constellation Sagittarius. Image copyright Serge Brunier; used with permission.

It is the brightest night-time celestial object you can see from a southern hemisphere location. The view of the Milky Way was so memorable from that time that every semester I taught introductory astronomy, I told my students that it is worth spending $1,500 to fly to Chile to be there on a moonless night in July, *just to see the Milky Way*. As an added bonus of such a journey, one would get to see the far southern Milky Way in Carina, which is spectacularly bright compared to the northern Milky Way in Cygnus and Scutum. The giant interstellar cloud known as NGC 3372 (the Eta Carinae Nebula), and the beautiful dust patch called the Coal Sack in Crux, are well-known showpieces of the far southern Milky Way.

My first run at CTIO was remarkably complex. I had applied to get images and spectroscopic observations[10] of six ringed galaxies, five (NGC 1433, 6300, 6744, 7531, 7702) in the southern sky, and one (NGC 5364), located near the celestial equator[11] (and still observable from Chile). NGC 1433 was not observable in July 1981, but was instead observed in a follow-up run on the CTIO 4-m in October

---

[10]Spectroscopy is the science of studying dispersed light, or light broken up into its component colors, as in a rainbow.

[11]The extension of the Earth's equator onto the sky, or celestial sphere.

1981. In September 1981, I also had an additional run at CTIO to use a photometer on the 36-inch telescope to obtain calibrating data for my images. These data were in the form of photoelectric multi-aperture photometry, a technique which is described further in Chapter 7.

Figure 4 shows modern blue light[12] digital images of my six sample galaxies, five of which are spirals and one (NGC 7702) which is not. The images I wanted to get with the 4-m telescope were

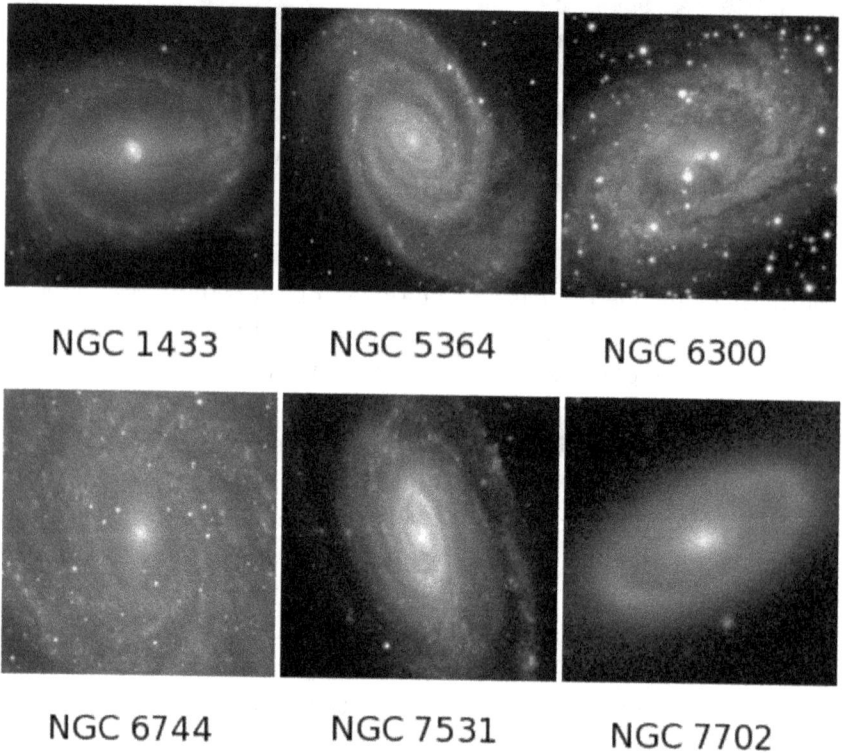

NGC 1433   NGC 5364   NGC 6300

NGC 6744   NGC 7531   NGC 7702

Fig. 4.   The six galaxies I observed with the CTIO 4-m telescope in 1981. NGC 5364, 6300, 6744, 7531, and 7702 were observed in July 1981, while NGC 1433 was observed during a separate run in October 1981. Images source: The de Vaucouleurs Atlas of Galaxies published by Cambridge University Press, 2007; reprinted with permission.

---

[12]Blue light refers to the part of the electromagnetic spectrum between about 400 nm and 500 nm wavelength, where 1 nm is a billionth of a meter.

needed to study the structure of the galaxies in different colors using special filters. The spectra were needed to study the rotation of the galaxies. All are rotating, disk-shaped systems that are tilted enough to the line of sight that Doppler wavelength shifts of the light of ionized clouds of hydrogen gas due to the rotation are easily detectable. All of the data I collected at the time were photographic; the digital imaging revolution in astronomy was still a few years in the future. My PhD dissertation was one of the last to come out of the photographic era in astronomy.

The imaging of the galaxies was carried out using Kodak 8×10-inch plates, which were relatively thin pieces of glass coated on one side with a light-sensitive (silver bromide-related) emulsion. To get the exposures needed required that the 4-m telescope be set-up for use in prime focus mode. The prime focus of a reflecting telescope is the point in front of the primary mirror where the light rays from an object are brought to a focus (Figure 5(a)). In small telescopes, a flat secondary mirror tilted at 45° is placed near this point on the front end of the telescope in order to deflect the focal point out the side of the tube into an eyepiece where an object can be viewed. In large (4-m class) reflecting telescopes, a person can observe at the prime focus directly without blocking much of the light, and as a result there is no need to deflect the light to the side of the tube. The CTIO 4-m telescope has a black metal can at the prime focus that is large enough for a person to get into, kind of like an astronaut in a space capsule (Figure 2). To observe at prime focus, a night assistant first moves the telescope until it is horizontal. The black can is held over a platform, a hatch is opened, and the astronomer gets in, carrying plates and plate holders as if in a darkroom. The situation was different for the spectra. The front end of the telescope was rotated to convert the design to Cassegrain (Figure 5(b)), where the light reflecting off a curved secondary mirror passes through a hole in the primary mirror into a spectrograph[13] attached to the back side of the telescope.

---

[13] An instrument designed to disperse light into its component colors and record the spectrum on a photographic plate or other detector.

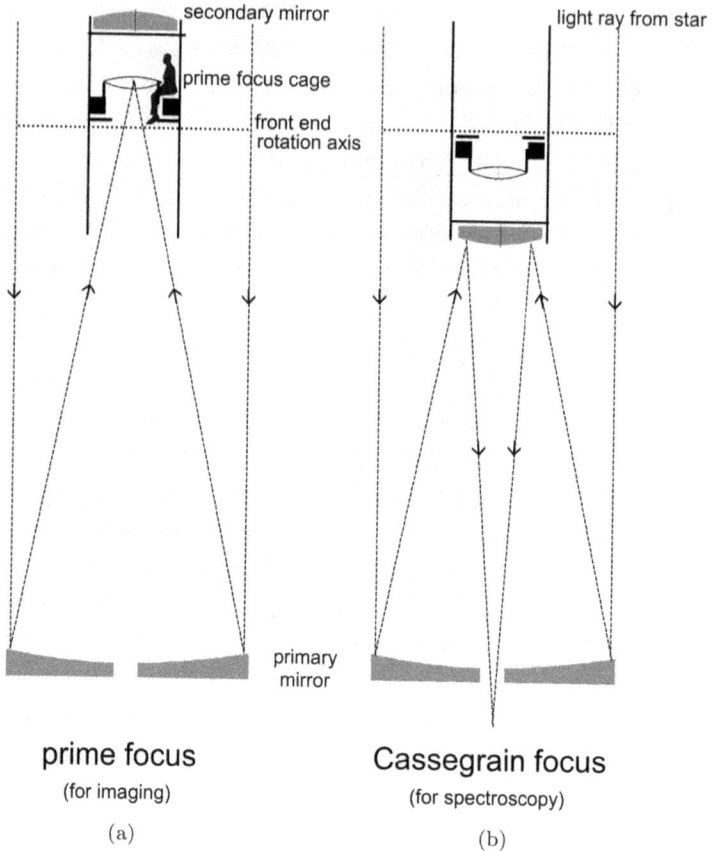

Fig. 5. Schematic diagrams showing what is meant by prime focus and Cassegrain focus observing with the CTIO 4-m telescope.

A major concern I had before I embarked on my Chilean observing experience was knowing how to properly develop my plates. I was worried about background gradients due to improper development that would complicate analysis of such large plates and make them less accurate for my study. At the time of my observations, the American Astronomical Society (AAS) published a journal called the AAS Photo-Bulletin. In this journal, I found several papers by William C. Miller, the chief photographer of Hale Observatories in California, that offered some useful advice. Of various methods that could be used to develop my prime focus plates, he recommended using a rotating tray rocker to get the best uniformity. I followed his recommendation, and of the 28 (8 × 10-inch) plates I eventually took during my July 1981 CTIO run, not one had a serious uniformity problem. I knew I had followed good advice when at a later CTIO run, I saw the plates another student had taken of some galaxies, plates that he had developed using a nitrogen burst technique applied probably to multiple plates simultaneously. All of this student's plates that I saw had background problems.

Getting good plates also required placing them correctly into the plate holder. During my September 1981 run, another 4-m prime focus observer at the time showed me a 90-min exposure she had taken of a globular cluster having a severe focus gradient across the field. From one side of the plate to the other, stars were increasingly out of focus, an effect which rendered the long exposure plate virtually useless. Apparently, she had misplaced the plate in its holder. Given how hard it would have been to get telescope time, such mistakes could be devastating.

Most of the nights I received at CTIO were perfectly clear and exceptionally dark photometric nights. When it was dark enough, I got into the prime focus can and asked the night assistant to take me to the zenith, i.e., to orient the telescope so that it is pointing straight up. After getting comfortable sitting in the can, I placed into the plate holder a slide with a wide field eyepiece having a sharp knife edge. The knife edge was used to visually focus the telescope. The telescope was in focus when the image of a bright star, being moved across the knife edge with a hand-paddle, vanished instantly. After this task, I was ready to observe. I asked the night assistant to point the telescope at the first galaxy on my list. I used the eyepiece slide for each galaxy to check that the telescope was pointed at the right object, and to make sure the galaxy was centered in the field. In so doing, I had the rare privilege of seeing my six sample galaxies with the light-gathering power of a 4-m (160-inch) telescope on an extremely clear night from one of the darkest observing sites on Earth. What did they look like with such power?

I only had a few seconds per galaxy to determine this, but the answer is, they looked a lot like the pictures in Figure 4. In NGC

1433, the bar and a faint trace of the large ring around the bar were visible, but most memorable in this case were a few star-like "knots" seen in the central region. These likely were star-forming regions that had formed as part of a small ring. In NGC 5364, I do not recall exactly if I saw the ring inside the main spiral pattern (I believe I did), but I do remember seeing the two spiral arms to the north (up in the image). NGC 6300 was very large in the knife edge eyepiece, showing a stunningly bright ring-spiral pattern. NGC 6744 overflowed the field of the eyepiece, but I remember two things about it: the vertically-oriented bar, and the sharp faint arc to the east of the bar region (left in image), part of a low-contrast ring. This galaxy was once thought to be similar in appearance to what the Milky Way might look like from the outside (Chapter 8). In NGC 7531, a highly-inclined spiral, the bright ring in its inner regions was easily seen, but in addition, I also noticed the outer disk as narrow and curved, due to the faint outer spiral structure. Finally, NGC 7702, although not a spiral, was the most memorable object I saw with the 4-m. The ring was diffuse and so bright that the galaxy resembled the planet Saturn. I was stunned when I saw it. I also noticed faint dust lanes on the south side as also seen in the image.

Mount Wilson and Palomar Observatories astronomer Allan Sandage (1926–2010) had a similar experience in viewing galaxies with a large reflector. In the Carnegie Atlas of Galaxies,[14] he comments on the view of the very bright spiral NGC 1566 with the Las Campanas 2.5-m telescope located in northern Chile, not far from CTIO. He comments: "NGC 1566 is an awesome sight. No written description is adequate to convey the scene when the galaxy is viewed through the eyepiece of a telescope rather than on an electronic screen." Like NGC 1365, NGC 1566 is a bright spiral that was too far south to be observed from Birr Castle. The galaxy lies 59 million light years away in the constellation Dorado, the swordfish.

My five CTIO nights in July 1981 went very well, and I acquired a nearly complete set of images in ultraviolet, blue, visual, and red light for each of my sample galaxies. Exposure times ranged from

---

[14] Carnegie Institution of Washington Publication No. 638, 1994.

10 min to 55 min. Ultraviolet images show where young stars and star-forming regions are located. Blue light images also reveal where young stars are located, but in addition such images show aspects of the dust morphology of spirals. Visual and red images progressively show older stars and minimize the effects of dust.

Once my observing run was over, I had to prepare to leave the mountain. "Packing" meant not only packing my luggage, but also having all my plates ready to carry. Since I had adopted the rotating tray rocker as the method for developing my plates, I could only develop one plate at a time. This became a problem on the last night — during the run, I had taken 28 plates, and all had to be developed, dried, placed in an envelope with observing details, and put into a sturdy wooden box as carry-on luggage. To achieve this state on my last night took so much time that I had no time to get any sleep before taking a taxi to the bus station in La Serena. This might not have been a problem but for the fact that Chile was governed by a military dictatorship at the time, and corruption was rampant. Visitors were warned to avoid black taxis and use only yellow taxis. When I arrived in Santiago after the unpleasant 7-h bus trip, I heard someone looking at me say "Aeropuerto?" I was so tired that in response I said "si," not knowing what kind of taxi was involved. When I got to the taxi, I saw that it was black, and realized I had made a mistake. I got into the taxi and the driver, instead of taking me to the airport, took me to a gas station and made me pay for filling up his tank. Then he charged an outrageous price for the ride, to the point that I felt I had been robbed. When I got to the airport, I had only a few dollars left and barely made it in time for my flight. After the July run, I was able to avoid the same thing happening again. Also, Chile switched to a democracy not long after, and things changed dramatically for the better. All of this was part of my foray into the world of the spirals.

Chapter 2

The Realm of the White Nebulae

Observers who have seen a large number of galaxies with a telescope, and who also have seen the Milky Way from a dark observing site, are likely to notice the resemblance of the Milky Way to some edge-on spiral galaxies (see Figure 1). In a way, it is almost obvious that the Milky Way is just another one of these objects, only much closer to us, like the Sun is just another star, only much closer to us as well. It was nevertheless not easy to prove the existence of other galaxies in the early days of visual astronomy. There was little to tell us about distances, and without such information, all that could be done was speculation.

How we came to know about the existence of other galaxies is one of the most interesting stories in the history of astronomy.[1] The Milky Way figures into the story because we live in it, which makes it the nearest spiral galaxy to us. It is a highly flattened, disk-shaped system of stars, that from the inside looks like a milky band going all around the sky. It has spiral arms, but these cannot be seen directly, only inferred from complicated and detailed studies (Chapter 8). The story of the spirals begins not with the Milky Way itself, but with objects that were discovered away from the Milky Way that came to be known as "white nebulae."

Nebulous objects in the night sky have been known since before the invention of the telescope. A few such objects are visible to

---

[1]For example, R. Berendzen, R. Hart and D. Seeley, *Man Discovers the Galaxies* (McGraw-Hill, 1976).

(a)

(b)

Fig. 1. (a) An all-sky, projected view of the Milky Way centered towards the constellation Sagittarius. The two small, diffuse objects to the right of the center and below the Galactic plane are the Magellanic Clouds. The smaller object at far left, below the Galactic plane also, is the Andromeda Galaxy, Photo: Serge Brunier; used with permission. (b) The edge-on spiral galaxy NGC 891 bears a strong resemblance to the Milky Way. The dark lane dividing the galaxy is obscuring matter in the form of interstellar dust that appears mostly confined to the plane of its disk of stars. Copyright: Adam Block/Mount Lemmon Skycenter/University of Arizona, reproduced with permission.

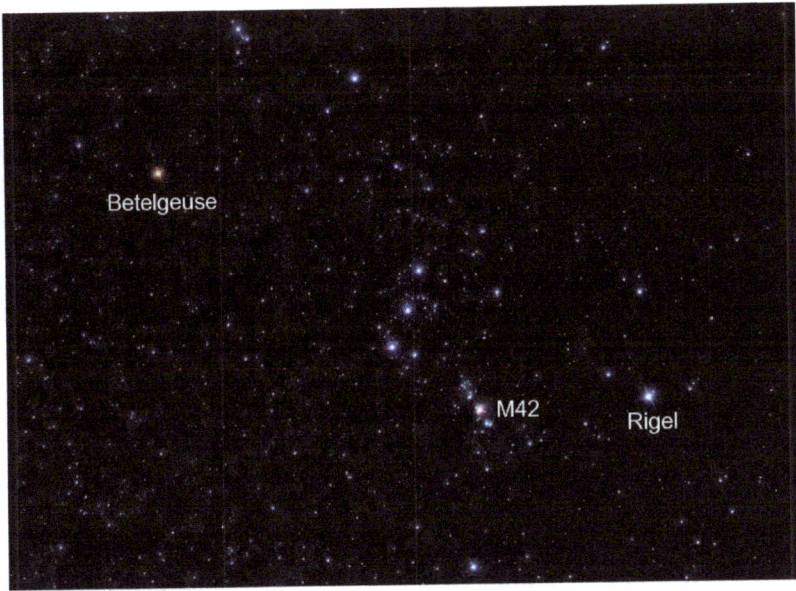

Fig. 2.   The Orion Nebula M42 is visible to the naked eye in the Sword of Orion, just north of the double star Iota Orionis. The bright stars Betelgeuse and Rigel are indicated. Astrophotography: Noel Carboni, with permission.

the naked eye. The Orion Nebula M42 is an example; a true cloud of glowing gas appearing as a "fuzzy" star in the Sword of Orion (Figure 2). The Orion Nebula is so bright that it can be seen without a telescope even in light-polluted sites. The Andromeda Galaxy can also be seen with the naked eye as a somewhat elongated and diffuse patch of light, but its visibility is more sensitive to light pollution or moonlight than is M42.

Some nebulous objects seen without a telescope are found to be nothing more than star clusters whose individual stars are too faint to be distinguished. The light of the faint stars blends together, so that one sees a cloudy glow of light rather than a clear cluster of stars. An example is the Beehive Cluster M44 in the constellation Cancer, the Crab. Its nebulous character can be easily seen with the unaided eye on dark, clear northern spring evenings, but through a telescope the Beehive is a swarm of moderately bright stars detectable with the slightest optical aid.

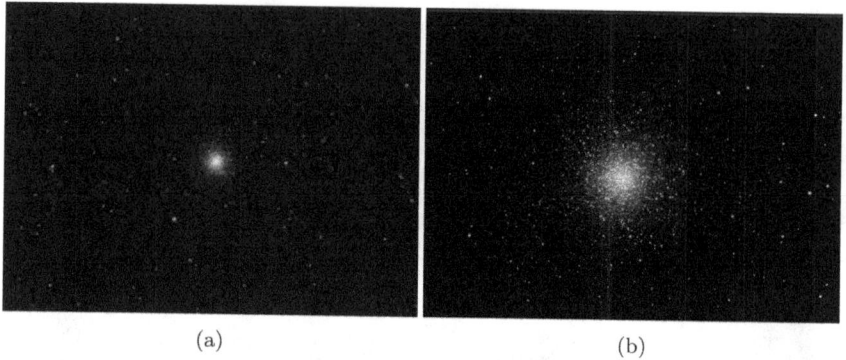

(a)   (b)

Fig. 3.   (a) Appearance of the globular cluster M13 with a small (approximately 2-inch) telescope, showing little resolution (image due to Filippo Bradaschia); (b) Image of M13 showing the kind of view that a larger (10-inch telescope) would give, showing full resolution of the cluster into stars. Image source: Astrophotography: Noel Carboni, with permission.

Some objects that appear nebulous with a small telescope can resolve into faint stars with a larger telescope. I recall observing the famous globular cluster M13 with a 2-inch refractor telescope in 1966. M13 appeared in that telescope as a round nebulous ball, bright near the center and gradually fading near the edges (Figure 3, (a)). Few, if any, individual stars were detected, and had I not known the true nature of the object, I would have had no reason to call it anything but a nebula. When I later obtained a 4-inch telescope, I quadrupled my light-gathering power, and M13 resolved into some of its stars. It became an obvious star cluster, although it was still largely unresolved. With a larger aperture telescope, M13 resolves into hundreds of faint stars (Figure 3, (b)), even in its bright inner section, and with the world's largest telescopes, one would see thousands of faint stars. Approximately 150 globular clusters have been catalogued in our Galaxy, and all resolve into stars with large telescopes. These objects, which appear mostly nebulous in small telescopes, are all distant systems of stars whose individual stars require larger telescopes to be seen.

Given what can be seen with globular clusters, one could imagine that a particular star cluster might be too far away for even a large telescope to reveal its individual stars. Indeed, this is what seemed to be the case for most of the nebulous objects catalogued by early visual astronomers. Recall that the first major catalogue of nebulous objects seen with a telescope was that of Charles Messier,[2] who made a list of 103 (later extended to 110) objects he considered a "nuisance" for comet hunting, which was his main interest. A comet is an icey planetesimal (called a nucleus) that usually follows a long elliptical orbit around the Sun. The fragment spends most of its time far from the Sun in the frigid expanse of the outer solar system, but for a short time is close enough to the Sun for the ices to sublimate, or convert directly from solid to gaseous form. The vapors interact with sunlight and the solar wind to produce a nebulous glow around the nucleus and, when closest to the Sun, tails. As a comet comes in but well before its closest approach to the Sun or the Earth, it can look like some "sidereal nebulae," or the nebulae always present among the "fixed stars." If you wanted to be sure that a newly viewed nebulous object was a comet, you first had to rule out that it was one of the sidereal nebulae that were always there. Messier did not discover all 110 of the objects with "M" numbers, but his catalogue is still popular because it includes many of the most spectacular deep-sky objects one can see with a small telescope.

It is a fact that any increase in telescopic light-gathering power can substantially increase the number of nebulous objects detectable. William Herschel (1738–1822) was the first to discover this. He perfected the construction of large mechanical telescopes whose light-gathering power greatly exceeded that of previous observers like Messier. Herschel's largest favored telescope used an 18.7-inch speculum metal mirror which, although no match for a modern 18.7-inch reflector, probably gathered 10 times as much light as Messier's telescopes. With such unprecedented light-gathering power, Herschel discovered literally thousands of nebulae, initially from his home in Bath, a town about 100 miles west of London, and then (after 1788) from Slough (pronounced sloo), a town about 20 miles west of London. He eventually inspired his son, John (1792–1871), to follow

---

[2]C. Messier, *Catalogue of Nebulae and Star Clusters* (Connoissance des Temps ou des Mouvements Célestes, 1784), p. 227.

in his foosteps, and also passed his enthusiasm for nebular studies to his sister, Caroline. Herschel searched for and compiled nebulae not necessarily to facilitate comet-hunting like Messier, but to learn about the nature of the universe itself. Ironically, a few comets infiltrated Herschel's catalogues. In his 1786 list,[3] he comments on two cases that he believes were comets that he did not recognize at the time.

Figure 4 shows sketches of the different forms of nebulae that Herschel found and catalogued.[4] The figure has been augmented with the NGC numbers[5] of the objects; examining these gives a good indication of what Herschel saw. The most typical nebulae

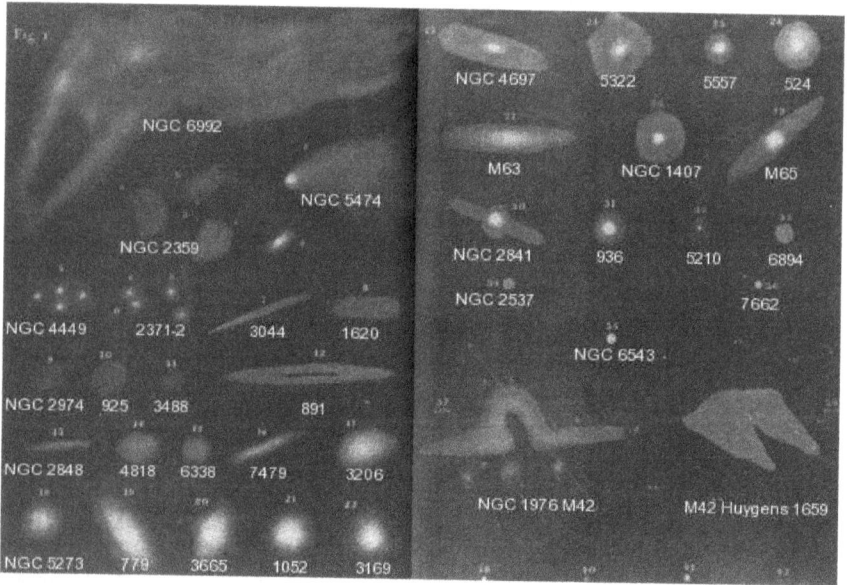

Fig. 4.   Sketches of nebulae made by William Herschel, published in 1811. The numbers above or next to the object are figure references in the original paper. A cross-index (www.messier.seds.org/xtra/similar/h2500a.txt) was used to get the NGC or Messier numbers, which are indicated below each sketch.

---

[3]W. Herschel, *Philosophical Transactions of the Royal Society of London*, **76**, 457 (1786).

[4]W. Herschel, *Philosophical Transactions of the Royal Society of London*, **101**, 269 (1811).

[5]More background on these numbers is given in Chapter 3.

that correspond to galaxies are the ones in the bottom three rows in the left half of the figure, and in the top three rows of the right half. Most are diffuse and featureless, regular in shape, whitish in color, sometimes with a bright center. Two of the illustrated cases are telling: No. 12 is NGC 891, the edge-on galaxy shown in Figure 1. Herschel's sketch shows that he saw the planar dust lane of this famous object. Also, No. 16 shows NGC 7479 as having a highly-elongated shape, but this shape is due to the presence of a strong bar, not to tilt. The galaxy is actually seen nearly face-on.

Of the 37 objects depicted in Figure 4, 28 are galaxies. The remainder are a supernova remnant (NGC 6992), two galactic nebulae (NGC 2359 and M42), and four planetary nebulae (NGC 2371-2, 6543, 6894, and 7662). These latter objects, which got their name through (in some cases) a superficial resemblance to planets, would soon play a role in showing that some nebulae are not unresolved distant star systems, but real clouds of glowing gas. Nevertheless, at least initially, Herschel believed that most, if not all, of the nebulous objects he catalogued were star clusters too remote for his telescopes to reveal the individual stars. He described this view in his monumental 1785 paper titled "On the Construction of the Heavens."[6] Over a period of nearly 20 years, he catalogued 2,500 star clusters and nebulae of various brightness classes through his famous "sweeps of the heavens" (referring to his method of systematically searching for nebulae).[7] The mystique of the nebulae was such that the bulk of the new ones found lay away from the band of the Milky Way. At the time of Herschel's observations, photography had not been invented, and Herschel recorded his observations of nebulae mostly as their positions on the sky, a brightness class, and a brief description of what the object looked like.

It is interesting that the concept of galaxies goes back to at least the early 18th century. From a purely speculative viewpoint, the Swedish philosopher Emanuel Swedenborg (1688–1772) imagined

---

[6]W. Herschel, *Philosophical Transactions of the Royal Society of London*, **75**, 213 (1785).

[7]F. Dyson, *Journal of the Royal Society of Canada*, **32**, 441 (1938).

in a 1734 publication[8] that our Galaxy is likely to be only one of many similar stellar systems scattered throughout space. Most of Swedenborg's work was theological in nature, and he gained a strong following during and after his lifetime, but the fact that he thought about other star systems like our own even before observers like Messier and Herschel began to find them in large numbers was remarkable.

Later 18th century philosophers followed a similar lead. In a 1750 publication,[9] Thomas Wright (1711–1786) speculated on the nature of the Milky Way and how what we see in its appearance might relate to its true (highly flattened) structure. He suspected that "cloudy spots" seen on the sky were likely distant star systems whose individual stars could not be reached with the telescopes of his day. In 1755, Immanuel Kant (1724–1804) more forcefully suggested that the faint nebulae seen in the sky, especially the ones that were regular elliptical patches of light, were not just star clusters too far away to be resolved by existing telescopes, but were separate, flattened star systems much like the Milky Way.[10] Later in his life, Kant became familiar with the nebular discoveries of Herschel and others, which convinced him even more of the existence of other galaxies.[11]

The reason for considering a different scale for such systems (that is, star clusters versus something more substantial) was that the unresolved nebulae were as bright as some nebulae that did resolve. For example, the Andromeda Galaxy is easily detectable as a nebulous object with the naked eye, and is in fact one of the brightest nebulae in the sky, but no 18th century telescope could resolve it. If the systems contained stars like the resolved nebulae, then in order to appear as bright as resolved nebulae but still far enough away to be unresolved, the systems had to be more significant than mere

---

[8]E. Swedenborg, *Basic Principles of Nature* (1734).

[9]T. Wright, *An Original Theory or New Hypothesis of the Universe* (H. Chapelle, London, 1750).

[10]I. Kant, *General Natural History and Theory of the Heavens* (Konigsberg and Leipzig, Johann Friederich Petersen, 1755).

[11]G. J. Whitrow, *Quarterly Journal of the Royal Astronomical Society*, **8**, 48 (1967).

star clusters. They had to be "island universes,"[12] which was the 19th century term for galaxies. It would seem that Kant's speculation did not necessarily include nebulae like the Orion Nebula, or most nebulae in the band of the Milky Way, but was an interpretation meant mainly for the regular nebulae seen away from the band. It was Herschel's discovery, in 1790, of the large planetary nebula NGC 1514 in Taurus that caused him to modify his view on nebulae. This object, which appears as a bright star surrounded by a glowing nebula (Figure 5), led Herschel to believe that real nebulae existed, that is,

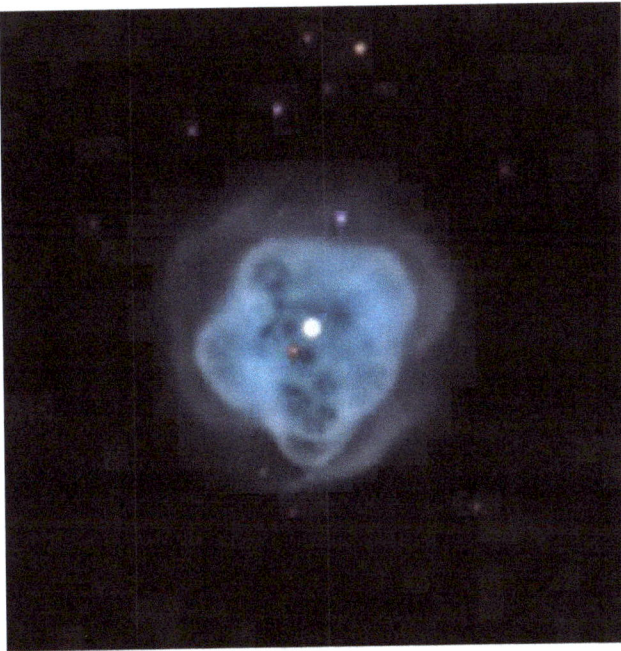

Fig. 5.   Color image of NGC 1514, the planetary nebula in Taurus that convinced William Herschel that some nebulae are gaseous "fluid" and not distant, unresolved star clusters. Image source: Don Goldman, used with permission.

---

[12] A term credited by Hubble and others to Alexander von Humboldt in 1850, but recently shown to be actually due to Ormsby MacKnight Mitchel, founder of the Cincinatti Astronomical Society, in 1846 (T. Siegfried, Science News, October 3, 2017). In a 1927 paper, Knut Lundmark argues that "The island-universe theory, has in fact its origin in the...paper of Herschel of 1785."

real clouds of glowing gas. In this case, the cloud glowed like an atmosphere around a bright central star.

William Herschel's son, John Frederick William, carried the mantle of his father's observations well into the 19th century and produced several monumental works. The first, the Slough catalogue,[13] included all the observations of nebulae and star clusters made by him between 1825 and 1833 at Slough, using the same kind of telescope, an 18.7-inch reflector with a speculum metal mirror and a focal length[14] of 20 feet, as was used by his father. The objects in this catalogue included not only about 2,000 of his father's objects but also 500 that the younger Herschel had discovered himself. The catalogue assigned numbers (1 to 2,306) to the entries listed in order of right ascension, a sky coordinate analogous to longitude.[15] This approach was novel and different from what his father used. The elder Herschel had previously published three long lists of nebulae and clusters in 1786, 1789, and 1802, amounting to 2,500 objects, and placed them into five "classes": I — bright nebulae; II — faint nebulae; III — very faint nebulae; IV — planetary nebulae; V — very large nebulae; and VI, VII, VIII — star clusters of different degrees of concentration. For example, the galaxy now known as NGC 891 has the W. Herschel number "V.19." The younger Herschel notes in his catalogue that of his 500 objects, only one was "very conspicuous and large," and that very few ranked in what his father referred to as "first class," or bright nebulae (Category I). He attributed this to the effectiveness of his father's approach to finding nebulae.

The latitude of Slough, 51.5 degrees north, imposed a significant limit on the catalogue's coverage of the southern sky. In 1833, the younger Herschel traveled to the Cape of Good Hope, South Africa, to gain access to the far southern sky, and catalogued more nebulae and clusters not previously seen by others.[16] The resulting catalogue

---

[13] J. F. W. Herschel, *Philosophical Transactions of the Royal Society of London*, **123**, 359 (1833).

[14] The distance from the primary mirror to the point where the image is formed.

[15] Right ascension is discussed further in Chapter 3.

[16] J. F. W. Herschel, *Results of Astronomical Observations Made During the Years 1834–1838 at the Cape of Good Hope* (Smith, Elder and Company, London, 1847).

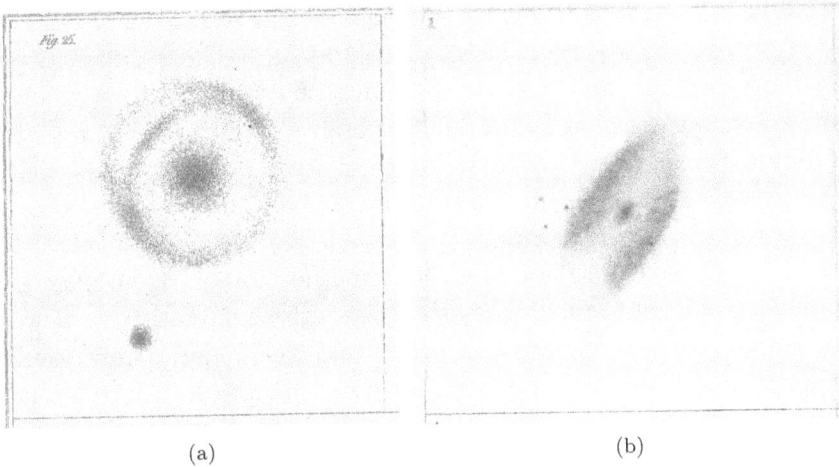

Fig. 6. Sketches of M51 (a) and NGC 1365 (b) made by Sir John Herschel. The M51 sketch is Figure 25 from his 1833 Slough catalogue, while the NGC 1365 sketch is Figure 1 from his 1847 Cape of Good Hope catalogue.

was published in 1847 and continued the 1833 numbering system, now including entries from 2308 to 4015. In both his 1833 and 1847 catalogues, Herschel included a number of impressive, well-reproduced sketches. Of these, Figure 25 in the 1833 catalogue depicted M51, while Figure 1 in the 1847 catalogue depicted NGC 1365. Both of these are shown in Figure 6. The younger Herschel's sketch of M51 shows two partially overlapping rings. Comparison with modern photographs shows that the split region is where the two main arms have the highest contrast on one side. It is interesting also that the younger Herschel's sketch of NGC 1365 shows an object having two enhancements flanking a bright center. There can be little doubt that these enhancements are what he saw of the two bright spiral arms. The bar was not seen, but it is likely that had the Leviathan been used to observe NGC 1365, the galaxy would have been seen as a barred spiral.

With these catalogues, names assigned in the younger Herschel's catalogues were sometimes prefixed with the letter "h," as in h 2301, while names taken from the elder Herschel's lists were prefixed by the letter "H," as in H III.230. The Birr Castle observers used "H" as the prefix for the Slough and Cape catalogue numbers, rather than "h."

The best way to test the idea that nebulae were mainly remote star systems was to try and resolve them as such. The main telescope William Herschel had used, an 18.7-inch reflector, could not resolve most nebulae. Although Herschel had built a 48-inch telescope (with funding from King George III[17]) after his 1785 paper had been published, the instrument had mirror problems and was hard to use. Even though it was the largest telescope in the world for 50 years (from 1789 to 1839), it added little to the store of knowledge on nebulae. This telescope had nearly 7 times the light-gathering power of Herschel's more favored 18.7-inch telescopes, and surely would have revealed details in nebulae not noted before. In principle, had it been effective, the elder Herschel might have been credited with discovering spirals. But as it turned out, this breakthrough was left to a talented Irishman who built an even bigger telescope from scratch with the main goal being, at least initially, to resolve the nebulae.

William Parsons, the Third Earl of Rosse, was a mechanically-inclined genius who was interested in astronomy and in particular the question of the nature of the nebulae. He had married Mary Field, the daughter of a wealthy landowner in Yorkshire, in 1836 and they were essentially a "team" from the beginning. In the 1830s, Parsons built a 36-inch telescope using a mirror made of speculum metal. In building this telescope, he was not able to benefit from the experiences of William Herschel, because the latter had not published details of his procedures for making large mirrors. The third Earl used his own ingenuity to build what certainly would have been an impressive telescope in his day. But the third Earl had bigger plans than this. He was excited about the possibility of resolving previously unresolved nebulae. It would be a major achievement to do so: it would verify a long unproven theory of the nature of the nebulae, and open nebular studies up to new understandings. The bulk of the nebulae discovered up to 1840 did not resolve with any of the Herschels' telescopes, although many were described in catalogue notes as "resolvable." It would seem, then, that objects on the brink of resolution with the

---

[17]Herschel came into the financial support of King George after his discovery of the planet Uranus in 1781.

(a)   (b)

Fig. 7. (a) One wall of the "Leviathan" Observatory on the grounds of Birr Castle, County Offaly, Ireland. (b) Close-up of the front end of the replica telescope, where the eyepiece was positioned. Images credit: R. Buta.

Herschels' telescopes might fully resolve with a substantial increase in light gathering power.

The "Leviathan of Parsonstown," as the new telescope came to be called (Figure 7),[18] had a 72-inch (6 feet) diameter speculum metal mirror that gathered 15 times as much light as the Herschel 18.7-inch telescopes, with which the greatest number of nebular discoveries had been made. The mirror had a focal length of 54 feet, making the telescope a massive instrument requiring the utmost in design to make it work. It had to be capable of pointing at objects of known position, and it had to be possible to move it easily enough to follow an object while an observer tried to study it. The 15 times increase in light-gathering power gave the telescope a limiting magnitude at least 3 magnitudes fainter than an 18.7-inch telescope.[19] It has been estimated that because speculum metal, even at its untarnished best, is not as reflective as a modern silvered mirror (as described in Chapter 1), the Leviathan had a light-gathering power equivalent

---

[18]The original Leviathan fell into disuse after 1890; in 1997, a fully operational replica of the old telescope was completed in the same place where the original had been located. This is described in M. Tubridy, *Reconstruction of the Rosse Six Foot Telescope*, B. Roden (ed.), (Brosna Press, Ferbane, County Offaly, 1998).
[19]The limiting magnitude of a telescope is the faintest apparent magnitude the telescope can reveal by eye.

to a modern 50-inch telescope.[20] Even so, the 15 times increase in light-gathering power over the Herschels' telescopes is valid because their telescopes used speculum mirrors as well.

Although the Leviathan had a familiar Newtonian design, i.e., used a curved primary mirror and a flat secondary to deflect the collected light to a focus near the front side of the tube, its mounting was nothing like that of the reflectors of modern observatories. The Leviathan was mounted between two massive stone walls 40 feet high and two feet thick, and was manipulated by swiveling the primary mirror end on a massive universal joint. The walls were oriented exactly north-south. Movement of the telescope up and down and left and right was achieved with chains and pulleys. It took a team of four people, in addition to the observer himself, to use the telescope effectively. Because of the walls, the telescope could not move very far east or west of the meridian.[21] In order to observe any specific object, the observer had to come prepared to "catch" that object just 30 minutes or so before it crossed the meridian. If the observers saw anything interesting, they would have at most about an hour to study it and sketch it at any given time. They would follow the object across the meridian until forced to stop by the west wall. Any object that was already past the meridian by more than half an hour was inaccessible the rest of the night.

The telescope was unlike any modern observatory because it was left out in the open air. The walls supporting the telescope did not protect it from rain, humidity, snow, and temperature changes. If a modern observatory were so exposed, it would not function for long. This kind of exposure caused the Leviathan's mirror to tarnish (lose its reflectivity) fairly quickly. To deal with this, the third Earl fabricated two specula, so that while one was in the mirror box of the telescope, the other was being refigured and polished. It has been

---

[20]D. W. Dewhirst and M. Hoskin, *Journal for the History of Astronomy*, **22**, 257 (1991).

[21]The meridian is the great semi-circle that divides the sky into rising and setting halves. An object is at its maximum point above the horizon when it is crossing the meridian.

said that each time the mirror was replaced, the Leviathan became a new telescope.

The third Earl encountered many impediments to his endeavor. First, a 72-inch speculum metal mirror is very heavy. The first successful one cast weighed 4 tons, and just getting it into the back end of the tube must have been a major undertaking. Everything about the great telescope was challenging and difficult. Second, at the time when the Leviathan was "stretching her legs," so to speak (that is, relatively brand new and ready for science), Ireland was hit by a devastating potato famine, a blight that killed by starvation up to 2 million people. During this time, from 1845 to 1847, astronomy at Birr Castle took a back seat to more urgent matters. Third, the Irish climate is not ideal for astronomy because of the large number of cloudy nights. As shown in the following section, the Birr Castle observers were able to effectively use the Leviathan for nebular astronomy on only about two and a half nights per month, or 30 nights per year. Compare this with a modern, well-placed observatory on a remote mountain top in a dry climate, like McDonald Observatory in west Texas, where it is possible to get more than 200 clear (or partially clear) nights per year.

Although the third Earl was interested in discovering new nebulae and fully expected that his telescope would reveal many because of its considerably greater light-gathering power, this was not his main priority. He wanted to mostly observe the brighter, previously discovered nebulae, to see what new information his telescope would provide. Specifically, could he resolve these nebulae given that people expected the nebulae were star systems? In March 1845, the third Earl invited two distinguished visitors, Sir Thomas Romney Robinson, Director of Armagh Observatory (pronounced "arm ah," located 72 miles north of Dublin), and Sir James South of Dunsink Observatory (located in Dublin), to observe with him and evaluate the performance of the Leviathan. Robinson wrote an account of the experience,[22] which included the observation of M51. When they all

---

[22] T. R. Robinson, *Proceedings of the Royal Irish Academy*, **3**, 114 (1845); T. R. Robinson, *Proceedings of the Royal Irish Academy*, **4**, 119 (1848).

viewed M51 at that time, they had in mind J. Herschel's "Fig. 25" impression (see Figure 6 (a)) of a double nebula where the primary "nucleus" had rings, one off center. Looking at M51 this way, they convinced themselves the nebula had indeed resolved, although South was doubtful. The spiral pattern in the primary "nucleus" of M51 was apparently first noticed in April 1845, and sketched as such by the third Earl. This intriguing sketch, on display at the Birr Castle museum, is shown in Figure 8. It was described by Hoskin[23] as the "first drawing of a spiral nebula," and indeed what makes this sketch interesting is the inscription that it was "Handed round the Section at the Cambridge meeting." This note refers to the June 19, 1845 meeting of the British Association for the Advancement of Science, and was the first circulation of the news of the discovery. As noted by Trevor C. Weekes (1940–2014),[24] the actual date when M51 was

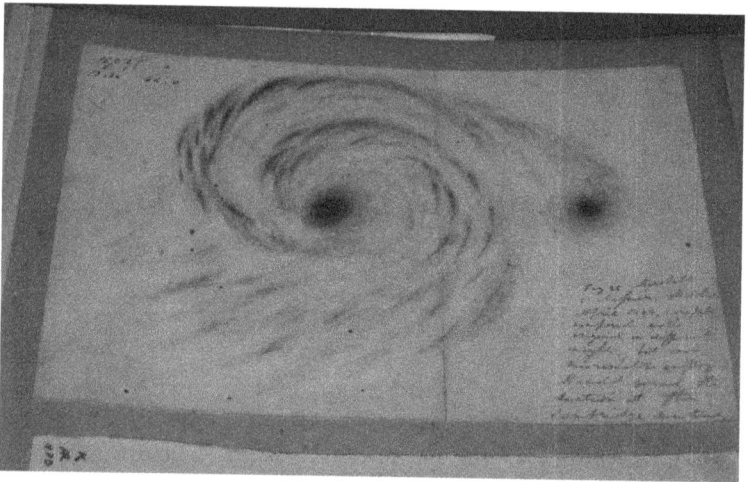

Fig. 8.   The first drawing of a spiral nebula sketched with the Leviathan telescope by the Third Earl of Rosse in April, 1845. The writing on the sketch identifies the object by its Herschel number, 1622, and points to a "Fig. 25," referring to a sketch of the same object made by John Herschel published in the Slough catalogue. This Figure 25 sketch is shown here in Figure 6. Image credit: R. Buta.

---

[23] M. Hoskin, *Journal for the History of Astronomy*, **13**, 97 (1982).

[24] T. C. Weekes, "The Nineteenth-Century Spiral Nebula Whodunit," *Perspectives in Physics*, **12**, 146 (2010).

noticed to be spiral as well as the circumstances of the discovery were never recorded, which unfortunately shrouds the discovery in mystery.

The Birr Castle Experience — Laurence Parsons (1840–1907), the eldest of William Parsons' sons, became the fourth Earl of Rosse in 1867, after the death of his father. Eventually becoming a well-known scientist in his own right, the fourth Earl continued his father's work with the Leviathan and made the decision to publish all of the visual observations in the 1879 and 1880 volumes of the Scientific Transactions of the Royal Dublin Society (STRDS II). The dates of observations listed in these volumes are telling. These highlight details of observing activities beyond the lists for single objects. It is interesting to ask: over a period of 30 years, how many nights were actually used to observe nebulae and clusters at Birr Castle? Based on the STRDS, 888 nights provided some useful observations. If we assume the observers tried to use any nights when the sky was moonless at least half the night, then technically they were able to observe during 16% of such nights from 1848 to 1878. Table 1 lists

Table 1. Number of nights used for observing nebulae and star clusters at Birr Castle, mostly with the "Leviathan" (1843–1878).[a]

Year 18__	No. of nights	Average No. nights/month	Earl	Assistant observer
43	2	0.2	3rd	none
44	1	0.1	3rd	none
45	1	0.1	3rd	none
46	1	0.1	3rd	none
47	0	0.0	3rd	none
48	34	2.8	3rd	William H. Rambaut (Jan.–June), George Johnstone Stoney (July–Dec.)
49	37	3.1	3rd	George Johnstone Stoney (Jan.–Dec.)
50	51	4.2	3rd	George Johnstone Stoney (Jan.–Aug.), Bindon Blood Stoney (Aug.–Dec.)
51	73	6.1	3rd	Bindon Blood Stoney (Jan.–Dec.)
52	36	3.0	3rd	Bindon Blood Stoney (Jan.–Apr.); George Johnstone Stoney (Aug.–Dec.)

(Continued)

Table 1.   (*Continued*)

Year 18__	No. of nights	Average No. nights/month	Earl	Assistant observer
53	2	0.2	3rd	none (Jan.–Nov.), Robert J. Mitchell (Dec.)
54	37	3.1	3rd	Robert J. Mitchell (Jan.–Dec.)
55	47	3.9	3rd	Robert J. Mitchell (Jan.–Dec.)
56	35	2.9	3rd	Robert J. Mitchell (Jan.–Dec.)
57	39	3.2	3rd	Robert J. Mitchell (Jan.–Dec.)
58	15	1.2	3rd	Robert J. Mitchell (Jan.–May), none (June–Dec.)
59	0	0.0	3rd	none
60	33	2.8	3rd	none (Jan.), Samuel Hunter (Feb.–Dec.)
61	39	3.2	3rd	Samuel Hunter (Jan.–Dec.)
62	26	2.2	3rd	Samuel Hunter (Jan.–Dec.)
63	16	1.3	3rd	Samuel Hunter (Jan.–Dec.)
64	11	0.9	3rd	Samuel Hunter (Jan.–May.), none (June–Dec.)
65	2	0.2	3rd	none (Jan.–Dec.)
66	30	2.5	3rd	none (Jan.), Robert S. Ball (Feb.–Dec.)
67	30	2.5	3rd	Robert S. Ball (Jan.–Aug.), none (Sept.–Dec.)
68	15	1.2	4th	none (Jan.), C. E. Burton (Feb.–Dec.)
69	4	0.3	4th	C. E. Burton (Jan.–Mar.), none (Apr.–Dec.)
70	2	0.2	4th	none (Jan.–Dec.)
71	16	1.3	4th	Ralph Copeland (Jan.–Dec.)
72	37	3.1	4th	Ralph Copeland (Jan.–Dec.)
73	37	3.1	4th	Ralph Copeland (Jan.–Dec.)
74	34	2.8	4th	Ralph Copeland (Jan.–May), none (June–July) John L. E. Dreyer (Aug.–Dec.)
75	42	3.5	4th	John L. E. Dreyer (Jan.–Dec.)
76	45	3.8	4th	John L. E. Dreyer (Jan.–Dec.)
77	39	3.2	4th	John L. E. Dreyer (Jan.–Dec.)
78	24	2.0	4th	John L. E. Dreyer

Note: [a]Based on information in the Scientific Transactions of the Royal Dublin Society, II, 1879, 1880.

the number of nights observed per year (or at least that provided information for the STRDS II) from 1843 to 1878 and the number of nights per month during each year. The table also shows the periods of time that the various assistant observers were in the employ of Birr Castle. Following are some notes:

(1) When there was no assistant observer, few or no recorded nebular observations were made. The lowest observing years were 1847 (famine), 1853, 1859, 1865, 1869, and 1870. The highest observing year was 1851, with an average of 6 nights per month. Bindon Stoney was the assistant observer during that time. The period 1848–1852 was very productive, as was the period 1872–1878. The fourth Earl was undoubtedly active during this latter time, and he had very effective assistants in Ralph Copeland and J. L. E. Dreyer (pronounced "Dryer"). In a letter from Ralph Copeland to the fourth Earl in 1871, we learn that the pay scale of an assistant observer was L200 per annum.

Table 2 lists the total number of observations by month from 1843 to 1878. The busiest months were March and April, which is

Table 2.   Observing frequency by month at Birr Castle (1843–1878).[a]

Month	No. nights observed
January	87
February	81
March	133
April	137
May	57
June	1
July	7
August	54
September	82
October	81
November	72
December	99

Note: [a]Based on information in the Scientific Transactions of the Royal Dublin Society, II, 1879, 1880.

not surprising since these are the optimum months for spiral nebular observations. Almost no observations were made in June or July, probably because the nights were short and the number of spiral nebulae to be found during these months was low. Also, the weather was probably more cooperative in March and April than in June or July.

(2) Based on diaries in the Birr Castle archives as well as several of the observing notebooks, one can deduce that 74 nebulae recognized as spiral in the Birr Castle observations are in fact spiral galaxies.[25] A few recognized cases are planetary nebulae while fewer still are diffuse galactic nebulae. Some of the spirals were only suspected, but the sketches and notes agree with available images.[26]

Figures 9–12 show some of the intriguing nebular sketches made at Birr Castle. While today such sketching would be done mainly for artistic or aesthetic purposes, in the 18th and 19th centuries sketching was the scientific way of recording details in nebulae. This is described in Omar Nasim's book *Observing by Hand: Sketching the Nebulae in the Nineteenth Century*.[27] Nasim discusses the drawings made by the Herschels, the third and fourth Earls of Rosse and their assistants, William Lassell, E. P. Mason, Wilhelm Tempel, and George P. Bond. Nasim describes how these drawings were made, what lighting was used, how sketching "evolved" from what was seen in the first of a set of observations to the last in that set, and how the drawings were actually prepared for publication.

One of the most talented visual artisans of Birr Castle was assistant Samuel Hunter, who made the remarkable drawing of the quintessential cosmic pinwheel M101 shown in Figure 9. This

---

[25]See Table 1, Appendix 2.

[26]Even a few globular clusters were characterized as spiral in the Birr Castle notes. In these cases, spirality is likely due to nothing more than random star patterns.

[27]O. Nasim, *Observing by Hand: Sketching the Nebulae in the Nineteenth Century*, (Chicago, The University of Chicago Press, 2013).

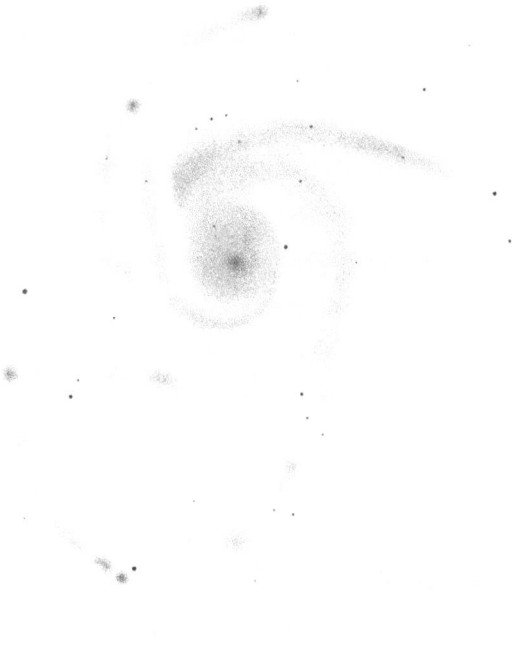

Fig. 9. Visual sketch of the famous "Pinwheel Galaxy" M101 (NGC 5457) in Ursa Major, as seen with the Leviathan. The observer was Birr Castle assistant Samuel Hunter, who had the job from February 1860 to May 1864. The sketch is based on a week's worth of observing and is remarkably accurate. Many of the "knots" in the spiral arms are star-forming regions that are so bright they each have their own NGC number. For a modern photograph of M101, see Figure 9 of Chapter 16.

complex object, located 22 million light years away in Ursa Major, showed considerable structure in the Leviathan eyepiece, including several bright knots which modern photographs show are massive clumps of newly-formed stars. M101 is a large star-forming spiral that has hosted at least four known supernovae.

Figure 10 shows Hunter's visual sketch of M51, made with the Leviathan starting in May 1863 and completed in May 1864 on what may have been one of his final nights according to Table 1. The drawing is astonishingly accurate and is clearly not just an awed impression as earlier sketches seemed to be.

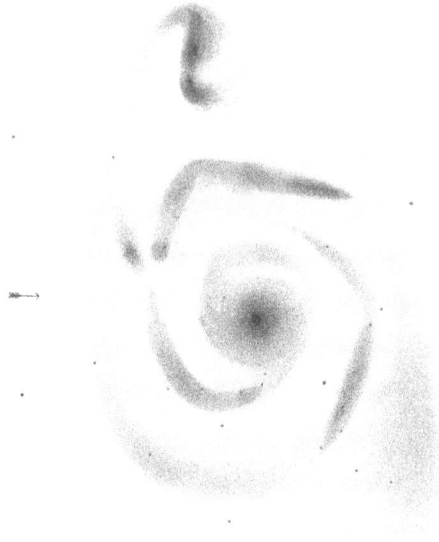

Fig. 10. Samuel Hunter's visual sketch of M51, one of the most accurate renditions of the great spiral ever made with the Leviathan.

Figure 11 shows a sampling of other Leviathan sketches made by the Birr Castle observers. All of the objects illustrated are galaxies except for the one marked with an "X," which is the diffuse nebula NGC 2245. The illustrations show a variety of visually detected structures including not only spiral arms, but a clear planar dust lane in NGC 3190 and aspects of several galaxy pairs.

(3) The observing notes show that not only was spiral structure detected in many nebulae, but rings and bars were detected in a few cases.[28] One sketch, highlighted in the upper left panel of Figure 12, shows about three-quarters of the large inner ring

---

[28]Earl of Rosse, *Philosophical Transactions of the Royal Society of London*, **151**, 681 (1862); Earl of Rosse, *Scientific Transactions of the Royal Dublin Society*, **2**, 1 (1879).

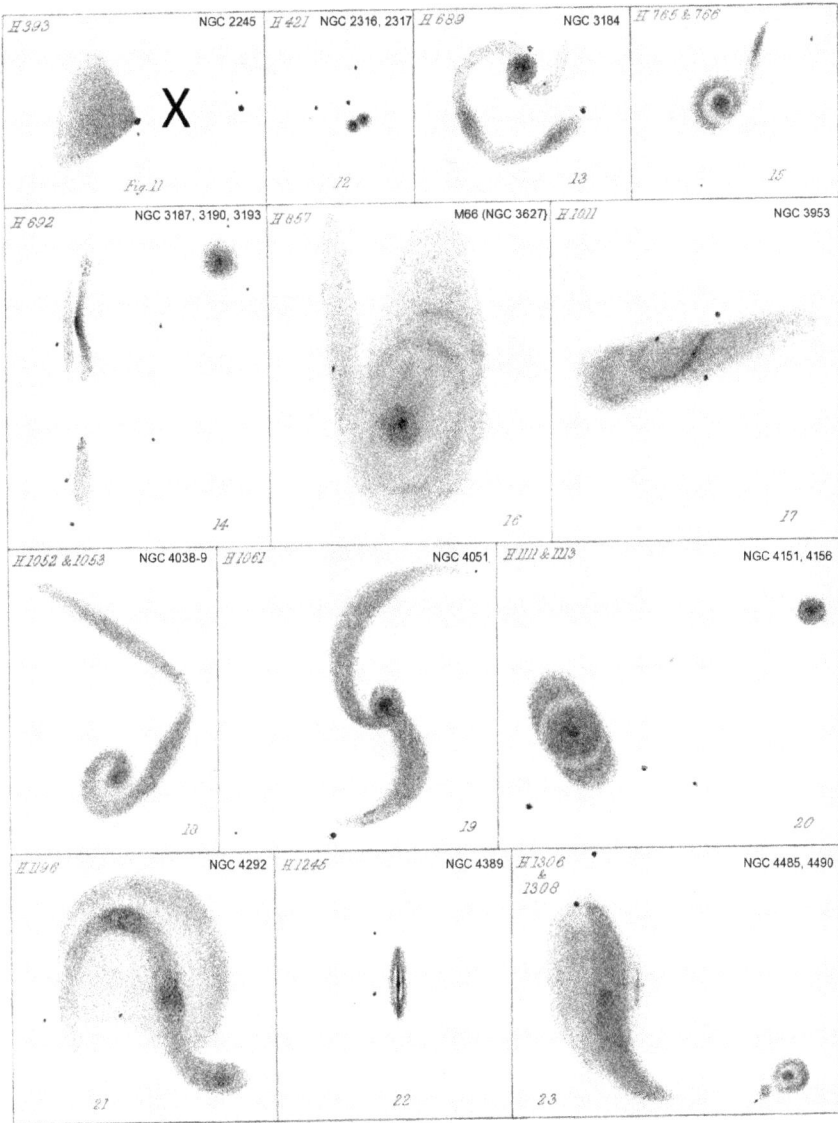

Fig. 11. A set of nebular sketches made by Birr Castle observers using the Leviathan. The galaxies are identified by their Slough and Cape catalogue numbers, which have been augmented by NGC numbers.

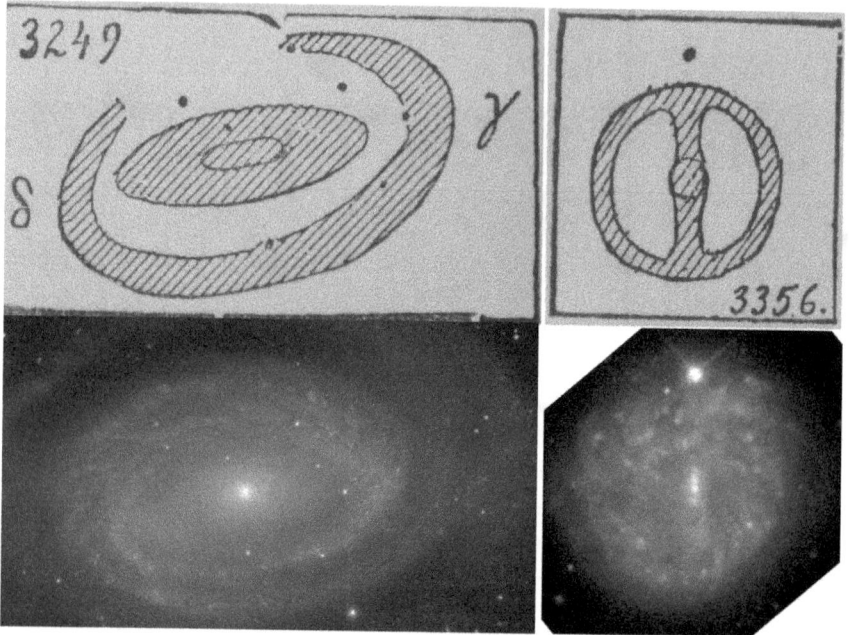

Fig. 12.   Visual sketches of two nebulae, both showing a bar and a ring rather than a clear spiral pattern, made with the Leviathan. The galaxies are GC 3249 = NGC 4725 and GC 3356 = NGC 4900. Images source: Sloan Digital Sky Survey Collaboration, www.sdss.org.

of NGC 4725 (GC 3249), a spiral 54 million light years away in Coma Berenices. The sketch is compared with a blue-green (*g*-band) image roughly oriented in the same manner. This image confirms the main features, including the bar. The upper right panel of Figure 12 shows a Leviathan sketch of NGC 4900 (GC 3356), a small bulge, face-on spiral located 40 million light years away in Virgo. As subtle as the features are in the accompanying *g*-band image, the Birr Castle observers were still able to see the large, ring-like enhancement in the outer regions as well as the thin bar. GC numbers are described in the next chapter.

In the Birr Castle notes, NGC 4725 was described as being "another spiral" by George Johnstone Stoney, based on an observation made in March 1850. Then his brother, Bindon Blood Stoney, described it as a "Spiral, 2 arms" based on a February 1851 observation. Based on an April 1858 observation, Robert J. Mitchell described NGC 4725 as being "very large and very bright. The centre itself is like an extended nebula with a nucleus; this centre is enveloped in an irregular ring or rings of nebulous light..." Then, based on an observation made about two weeks later, Mitchell recorded that "The surrounding ring of nebulosity is of irregular shape." The final observation, recorded in March 1867 and possibly made by either the fourth Earl or assistant observer Robert S. Ball, characterized the appearance of NGC 4725 as "somewhat like an anchor," which is the most accurate description of its appearance in modest-sized telescopes.

NGC 4900 was discovered to be an interesting case in April 1855 by Robert J. Mitchell, who wrote "Looks sometimes like the owl nebula [h858] when badly seen, with a bright, extended patch in the centre and dark spots on each side of this; sometimes dark ring is seen all the way around." The Owl Nebula (M97) is a large planetary nebula in Ursa Major having two dark regions that make the nebula vaguely resemble an owl. NGC 4900 apparently looked like this also. In both the NGC 4725 and 4900 descriptions, a bar is described as an extended nebula or patch of light.

(4) The observing notebooks indicate that more emphasis was placed on micrometer measures in later years. The notebooks of J. L. E. Dreyer are covered with calculations, while those of W. H. Rambaut and G. J. Stoney are focussed more on visual impressions. Micrometer observations of field stars were used to make visual sketches and positioning more accurate.

(5) The Birr Castle observations took place during a major development: the birth of astrophysics. This began when spectroscopy started to be used for both stellar and nebular observations. The field was pioneered by William Huggins (1824–1910) beginning in 1864 and was carried further with his wife, Margaret Lindsay Huggins (1848–1915), whom he married in 1875. Huggins work captured the interest of the fourth Earl. Huggins had determined that planetary nebulae and the Orion Nebula had strong emission lines in their spectra, signaling that these objects were in fact

real glowing clouds of very low density, gaseous matter rather than the unresolved light of very faint stars. In contrast, regular nebulae like the Andromeda Nebula showed mainly a continuous spectrum, as if their light was due to the unresolved glow of very faint stars. It is interesting that many of the brightest planetary nebulae observed with the Leviathan are described in the STRDS notes as being bluish in color, while the regular nebulae generally showed no color but white. This is how these latter objects became known as the "white nebulae."

Although the Leviathan failed to resolve the Herschel nebulae, the third Earl had the right idea: build bigger and better telescopes. The key to understanding the nature of the nebulae was to resolve them into their individual stars. Photography eventually made this possible.

Chapter 3

Spiral Nebulae and the
Photographic Revolution

The discovery of spirals was made at a time when Herschel (H or h) numbers reigned supreme as the naming convention of nebular objects. However, as more and more nebulae and star clusters were discovered, the more cumbersome and incomplete existing catalogues became. This is what led to Sir John Herschel's *General Catalogue of Nebulae and Clusters of Stars* published in 1864.[1] All the previously published lists with their own numbering systems were combined into a single list with a new numbering system, the GC number. Thus, the bright spiral known as M101, which was h 1744 in John Herschel's 1833 Slough catalogue, became GC 3770. About 90% of the 5,079 entries in the GC were found by the two Herschels; the rest were found by other observers.

The time of the GC numbers was short-lived. In the mid-1870s, John Louis Emil Dreyer (1852–1926), then an assistant observer at Birr Castle, decided that a new GC had to be compiled in order to account for all the nebular discoveries made at Birr Castle and elsewhere. Although the focus at Birr Castle was on observations of bright, well-known nebulae, the Leviathan often revealed new nebulae in the vicinity of already catalogued objects. For example, Heinrich D'Arrest (1822–1875) discovered the nebula that became

---

[1] J. Herschel, *Philosophical Transactions of the Royal Society of London,* **154**, 1 (1864).

GC 1 with the 11-inch refractor of the Copenhagen Observatory in 1860. The fourth Earl later observed GC 1 with the Leviathan, and found that it had a faint companion that had not been recorded by others. This companion was not noticed soon enough to make it into the published GC, but was assigned the number GC 6246 in a supplement Dreyer made to the GC that was published in 1878.[2]

Dreyer proposed a revision to the GC because, by the 1870s and 1880s, many other observers had taken on the task of searching for new nebulae,[3] and especially of measuring reliable positions of many previously discovered cases. Also, trying to determine if an object was "new" or previously seen and published by others again proved cumbersome. The "New General Catalogue of Nebulae and Clusters of Stars" (NGC) was a major undertaking and was published in 1888 as a Memoir of the Royal Astronomical Society.[4] Like the GC, the NGC involved the assignment of new numbers, which Dreyer clearly did not relish as he advocated against using the numbers over the old Herschel numbers. But while Dreyer was reluctant to use his NGC numbers, which ran from 1 to 7840, the astronomical community embraced them, such that NGC numbers are still used today for the majority of bright galaxies. The wisdom of this is shown by GC 1 and GC 6246. In Dreyer's catalogue, they became NGC 1 and NGC 2, respectively. The two galaxies are shown in Figure 1. NGC 1 is an excellent, somewhat asymmetric spiral 210 million light years away, while NGC 2 is also a spiral, but 345 million light years away. The two objects appear to be unrelated and form what astronomers call an optical double galaxy, that is, a close pair formed by chance alignment.

---

[2] J. L. E. Dreyer, *Transactions of the Royal Irish Academy*, **26**, 381 (1878). This supplement assigned additional GC numbers from 5080 to 6251.

[3] The fascinating history of the observers behind the NGC is recounted in W. Steinicke, *Observing and Cataloguing Nebulae and Star Clusters: From Herschel to Dreyer's New General Catalogue* (Cambridge, Cambridge University Press, 2010).

[4] J. L. E. Dreyer, *The New General and Index Catalogues* were reprinted in 1962 in a single volume of the *Memoirs of the Royal Astronomical Society* (Royal Astronomical Society, London).

Fig. 1. Two spirals forming an optical double galaxy: (Top) NGC 1, a spiral 210 million light years away. (Bottom) NGC 2, a spiral 345 million light years away. Both are in the constellation Pegasus. Image source: Sloan Digital Sky Survey Collaboration, http://www.sdss.org.

Figure 2 shows the page of the NGC that includes M51 (NGC 5194–5195). Except around the sky near the Large Magellanic Cloud, most of the objects on any page of the catalogue are galaxies. In fact, on this particular page, all but one, NGC 5189, is a galaxy. NGC 5189 is a planetary nebula in the southern constellation of Musca, the fly. Ironically, NGC 5189 is unusual in that it strongly resembles a barred spiral galaxy even though it is not a galaxy. One of the galaxies on the page, NGC 5219, is actually a deleted entry because it was found

No.	G. C.	J. H.	W. H.	Other Observers.	Right Ascension, 1860·0.	Annual Precession, 1880.	North Polar Distance, 1860·0.	Annual Precession, 1880.	Summary Description.	Notes.
					h m s	s	° ′	″		
5184	3566	1618	II 680	...	13 23 0	+3·08	90 56·1	+18·7	pF, pL, iR, bM, f of 2	
5185	3567	1619	III 642	...	13 23 9	2·95	75 53·5	18·7	vF, S, iR	
5186	...	...	...	Hartwig	13 23 10	2·96	77 6·0	18·7	No description	
5187	3568	1620	III 652	...	13 23 18	2·77	58 8·6	18·7	vF, vS, R, glbM	
5188	3569	3515	...	...	13 23 31	3·40	124 3·8	18·7	F, pL, vlE, vglbM	
5189	3570	3514	...	Δ 252?	13 23 41	4·13	155 15·2	18 7	!, B, pL, cE, bM curved axis, 4 st inv	†
5190	3571	1621	...	...	13 23 50	2·91	71 8·3	18·7	cF, S, R, bM, * * f	
5191	...	...	...	Hough, T VIII	13 23 50	2·98	78 4·1	18 7	cF, * 9 f 57′	
5192	5740	...	...	m 259	13 23 56	3·08	91 1	18·7	vF	
5193	3573	3516	...	...	13 23 57	3·38	122 30·5	18·7	pB, S, R, g, psbM	
5194	3572	1622	...	M 51	13 23 58	2·54	42 4·9	18·7	!!!, Great Spiral neb	†
5195	3574	1623	I 186	...	13 24 5	2·54	42 0·6	18·7	B, pS, lE, vglbM, inv in M 51	†
5196	5741	...	...	m 260	13 24 5	3·08	90 54	18·7	vF	
5197	5742	...	...	m 261	13 24 9	3·08	90 59	18 7	vF	
5198	3576	...	II 689	...	13 24 16	2·55	42 36·3	18·7	pF, pS, R, mbM	
5199	3577	1624	III 406	...	13 24 22	2·73	54 26·5	18·7	vF, vS, lE	
5200	5072	...	...	S Coolidge	13 24 32	3·07	89 18·5	18·7	* 12 in F neb	
5201	3578	...	II 797	...	13 24 40	2·41	36 12·2	18·7	pF, cS, R, rglbM	
5202	5743	...	...	m 262	13 24 45	3·08	90 59	18·7	vF	
5203	3579	3517	III 507	...	13 24 52	3·14	98 3·3	18·7	vF, eS, R, gbM, r	
5204	3575	1625	IV 63	d'A	13 24 53	2·26	30 51·5	18 7	pB, cL, iR, gmbM, r	
5205	...	...	...	Sw VI	13 25 2	2·11	26 46·4	18·7	vF, pS, R, bet 2 vF st	
5206	3580	3518	...	...	13 25 19	3·61	137 24·7	18 6	F, pL, R, vgbM	
5207	3581	1626	III 643	...	13 25 22	2·95	75 22·9	18·7	F, S, cE, * 11 att np	
5208	3582	1627	III 9	...	13 25 27	3·00	81 57·7	18·6	F, vS, R, psbM, p of 2	
5209	3583	1628	III 10	...	13 25 43	3·00	81 57·4	18·6	F, vS, R, stellar, f of 2	
5210	3584	1629	III 99	...	13 25 50	3·00	82 6·7	18·6	F, S, R, psbMN	
5211	3585	1630	...	...	13 25 54	3·08	90 18·9	18·6	pB, S, R, psmbM	
5212	3586	1631	...	...	13 26 22	3·00	81 59·1	18·6	eF	
5213	5744	...	...	m 263	13 26 35	3 03	85 10	18 6	vF, S, lE	
5214	3587	1632	III 656	...	13 26 45	2·62	47 24·6	18·6	vF, S, R, lbM	
5215	3589	3519	...	...	13 27 10	3·40	122 45·7	18·6	eF, eS, * s and * p	
5216	3590	1635	II 841	...	13 27 15	2·08	26 33·9	18·6	pB, S, vlE	
5217	3591	1634	...	...	13 27 18	2 90	71 25·4	18·6	vF, S, R, bM	
5218	3592	1636	II 842	...	13 27 21	2·07	26 30·7	18·6	pB, pL, R, gbM	
5219	3593	3520	...	...	13 27 48	3·58	135 11·3	18·6	vF, S, R, * n, nr	
5220	3594	3521	...	...	13 28 0	3·40	122 44·2	18 6	vF, S, R, * 10 f	
5221	3595	1637	III 86	...	13 28 4	2·94	75 27·8	18·6	vF, S, vlE, 1st of 3	
5222	3596	1638	III 85	...	13 28 4	2·94	75 32·3	18·6	cF, S, R, bM, 2nd of 3	
5223	3598	1640	III 407	...	13 28 6	+2·71	54 34·8	+18·6	F, cS, R, * 10 p, p of 2	

Fig. 2.   The page in the NGC that includes M51, NGC 5194-95. Image credit: R. Buta.

long after the publication of the NGC to be a duplicate of NGC 5244. In spite of Dreyer's heroic effort to make the New General and Index Catalogues as accurate and reliable as possible, numerous errors like this crept in, often in the form of inaccurate coordinates that would lead to an object either not being found or assigned more than one

Fig. 3. Two galaxies on the same NGC page as M51 that show how well an NGC description (Figure 2) can match to an actual image. Images source: Sloan Digital Sky Survey Collaboration, http://www.sdss.org.

identification number.[5] Thus, out of 40 objects on the page, 38 (95%) are distinct galaxies.

NGC descriptions have a well-defined arrangement: separated by commas are a brightness class, a size class, a shape class, and a central concentration class, followed by other details relevant to a specific object. For example, the object listed as NGC 5204 in Figure 2 is described as "pB,eL,iR,gmbM,r" which translated is to read "pretty bright, excessively large, irregularly round, gradually much brighter middle, resolvable." The next one in the list is NGC 5205 described as: "vF,pS,R,bet 2 vF st" which translates to "very faint, pretty small, round, between two very faint stars." When one looks into these objects more closely using modern images (Figure 3), both are found to be spiral galaxies, although NGC 5204 has only weak spiral structure. Interestingly, the image of NGC 5204 shows many scattered blue star-forming regions. Could these account for the "resolvable" part of this object's description in the NGC?

---

[5]The NGC/IC Project is a group of amateur and professional astronomers with the long-term goal of identifying as many of these kinds of discrepancies as possible, and of using the historical record to correct the catalogues. The project was conceived and organized by Harold G. Corwin, Jr., and is described on the webpage (http://haroldcorwin.net/ngcic/index.html).

Planetary nebulae and globular clusters are highlighted in the descriptions with special symbols when recognized as such. Galaxies and diffuse nebulae are not distinguished in the descriptions. Diffuse nebulae are real interstellar clouds, like the Orion Nebula (M42 or NGC 1976) and the Great Looped Nebula (NGC 2070) in the Large Magellanic Cloud. Visually, diffuse nebulae lack the regularity of typical white nebulae. Some diffuse nebulae are complex clouds of ionized hydrogen gas while others are dust clouds that are near one or more hot stars and reflect some of the light of these stars. The latter are known as reflection nebulae. The classification of some nebulae as galaxies had to wait another 30 years or so after the publication of the NGC, although exceptional spirals (such as M51) were recognized in the catalogue.

Figure 2 also shows the extensive cross-indexing the NGC has. In addition to the corresponding GC numbers, Dreyer included cross-indexing to John Herschel's 1833 and 1847 catalogues, William Herschel's various object lists (symbolized by a roman numeral followed by a number), and the names and numbers of objects in other observers' lists (for example, the Messier catalogue). The remaining columns in the catalogue give the location of the object on the sky. The *right ascension* is a sky coordinate analogous to longitude on Earth; it increases eastward from the direction of the vernal equinox[6] and is often specified in time units (hours, minutes, seconds). The North Polar Distance is the angle of the object relative to the north celestial pole. This type of coordinate is no longer used; instead, modern catalogues specify this coordinate as the *declination*, or angle (in degrees, arcminutes, and arcseconds) north or south of the celestial equator. This makes declination analogous to latitude on Earth.

In astronomy, coordinates like these always have a date attached because the Earth's rotation axis wobbles over time. This wobble, known as *precession*, continually changes the direction of the north celestial pole, which as a result changes the right ascension and

---

[6]The point on the celestial sphere where the Sun crosses the celestial equator from south to north.

declination coordinates of all celestial objects. Fortunately, precession is fairly slow. The coordinates listed in the NGC are for the year 1860, which was already long past by the time the catalogue was published. For this reason, the catalogue provides in columns 7 and 9 the yearly precessional correction that would have to be made to get the actual coordinates for any other year.

Dreyer's interest in keeping track of nebular discoveries did not end with the publication of the NGC. New nebular discoveries continued to be made, although most post-NGC finds involved rather faint objects. The first Index Catalogue of Nebulae and Clusters of Stars was published in 1895 and comprised IC numbers from 1 to 1529. It was followed in 1908 by the Second Index Catalogue, which comprised IC numbers from 1530 to 5365. Although the first IC was still based largely on visual discoveries, the second IC included a significant number of objects discovered on photographic plates. Because even these early plates could reveal some details of the morphology of faint nebulae, Dreyer notes in the introduction to the second IC that he added the term "spir" to the visual descriptions of objects to indicate that spiral structure was seen in the photographic image.

3.1. End of an Era

It was just around the time of the publication of the index catalogues that the era of visual discovery of nebulae largely came to an end. No large catalogues of visually-discovered nebulae and clusters would be published again. Photography became the prime method for both finding and studying nebulae. Silver bromide dry-emulsion plates started being used in the 1870s, and even though this was well within the era of the Leviathan, the mechanical requirements for astrophotography demanded a different kind of telescope, one that could automatically stay directed at the same point in the sky for hours at a time. This is called "tracking," which was something the Leviathan could not do accurately enough. Tracking was needed because the sensitivity of photographic plates was so low, and the brightness of celestial objects so dim, that "snapshots" were impractical, especially in the 19th century. As effective as photographic plates were for recording information, only about 1% of the light falling on a plate could be captured, meaning that to get a good quality photograph of something like a nebula required long exposures. It was essential to have a way of keeping a telescope pointed in a given direction. If accurate, images of stars would come out as round dark spots on a developed plate. More typically, however, tracking errors would make star images into slightly elongated spots. Image quality could be improved by attaching a "guidescope" to the main telescope, in order to allow an observer to correct for tracking errors due to imperfections in mountings and tracking motors, whether mechanical or (eventually) electrical.

Because so many of the objects in the New General and Index Catalogues are described as being rather faint, in the years after the catalogues' publication there was a great deal of mystique about what the catalogues contained. One might wonder: what are these objects, and who could ever see them? Even by the 1970s, very little was known about most NGC objects, and much less was known about most IC objects.

Photography added two major new dimensions to nebular studies. First, photography could reveal details in a long exposure that would be difficult or impossible to see visually with the same telescope. Second, photography could reveal large numbers of new nebulae that either were missed by visual observers, or were too faint to be easily seen with any visual telescope.

In 1888, Isaac Roberts (1829–1904) obtained the first photograph of the Andromeda Nebula, taken with a 20-inch telescope in his private observatory. This small telescope had two advantages over the Herschel and Leviathan telescopes: the glass mirror used a highly reflective silver coating, and the tube was mounted equatorially in order to reliably track. An equatorial mounting is a type where the pointing of the telescope is aligned with the rotation of the Earth. The telescope is adjusted to the observer's latitude, and the right ascension and declination coordinate system, through polar axis alignment, which sets the direction of the north celestial pole. The celestial equator is at an angle of 90° to the pole. Positive declinations are in the northern celestial hemisphere and negative declinations are in the southern celestial hemisphere. With an equatorial mounting,

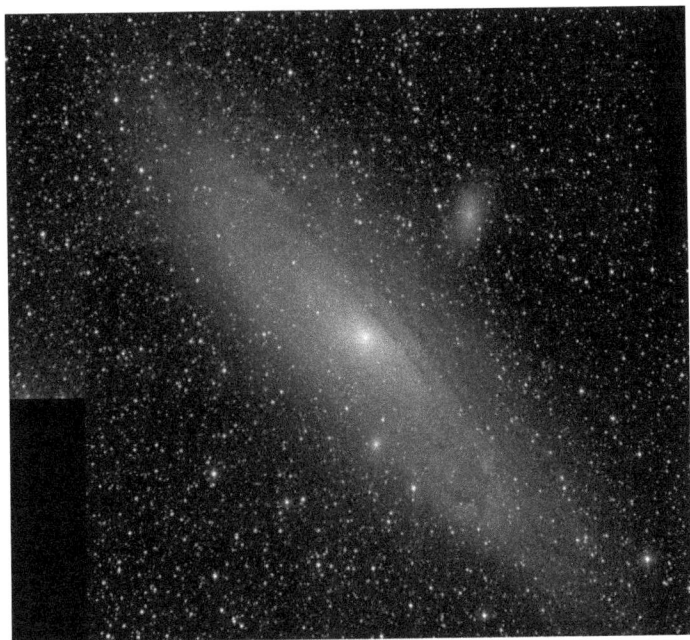

Fig. 4.   A modern black and white image of the Andromeda Galaxy, M31, and its two small companions M32 (below center) and NGC 205 (to upper right of center). The main aspect of the disk structure of M31 that can be seen visually is the innermost of the two sharp dust lanes on the west (right) side. M31 is inclined 78° to the line of sight (that is, close to edge-on), which accounts for some of the difficulty in seeing its spiral structure. Image source: The de Vaucouleurs Atlas of Galaxies published by Cambridge University Press, 2007; reprinted with permission.

a tracking mechanism only had to be applied to the right ascension coordinate and not the declination coordinate.

The silver coating improved the reflectivity of the 20-inch telescope's mirror over speculum metal mirrors, while the tracking capability facilitated long exposure guided photographs. Roberts' photograph of the Andromeda Nebula revealed the spiral structure of the galaxy for the first time, even though Andromeda is the visually largest and brightest white nebula in the northern sky. This characteristic of the Andromeda Nebula eluded early visual observations in part because the galaxy is tilted towards an edge-on view, and because the structure is rather faint (Figure 4). Roberts

Fig. 5.   A modern black and white image of M33, the Great Spiral in Triangulum. This one is more face-on than the Andromeda Galaxy, but its brightness per unit area is low. The spiral arms of M33 can be seen with a 12.5-inch telescope on dark, moonless nights, but the slightest light pollution can make the galaxy very difficult to see. Image source: The de Vaucouleurs Atlas of Galaxies published by Cambridge University Press, 2007; reprinted with permission.

compiled a collection of his photographs in a two-volume book set.[7] One of these includes the Triangulum Spiral M33, a modern image of which is shown in Figure 5.

3.2. Nebulae: Ad Infinitum

In 1898, James Keeler (1857–1900) carried out one of the first photographic surveys of bright nebulae using the 36-inch Crossley reflecting telescope of the Lick Observatory in California. The telescope[8] was especially effective for the task. It was mounted equatorially, and photographic plates were initially exposed at the Newtonian focus. Tracking was accomplished using a weight-driven

---

[7]I. Roberts, *A Selection of Photographs of Stars, Star-Clusters, and Nebulae* (The Universal Press, London, 1899).

[8]J. E. Keeler, *Publications of the Lick Observatory*, **8**, 11 (1908).

"clock drive" that had to be periodically rewound. Photographs were taken using a double-slide plate holder, a type of plate holder that allowed small adjustments in the positioning of the plate in two perpendicular directions.[9] Guiding (i.e., correcting for errors due to tracking and other effects) was done using a high magnification eyepiece with illuminated crosshairs at one edge of the field. After selecting a bright "guide star," an observer could guide by adjusting the position of the plate using a pair of set screws. This was only one approach to guiding. Another popular approach, as already noted, was to attach a "guidescope" parallel to the main telescope's optical axis. Like most telescopes built after 1880, the primary and secondary mirrors used a metal coating on glass for maximum reflectivity.

Keeler's program involved getting plates of 104 objects selected mostly from the NGC. Unfortunately, he died after observing only 68 (2/3) of the sample and the rest of the program was completed by an assistant astronomer, Charles D. Perrine (1867–1951). From the 68 objects Keeler did observe, he noticed that while spirals appeared to be relatively rare when recognized only by eye, in photographs many more nebulae were found to be spiral, to the extent that non-spiral nebulae became the exception rather than the rule. This caused Keeler to consider nebulae that did not reveal any spiral form as being of "greater interest."[10] He also noticed that spirals had to be flattened disk-shaped objects because his photographs showed how the appearance of spirals changed as the tilt changed from face-on to edge-on. This is shown with modern photographs by the four galaxies in Figure 6. Spirals are most clearly seen when the tilt is low, as in NGC 628 (M74). This is true also of the second galaxy, NGC 3370, even though it is inclined nearly 55°. The third galaxy, NGC 3877, is inclined almost 80°, and while it is still recognizable as a spiral, the arms are less distinct and the planar dust lane is starting to show. Finally, the last galaxy in Figure 6, NGC 5907, is inclined almost 90°, and is sufficiently edge-on that its spiral structure is not detectable at all. This relates directly to the appearance of the Milky

---

[9] A. A. Common, *Monthly Notices of the Royal Astronomical Society*, **49**, 297 (1889).
[10] J. E. Keeler, *Astronomische Nachrichten*, **151**, 1 (1899).

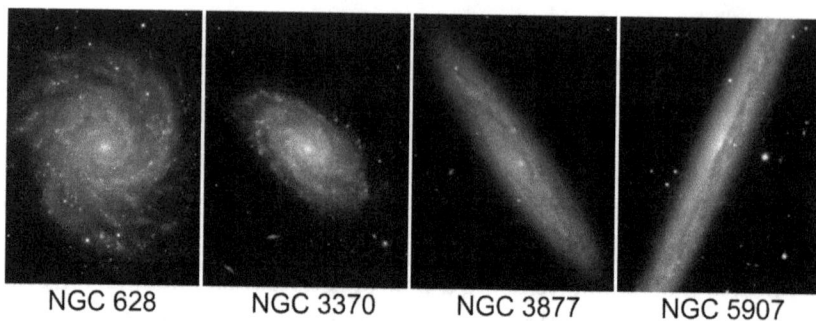

NGC 628   NGC 3370   NGC 3877   NGC 5907

Fig. 6. Four galaxies showing how the appearance of a spiral galaxy depends on how much it is inclined to our line of sight. The inclination increases from left to right. Images source: the de Vaucouleurs Atlas of Galaxies published by Cambridge University Press, 2007; reprinted with permission.

Way: our perspective on the Milky Way is an edge-on view, and thus we do not see the spiral arms of our Galaxy directly.

Keeler's most important discovery was the large number of previously uncatalogued nebulae that appeared on most any plate he took with the Crossley reflector, especially away from the band of the Milky Way. Just as for the brighter NGC nebulae, most of these fainter cases were spirals. Based on what his plates revealed within the small areas they covered, he deduced that the number of nebulae extrapolated over the whole sky must be huge, on the order of 120,000.[11]

Keeler's work with the Lick Crossley telescope was followed up by a very large survey with the same telescope by Heber D. Curtis (1872–1942). Curtis published descriptions of 762 nebulae and clusters as seen on Crossley plates, including 513 spiral nebulae.[12] This turned out to be a voyage of discovery for Curtis, for several reasons.

First, the survey provided a wealth of information on nebular morphology. Not only were some nebulae noted to be spiral, but subtle distinctions among these nebulae were also noted. Curtis called some nebulae "phi-type" (or ϕ-type) spirals, because these bore a

---

[11] J. E. Keeler, *Monthly Notices of the Royal Astronomical Society*, **60**, 128 (1899).
[12] H. D. Curtis, *Publications of the Lick Observatory*, **13**, 9 (1918).

Fig. 7. Blue-green (*g*-band) image of M95 (NGC 3351), located 30 million light years away in Leo. M95 is an original Curtis ϕ-type spiral, but with its bar and ring, it resembles the Greek letter θ more. Image source: the de Vaucouleurs Atlas of Galaxies published by Cambridge University Press, 2007; reprinted with permission.

resemblance to the Greek letter ϕ. An example is the bright spiral M95 (NGC 3351) in Leo (Figure 7). Other examples are NGC 5921 (Figure 7 of Chapter 8) and NGC 1300 (Figure 1 of Chapter 15). Of these, NGC 3351 and 5921 show a bar crossing a ring. It appears Curtis saw the impression of a bar crossing a ring as being like the

Greek letter ϕ. The only problem is that the bar in general does not overfill the ring in the manner of an actual ϕ. In this sense, the morphology is more like the Greek letter "theta" (θ) than it is ϕ.

The fact that spirals were so common in the sample also led Curtis to make an ultimately incorrect claim: that any nebula that did not show spiral structure on his plates would show such structure if photographed with better equipment. That is, he did not believe there was such a thing as a genuine non-spiral nebula.

Second, just as Keeler did with his plates, Curtis counted nebulae on his much more extensive set of plates and concluded that the number of nebulae over the whole sky must be on the order of a million, much more than Keeler had estimated. Even after correcting for a possible bias in his sample towards north Galactic pole fields, he still estimated that there must be more than 700,000 nebulae.

Third, just as for Keeler's work, Curtis's sample included spirals over a wide range of tilt angles, from roundish-looking face-on cases to ray-like edge-on cases. He noticed that spirals of higher tilt would often show a dark band either through their midsection or dark lanes on one side of the major axis (Figure 6). This band would show irregular structure, and appeared to be made of clouds of obscuring matter. In the edge-on view, the appearance is like the Milky Way with its absorbing clouds of interstellar dust. It was known since the time of the Herschels that most nebulae avoided the band of the Milky Way. This suggested to Curtis that the Milky Way would look very similar to the ray-like nebulae if it could be viewed from a great distance.[13]

Lastly, in 1917 Curtis noticed the appearance of a nova in the nebula known as NGC 4527, on a plate taken in 1915.[14] In astronomy, "nova" means "new star," a few dozen of which had been discovered in our Galaxy by 1917. Galactic novae would be discovered as a bright star that suddenly appeared and would subsequently fade less rapidly. They would often be discovered as naked eye stars since these were most noticeable.

---

[13]H. D. Curtis, *Publications of the Lick Observatory*, **13**, (2), 45 (1918).
[14]H. D. Curtis, *Lick Observatory Bulletin*, **9**, 108 (1917).

At nearly the same time as Curtis discovered his nova in NGC 4527, G. W. Ritchey (1864–1945) discovered a nova in the nearby spiral NGC 6946 with the 1.5-m telescope of Mt. Wilson Observatory. These discoveries led Curtis to examine past plates for similar phenomena, a few more examples being found. The discoveries also had a significant impact on Curtis's view of the nature of the nebulae, and he became well-known as an advocate for the "island universe" interpretation of the nebulae. This was because the novae he found, which were six total by 1917, were much fainter than a typical Galactic nova. Curtis deduced that if the novae he found were the same as those seen in the Milky Way, then they would have to be on the order of 100 times further away than typical Galactic novae.[15]

The photographic discovery of novae in spirals set the stage for establishing the existence of galaxies. In 1919, Swedish astronomer Knut Lundmark (1889–1958) analyzed the novae discovered in the Andromeda Galaxy and obtained a distance to the galaxy of 650,000 light years,[16] thus already proving that Andromeda was an external galaxy (that is, a system like the Milky Way but outside its boundaries) before Hubble's famous work on Cepheid variables did the same a few years later.

Figure 8 shows how a typical nova in spiral nebulae would have looked to Ritchey and Curtis. A picture of a spiral nebula taken one night is compared with a picture of the same object taken on an earlier night. The later picture shows a star that was not present on the earlier picture. If the new star is followed over a period of time, it will generally appear to fade until it disappears altogether. The aspect that was most puzzling was why these novae were so bright in nebulae that seemed faint compared to the Andromeda spiral. There was another scale of nova that appeared in spirals too far away for conventional galactic novae to appear in. These became known as *supernovae*, and are discussed further in Chapter 13.

---

[15]H. D. Curtis, *Publications of the Astronomical Society of the Pacific*, **29**, 206 (1917).
[16]K. Lundmark, *Astronomische Nachrichten*, **209**, 369 (1919).

Fig. 8.   Images of the spiral galaxy NGC 6946 revealed a supernova in 2017, named SN 2017eaw, the 10th supernova discovered in that galaxy since G. W. Ritchey discovered one in 1917. In Ritchey's time, an object like SN 2017eaw would have been called a nova. Images source: W. C. Keel and R. Buta.

3.3.   Cataloguing Nebulae from Photographic Plates

The first half of the 20th century brought long lists of new nebulae discovered photographically. Harvard College Observatory was one of the leaders in the early compilations of such lists, based on photographs taken with the Bruce 24-inch refractor[17] at Arequipa, Peru. An example of one of these lists was "A Catalogue of 2778 Nebulae, including the Coma-Virgo Group," by Adelaide Ames (1900–1932) and published in 1930 as Harvard Annals, Vol. 88, No. 1. In Ames's catalogue, 622 nebulae were previously catalogued in the part of the sky she studied, while 2156 were newly identified.

---

[17]The Bruce refractor was funded by and named after Catherine Wolfe Bruce (1816–1900), an American philanthropist and significant benefactor of astronomy. The refractor was designed to be an astrograph, intended solely for photography and not visual observing. Bruce funded other telescopes, and is the namesake of the Bruce Gold Medal, which is regularly awarded by the Astronomical Society of the Pacific to an astronomer with a lifetime of outstanding achievement.

Photography with the Bruce refractor provided a number of advantages over visual cataloguing of nebulae:

(1) The telescope was designed to have a large photographic field of view, allowing large numbers of nebulae to be seen on a single plate.

(2) The telescope was located at a dark observing site, which facilitated long (4–5 h) exposures. Long exposures at a dark site could reveal many faint nebulae that would be hard to detect by eye with the same telescope.

(3) The right ascension and declination coordinates of nebulae could be more accurately measured using plates rather than by eye. The wide field of view generally allowed a few stars of known coordinates to be present on the plates. The coordinates of nebulae could then be determined relative to these stars.

(4) Details in some nebulae could be seen in photographs that would not be visible easily by eye. For example, photographs revealed that spirals were much more commonly present among nebulae than could be detected by eye.

(5) For the first time, photographs allowed the possibility of measuring magnitudes, angular sizes, and mean surface brightnesses[18] of nebulae, allowing nebular catalogues to take on an astrophysical character. Consistent classification of nebulae was also made possible by photography.

Another major list of nebulae was produced by Anders Reiz (1915–2000), who used plates taken with the Bruce double 16-inch astrograph telescope at the Observatory of the University of Heidelberg.[19] In a detailed 1941 study,[20] Reiz compiled information

---

[18]Surface brightness is apparent brightness per unit area and is often expressed in terms of magnitudes per square arcsecond. See Chapter 7 for further discussion of this important characteristic of galaxies.

[19]A double astrograph is a binocular-like pair of refracting telescopes designed to photographically survey even larger areas of sky than could be achieved with a single astrograph. Like the Harvard telescope, the Heidelberg double astrograph was funded by Catherine Wolfe Bruce.

[20]A. Reiz, "A Study of External Galaxies with Special Regard to the Distribution Problem," *Annals of the Lund Observatory*, **9** (1941).

on 4666 mostly new galaxies identified on Heidelberg Bruce plates for the purpose of studying their distribution. Like Ames, Reiz emphasized the importance of measuring accurate magnitudes for an astrophysical study. He argued that "For a thorough exploration of the metagalactic system, its structural and spatial arrangement, we must have access to accurate magnitudes." Studies like those of Ames and Reiz laid the groundwork for much astrophysical research to follow. The derivation of galaxy magnitudes is discussed further in Chapter 7.

Chapter 4

From Nebulae to Galaxies

In the late 19th century, the idea that all nebulae were distant star systems too far away to be resolved took a blow from the development of spectroscopy. Many bright nebulae, when examined spectroscopically, revealed bright emission lines, indicating that they were nothing more than glowing clouds of interstellar gas that were likely part of the Milky Way system. All nebulae clearly were not distant star systems too remote for the individual stars to be detected. Some saw this to mean that there was only one Universe, the Milky Way, and everything was a part of that universe, including the spirals.

The chief proponent of the one-universe idea was Harlow Shapley, who had made what he thought were the most reliable measurements of the size of the Milky Way. The basis for Shapley's view was his use of variable stars in globular clusters and other techniques to judge the actual extent of the Milky Way. He believed the Milky Way to be 10 times larger than had been previously deduced, with the globular clusters extending out far enough to make the Galaxy about 300,000 light years across. This was so huge that Shapley felt that the spiral nebulae had to be part of the Galactic system. If they were simply distant versions of the Milky Way, then to appear as faint and small as they do, they would have to be enormously far away. Shapley at that time may not have been prepared to fathom the idea of a universe extending millions of light years away from us.

Another issue that affected Shapley's view was the results of rotation studies of spirals. All astronomical objects are moving, and that movement can be broken into a component along the line of sight (the radial velocity) and a component perpendicular to the line of sight (the tangential velocity). The radial velocity is only measureable using spectroscopy through the Doppler effect, the shift in the wavelengths of spectral absorption or emission lines due to the relative motion of the light source along the line of sight.

In 1916, Francis G. Pease (1881–1938) used the Mount Wilson Observatory 60-inch telescope to detect, through the Doppler effect, the rotation of the "Sombrero" Nebula.[1] This object, also known as M104, is a very bright galaxy lying 40 million light years away in the constellation Virgo. As can be seen in Figure 1, the Sombrero is a disk-shaped system tilted almost edge-on to the line of sight. The disk is embedded within a bright, rounder-shaped bulge that together give the object the shape for which it is named. The nearly edge-on orientation of the disk is ideal for detecting rotation because the rotation speed along the major (long) axis projects almost entirely

Fig. 1.   The Sombrero Galaxy M104 (NGC 4594) in Virgo, based on multi-filter imaging with the Hubble Space Telescope. Image credit: NASA/ESA and the Hubble Heritage Team (STScI/AURA).

---

[1] F. G. Pease, Publications of the *Astronomical Society of the Pacific*, **28**, 191 (1916).

into the line of sight. Rotation of the highly-flattened, disk-shaped part of the Sombrero manifested itself in the slanting of absorption lines along the major axis. The slanting occurs because one side of the galaxy moves away from us while the other side approaches us. It took Pease $3\frac{1}{3}$ days worth of total exposure time of a single photographic plate to detect the slanted lines well enough to measure the rotation of the disk of the galaxy. The amount of Doppler wavelength shifting is directly proportional to the line of sight speed. At the farthest detectable point along the major axis of the Sombrero Galaxy (2 arcminutes from the center), Pease derived a rotation speed of 330 km/s.

The tangential velocity of a distant celestial object can be detected in a different way. Provided the object is not too far away, tangential motion can be manifested as *proper motion*, or angular movement along the sky. Spectroscopic observations led to the obvious question: if spirals are rotating, and if they are relatively nearby, could this rotation be detected directly in images of face-on spirals, for which the angular movement expected would be most favorably seen? Or if not pure rotation, could another type of motion be detected directly on plates? If all spirals rotate with a comparable speed to the Sombrero, and if all spirals are within a few thousand light years distance, then plates taken several years apart might show evidence of this motion.

Mount Wilson astronomer Adrian van Maanen (1884–1946) realized that searching for such movement would be a test of the "island-universe theory," and being an expert in the measurement of proper motion, he undertook an investigation of several of the brightest nearly face-on spirals in the sky: M33, M51, M63, M81, M94, M101, and NGC 2403. For each object, he had at least two plates of comparable quality taken 8–12 years apart. The change in position of objects in the spiral arms relative to local field stars is what gave the proper motions, if any. In every case, van Maanen's measurements pointed not only to the real detection of movement of discrete objects in the arms of these galaxies, he also concluded that these motions were outward along the spiral arms, not just pure rotation. If correct, the spirals had to be nearby because if they were at distances of millions rather than thousands of light years, the implied speeds would have been close to the speed of light, which was unlikely. In the 10th and final paper in a series that ran for 7 years, van Maanen concluded: "All this material seems to point to [distances] for the larger spiral nebulae between a few [thousand] and a few [hundred light years]. With such values the diameters of the spirals range from a few light years to several hundred light years. Since our present estimates of the Milky Way system vary, according to different authorities, from 20,000 to 300,000 light years, it is clear that the present material indicates that the spirals, while enormous in size as compared with our solar system, are not at all comparable with the Milky Way system."[2]

Nevertheless, as thorough and conclusive as van Maanen's results seemed to be, his measurements of the rotation of spirals were flawed. The spirals he observed were indeed rotating, but the rotation was too slow to be detectable on timescales of mere years. Even centuries likely would not have been enough to conclusively detect this motion, at least with the equipment van Maanan had to use. The problem was that he was trying to measure something that was far below his ability to detect with the plate material and measuring techniques of his time, and therefore his results were prone to significant systematic errors.

Novae figured more into resolving the issue because several examples had been well-observed in the Milky Way. By today's understanding, novae represent a transient phenomenon involving mass transfer onto a white dwarf[3] whose increase in luminosity results from the explosive fusion of hydrogen on its surface. For a few weeks to months, a nova can outshine most of the individual stars in a galaxy, the object's light following a characteristic rapid rise to maximum brightness followed by a slower decline. Novae occur probably in all spirals, and indeed photographic observations of bright spirals like M33 in Triangulum and M31 in Andromeda

---

[2]A. van Maanen, *Astrophysical Journal*, **57**, 264 (1923).
[3]White dwarfs are dead, sun-like stars that, at some point in their evolution, ejected their outer layers of gas and exposed their hot core.

revealed transient stars having light curves similar to Galactic novae. However, in 1885, a nova appeared close to the center of M31 that was much brighter than the typical novae seen in that galaxy. This confused the whole nova idea. In evaluating the significance of novae to resolving the "island universe" idea, Curtis wisely rejected Nova 1885 as being something different and excluded it from his analysis.

The key to establishing the existence of galaxies was to not only resolve the spirals into stars, but also to resolve them into recognizable types of stars, such as those found in the Milky Way. In the 1920s, the Mount Wilson 100-inch Hooker Telescope was the largest functioning telescope in the world and the only one that could do this well. Its construction in 1917 ended the reign of the Leviathan of Parsonstown as the world's largest telescope. With a highly reflective silver-coated glass mirror, tracking, and a capability for guided long exposures of photographic plates, it became the principal tool for nebular research a century ago. This was a telescope that could definitively resolve the nearest white nebulae, the Andromeda and Triangulum spirals, into stars.

The most resolvable star types in any galaxy will tend to be those of high luminosity. Studies of star colors and brightnesses, together with spectroscopy and distance estimates, have shown that the most luminous stars in the Milky Way tend to be the largest and often the hottest stars (Chapter 10). With knowledge of the luminosities of such stars, identifying them in another galaxy would allow the distance of the galaxy to be determined.

The most accurate way of deriving the distance to any star is trigonometric parallax, the small, periodic shift in position of a nearby star relative to background stars due to the Earth's orbit around the Sun. The background stars are also shifting in position, but are assumed to be so far away that their shifts are very small. The method is such that, in Curtis's time, it became mostly inaccurate beyond 50–100 parsecs.[4] This would not have been a problem had

---

[4]A parsec is a unit of distance. An object 1 parsec away would show a trigonometric parallax of 1 arcsec, which also corresponds to 3.26 light years. A kiloparsec is 1,000 parsecs, or 3,260 light years. A megaparsec is a million parsecs, or 3.26 million light years.

the kinds of luminous stars that could be seen in other galaxies been abundantly found within 50–100 parsecs distance from the Sun. Then we would accurately know their luminosities. Instead, the kinds of luminous stars seen in other galaxies are very rare and almost none are within 100 parsecs of the Sun. A variety of methods were developed to get distances and luminosities of these rare luminaries, one involving the "bootstrapping" of the brightnesses of stars in a distant star cluster with the brightnesses of similar stars in a nearby cluster of known distance.

4.1. "Breathing" Stars

One easily recognizable type of star that was resolvable in nearby galaxies was Cepheid variable stars. These interesting stars, which vary in apparent brightness in a well-defined, periodic way, were critical to proving that the white nebulae are galaxies. Cepheids are radially pulsating stars, meaning they vary in size. The variation in size causes the surface temperature to change, which we see as a variation in apparent brightness. The brightness changes are today understood to be caused by variations in the opaqueness (transparency to light) of the gases deep inside the stars. This would all be academic were it not for the fact that Cepheid variable stars have a well-defined period-luminosity relation, in the sense that the longer the period of the variation, the more luminous the star. The Cepheids with the longest periods can be tens of thousands of times as luminous as the Sun. As a consequence, they can be seen to great distances, well beyond the Milky Way.

Although the first Cepheid variable star was discovered in the 1780s, the Cepheid period–luminosity relation was not recognized until the early 20th century. The catalyst was a photographic search for variable stars in the Small Magellanic Cloud (SMC), a "cloud" of faint stars in the far southern sky now recognized to be a nearby dwarf companion galaxy to the Milky Way (Figure 2). Beginning in 1896, the 24-inch Bruce refractor of the Harvard College Observatory, Boyden Station, Arequipa, Peru, was being used to take plates of the

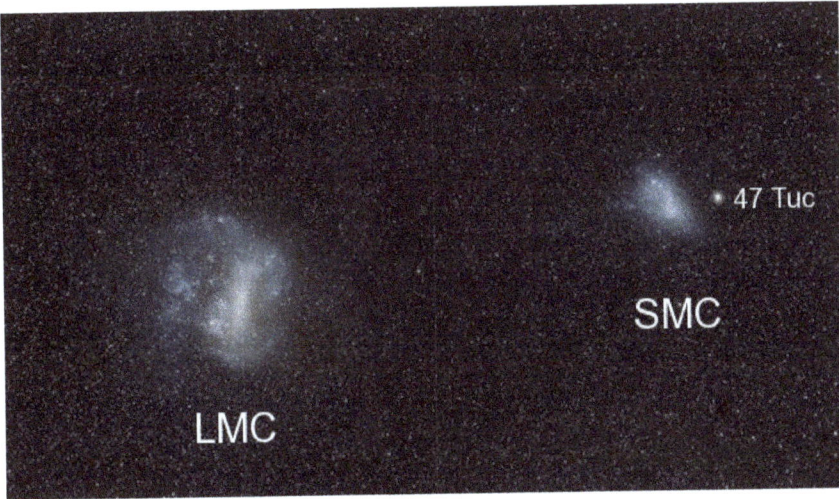

Fig. 2. The Large and Small Magellanic Clouds (LMC and SMC, respectively), two of the nearest external galaxies. The LMC is 158,200 light years away in the constellation Dorado, the Swordfish. The SMC is 199,000 light years away in the constellation Tucana, the Toucan Bird. Both galaxies are too far south to be seen from most of the US. The image also includes the globular cluster 47 Tucanae, which appears as a fuzzy star to the naked eye. Image source: Andrew Lockwood, Mount Pleasant Observatory, Western Australia, with permission.

SMC over extended periods, for the purpose of identifying variable stars of any kinds. Astronomy "computer" Henrietta S. Leavitt was assigned by her boss, Edward C. Pickering, to search for previously unidentified variable stars on these plates. The key to identifying variables efficiently on such plates was to make a glass positive of one plate, and then superpose other plates on top of this positive. If no variable stars are in the field, then the glass positive would cancel the light of all the stars, but if some are variable, they would be immediately recognized since they would not be canceled out by the positive plate.

In Harvard College Observatory Circulars in 1904 and 1905, Pickering announced the discovery of 900 variables in the SMC, indicating a rich area for follow-up study. He notes in the 1905 circular that the plates were taken in the autumn of 1904, were delivered to Cambridge, MA, in January 1905,[5] and that "an examination of them by Miss Leavitt led to the surprising discovery that hundreds of variable stars were present in this region." Based on the number density of stars on these plates in the central portion of the region, where the SMC is dominant, Pickering deduced that one out of every 300 stars is variable, while in the Galactic stellar region surrounding the SMC, only one out of every 3,300 stars is a variable. Thus, the variables had to be mostly associated with the SMC. He also noted that few of the variables, when at mininum light, became as faint as the barely resolved stellar background of the SMC, implying that the variables tended to be among the brighter stars in the system.

In another paper, Leavitt further discusses the SMC discoveries and also her examinations of plates of the Large Magellanic Cloud, a second "cloud" of faint stars in the far southern sky that is also now recognized to be a nearby companion galaxy to the Milky Way (Figure 2).[6] Both clouds look to the naked eye like detached portions of the Milky Way, and between the two objects, Leavitt reported that 1777 variables had been found. Most of the variables were initially found by comparing only two plates taken a few days apart. This readily identifies variables, but tells little about the nature of the variability. By the time of Leavitt's observations, several different types of variables were known, including Mira (or long period) variables, which tend to be red stars varying by many magnitudes on a time scale of months to years; δ Cephei or Cepheid variables, which tend to be yellowish stars varying by approximately one magnitude on timescales of a few to 100 days; and eclipsing variables, where the stars in a very close binary system periodically (on timescales of hours to days) eclipse each other, causing a brightness change. The light variations of each type of variable star tend to be distinctive, and thus the most important observation that must be established for any variable star is the *light curve*, or graph of the apparent

---

[5]H. S. Leavitt and E. C. Pickering, *Harvard College Observatory Circulars*, **96**, 1 (1905).
[6]H. S. Leavitt, *Annals of Harvard College Observatory*, **60**, 87 (1908).

brightness versus time. If the variations are periodic (that is, repeat after a certain amount of time), then the light curve would also tell the period. To get light curves, Leavitt examined plates taken with the 24-inch Bruce refractor from 1896 to 1906. Each plate represented an observation date or epoch, which was known accurately.

By 1908, Leavitt had derived reliable light curves for 17 SMC variable stars. She noticed that the measured light curves of these variables were of a characteristic form where the star rises quickly from minimum brightness to maximum, then fades from maximum to minimum brightness much more slowly. This behavior was very similar to that of variable stars found in globular clusters and also to known Cepheid variable stars (Figure 3). The most important characteristic she noticed was that "brighter variables have the longer periods." This is easily seen in Figure 4, where Leavitt's data for the 17 variables are plotted as the average of the maximum and minimum brightness versus the logarithm of the period in days (solid circles). Surprisingly, as clearcut as the correlation is, Leavitt did not include a graph of it in her paper, believing it needed more data. In a

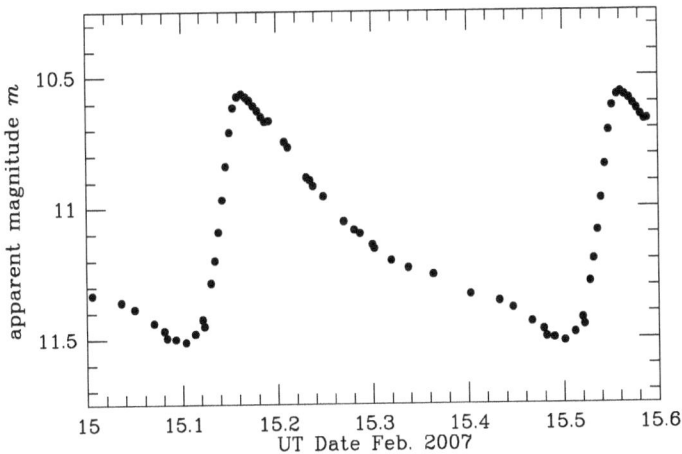

Fig. 3. Light curve of RR Geminorum, a cluster-type radially-pulsating variable star found in the field (RR Lyrae type). The apparent magnitude in this case is the visual magnitude. The UT date is the Universal Time date or Greenwich, England, date. The pulsation period is only 0.4 days, and the rise from minimum to maximum brightness took only 45 min. This light curve was obtained with the University of Alabama 0.4 m reflector. Illustration credit: R. Buta.

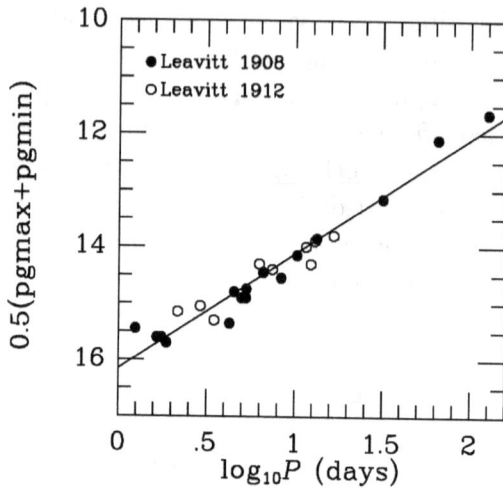

Fig. 4. The Leavitt Law, or Cepheid period–luminosity relation. Illustration based on data from Leavitt (1908) and Leavitt (1912) (see footnotes 6 and 7 respectively). The y-axis is the average of the photographic magnitude at maximum and minimum apparent brightness.

1912 paper,[7] she added eight more variables to the dataset, and these verified her 1908 finding (open circles in Figure 4). In this paper, she plots the correlation between the apparent brightness of SMC variables and the log of the period, because the eight additional variables falling on the relations bolstered her confidence in her result. She noted that her graphs showed a "remarkable relation between the brightness of these variables and the length of their periods." This was humanity's first encounter with the Cepheid period–luminosity relation, now known as Leavitt's Law. It is astounding that Leavitt's first derivation of the law had such a small scatter: the dispersion around the line shown is only 0.23 magnitudes over more than four magnitudes range in brightness.

Why did the SMC variables reveal the Cepheid period–luminosity relation so clearly? The reason is that the SMC, being a separate stellar system from the Milky Way, is far enough away from us that

---

[7]H. S. Leavitt and E. C. Pickering, *Harvard College Observatory Circular*, **173**, 1 (1912).

its three-dimensional structure is small compared to its distance. This means the variables are all being seen from about the same distance, such that differences in apparent brightness translate directly into differences in luminosity.

The Leavitt and Parsons' experiences are perfect examples of the nature of scientific discovery. This nature was captured beautifully in the 1963 television series titled "The Outer Limits," one of the most ground-breaking and influential TV series of the 1960s. In the first episode of the series, which aired in September 1963, a scientist, Alan Maxwell, owns a radio station but is less interested in broadcasting music and commercials and more interested in using the station for research. The episode was set in a time when research in microwaves was a field wide open for discovery. At one point the scientist is using power from the station's antenna to scan the microwave sky, but he uses so much power that the station's normal range is greatly reduced. His brother, Gene "Buddy" Maxwell, who acts as the station's DJ, notices this and asks the scientist's wife, Carol, to go to his lab and get him to raise the station's power, noting that the station's sponsors expect them to reach a certain-sized audience. When Carol confronts Alan about this, he tells her not to worry, the station's bills will be paid. He tells her what he is up to and that he was puzzled about a "static" he had noticed. He says that by scanning the sky in three dimensions, he can give "solidity" to this particular static, which she sarcastically refers to as "solid static." Then she asks him something she feels makes him uncomfortable: what good is the research he is doing? He tells her it is "interesting" and also that it is "important" because research in microwaves the past 30 years led to important discoveries. Then she asks: "What makes you think you can discover something? Who are you? He says: "Nobody. Nobody at all. But the secrets of the Universe don't mind. They reveal themselves to nobodies who care."

Henrietta Leavitt's discovery took place at a time when the photographic sky was wide open for discovery. The 24-inch Bruce refractor was a powerful tool for photographic research a century ago, and being located in the southern hemisphere, the SMC was a logical photographic target. Leavitt was only one of many women that Edward Pickering hired at the time to do the tedious aspects of research astronomy.[8]

4.2. A "Great Debate"

By 1920, little progress had been made in resolving the issue of whether spiral nebulae really are galaxies external to the Milky Way, or were local phenomena, like solar systems in formation, a popular alternate idea at the time. The stalemate came to a head in May 1920, when Curtis and Shapley met for the "Great Debate" about the scale of the universe.[9] Curtis had some powerful arguments in favor of the "island universe" idea: the faintness of the novae seen in spirals compared to those seen in the Milky Way; the fact that the Milky Way appeared to be highly flattened as did many edge-on spiral nebulae; that highly inclined spirals resembled the Milky Way, with dark lanes or clouds of absorbing material in their disk planes (later determined to be interstellar dust); and that no spiral nebulae appeared towards the band of the Milky Way, as if this absorbing material was blocking their view. If the spirals were actually part of the Milky Way, they would be avoiding being seen in the main part of the Milky Way where most of the Galaxy's stars are found.

In contrast, Shapley focussed on his work on the size of the Milky Way. He outlined his judgment of this size using globular star clusters having "cluster variables" (for example, Figure 5),[10] which are radially-pulsating variables with periods less than a day. Globular clusters can be used for this purpose because they are resolvable major systems of stars that can be seen out to great distances. All 150 of the globular clusters associated with the Milky Way are

---

[8]For the interesting story behind this, see Dava Sobel's *The Glass Universe: How the Ladies of the Harvard Observatory Took the Measure of the Stars* (New York, Viking, 2016).

[9]H. Shapley and H. D. Curtis, *The Scale of the Universe*, Bulletin of the National Research Council, **2**, 171 (1921).

[10]A term due to Harvard College astronomer Solon I. Bailey (1854–1931), who made some of the first studies of variable stars in globular clusters, especially finding that variables with periods less than a day are much more common in globular clusters than in the field.

resolvable with large modern telescopes, but none is near enough to have its distance directly determined by trigonometric parallax. Shapley noted that "The parallaxes of stellar systems that are too remote for direct trigonometrical measurement are best determined from the luminosities of the stars they contain." The key to getting distances to globular clusters, just as for spiral nebulae, is to identify the kinds of stars they contain and to determine the luminosities of stars that might act as "standard candles." A standard candle is a type of star (like a Cepheid) that has a known luminosity with a relatively small dispersion (or scatter). Once identified, a standard candle becomes a powerful indicator of distance. Also, the more luminous the standard candle, the greater the distance from which it can be seen. To determine the scale of the universe, it is essential to identify standard candles that can be used to as great a distance as possible.

Just as for the spiral nebulae, the most resolvable stars in globular clusters are their most luminous stars, which in this case are *red giants* (Figures 5 and 6). A red giant is a former sun-like star that, through the normal course of stellar evolution, used up all of its hydrogen fuel in the center, causing the star to expand drastically in size and to a cooler temperature on its surface. Some globular clusters have hundreds to thousands of such stars, as in Omega Centauri (Figure 5), each of which can be more than a thousand times the luminosity of the Sun. Thus, red giants can be seen from great distances. Red giants are also not only found in globular clusters. They are also found in any old, general field galactic stellar population.

Recent studies have shown that red giants can be used effectively as extragalactic distance indicators (Chapter 23), but a century ago steady stars like these were not necessarily the best choice for getting distances to globular clusters, at least with the plate material available at that time. In Shapley's day, cluster variable stars were found to work better. In the wake of Leavitt's monumental discovery, which clearly offered a way to get distances to extremely remote stellar systems, astronomers attempted to "calibrate" the relation, meaning they wanted to know what the actual luminosity a Cepheid

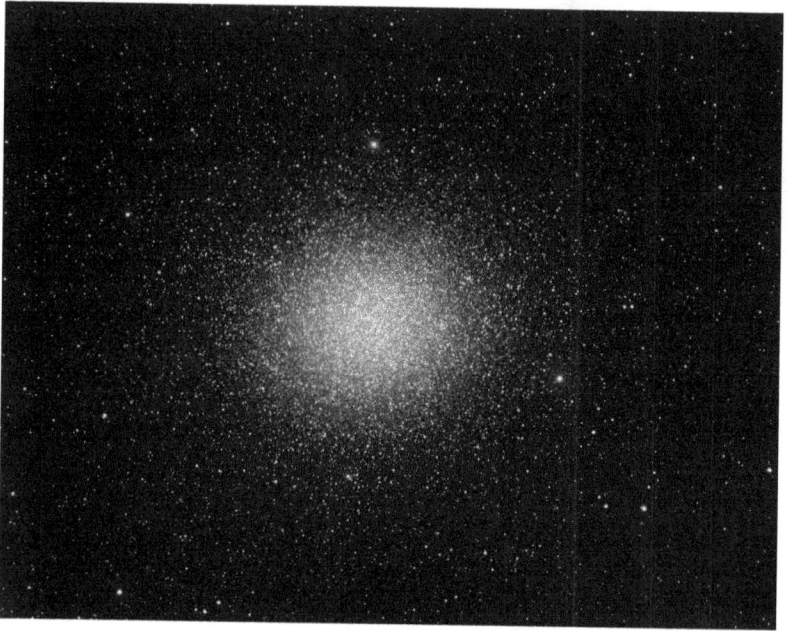

Fig. 5.   The globular cluster Omega Centauri, located 15,800 light years away, has more than a hundred "cluster variables" and likely also more than a thousand red giants. Image source: Andrew Lockwood, Mount Pleasant Observatory, Western Australia, with permission.

with a given period had. The most effective way of representing the luminosity was to use absolute magnitudes. The absolute magnitude of any star is the apparent magnitude it would have if it could be seen from a standard distance of 10 parsecs, or 32.6 light years. Thus, if you know the distance to a star, you can imagine moving it from its actual distance to the standard distance of 10 parsecs and asking how bright it would appear. When viewed in absolute magnitudes, differences in brightness translate into differences in luminosity.

The way to get the absolute calibration of the P–L relation is to derive distances to nearby Galactic Cepheids of different periods. Galactic Cepheids are part of our Galaxy and generally much closer to us than those seen in distant stellar systems, like globular clusters, but are also generally too far away for direct trigonometric parallaxes to be measured. Statistical methods can be used for

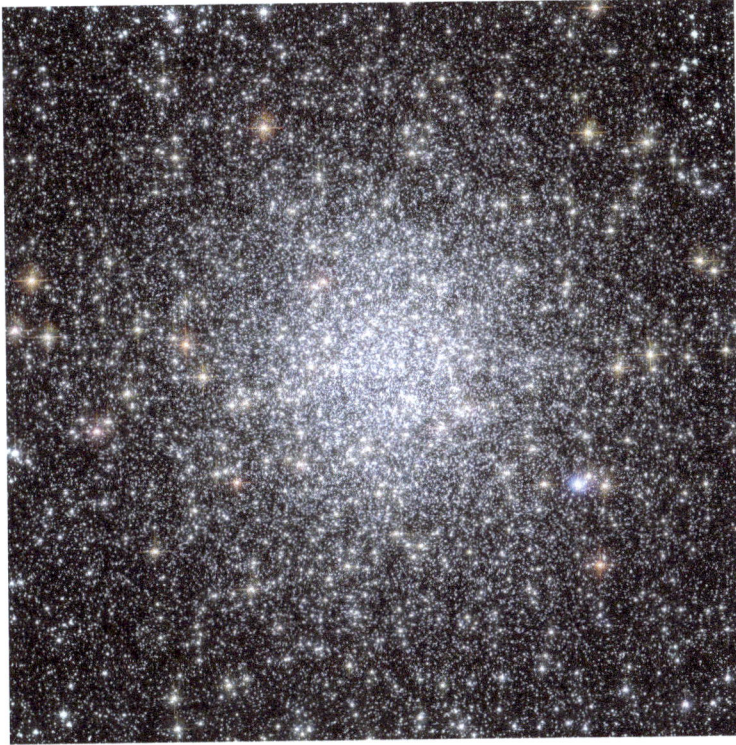

Fig. 6. The globular cluster 47 Tucanae, located 13,000 light years away in the constellation Tucana, the Toucan Bird. The reddish-tinted stars are red giants, and the picture shows that these are among the brightest stars in the cluster. Image source: NASA/ESA and the Hubble Heritage (STScI/AURA)-ESA/Hubble Collaboration.

Galactic Cepheids, and these led to Shapley's adopted calibrated P–L relation shown in Figure 7.[11] Shapley's relation is more extensive than what Leavitt found because it covers nearly the full range in luminosities and periods of Cepheids. Also, Shapley's calibration is in terms of visual absolute magnitudes, not photographic magnitudes, which became the standard way of illustrating the P–L relation.

The most luminous, long-period Cepheid known in our Galaxy is a star called S Vulpeculae. It has a period of 68 days, a

---

[11] H. Shapley, *Astrophysical Journal*, **48**, 89 (1918).

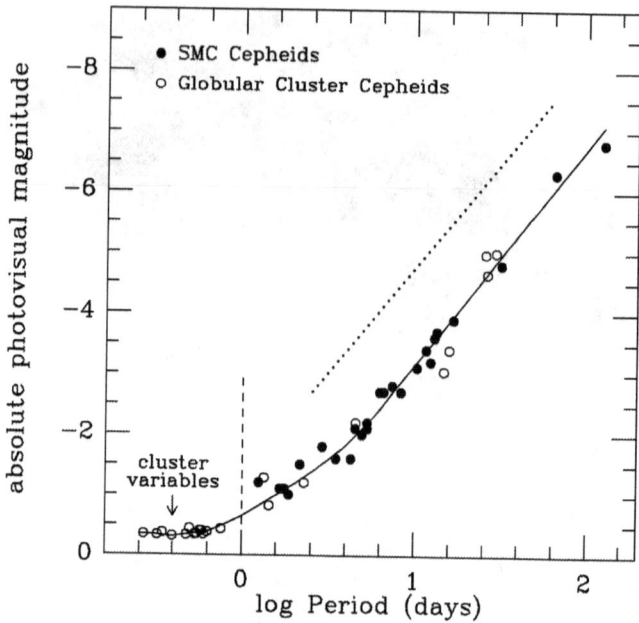

Fig. 7. Shapley's Cepheid P–L relation based on a combination of globular cluster Cepheids and SMC Cepheids, using the "cluster variables" as the "anchor" to join them. The dotted line shows the approximately corrected P–L relation for SMC Cepheids.

luminosity of 28,840 solar luminosities, and a radius nearly 400 times that of the Sun.[12] Such a Cepheid would be easily visible at the distances of the Magellanic Clouds and would be detectable to much greater distances. Cepheids tend to have negative absolute visual magnitudes, the more negative the magnitude the more luminous the star. S Vulpeculae has a visual absolute magnitude of −6.5 compared to the least luminous Cepheids which have an absolute visual magnitude close to 0. For comparison, the Sun's absolute visual magnitude is +5. At a dark observing site, the faintest naked eye stars have a visual magnitude of +6. This means that from a distance of 10 parsecs, the Sun would be just bright enough to be seen with the

---

[12]D. G. Turner, arXiv 1403.1968 (2014); W. P. Gieren *et al.*, *Astrophysical Journal*, **496**, 17 (1998).

naked eye, while S Vulpeculae would appear 5–10 times as bright as Venus when at its maximum angle from the Sun.

Cluster variables can represent about 90% of all the variables in a cluster,[13] if any are present. Some globulars do not have any such variables, or at least have very few. For the ones that do, Shapley's P–L relation shows that at the faint end of the relation, where the absolute visual magnitudes are close to 0 and the periods are about half a day, the luminosity is nearly constant independent of the period. This is the domain of the cluster variables, and it means that for globular clusters one can observe a set of variables, estimate their mean apparent magnitudes (average of maximum and minimum brightness), and then average these averages to get a mean apparent brightness for all the variables in the cluster. Comparing this mean apparent brightness with the known absolute brightness of such stars then gives the distance to the cluster.

The color and size of Cepheids depend on their luminosity. Cluster variables are whitish in color and on average a few times the size of the Sun. RR Lyrae, a field variable star that is almost identical in properties to a typical cluster variable (as is RR Geminorum in Figure 3), has been estimated to be about 5 times the size of the Sun. In contrast, the most luminous Cepheids like S Vulpeculae are orange-reddish in color and hundreds of times the size of the Sun.

The results of Shapley's analysis were several-fold: (1) he found that globular clusters are excentrically-situated with respect to the Sun, in the sense that they are concentrated roughly towards the constellation Sagittarius; (2) globular clusters avoid locations within ±1,000 parsecs of the Galactic plane; (3) if the clusters are distributed around the Galactic center, then that center is about 40,000 light years away; and (4) the Galactic system is about 300,000 light years across.

---

[13]H. S. Hogg, *Publications of the David Dunlap Observatory*, **2**, 33 (1955).

4.3. Why is there a Cepheid P–L Relation?

Cepheids are not just important for distance determinations, they are also fascinating physical laboratories for understanding stellar astrophysics. Not only do Cepheids tell us what goes on inside some stars, they also point to how variability connects to stellar evolution. Although Cepheid variables had been known since the 1780s, the cause of their variability was not understood until the early 20th century. For a while Cepheids were thought to be binary stars, but astronomers like Harlow Shapley were suspicious that Cepheids were genuine radially pulsating stars (meaning they varied due to periodic size changes, not to issues connected with orbital motion).[14] The cause of Cepheid variability only became fully appreciated when people started to make theoretical models of stars, specifically models which described the interiors of stars with a set of equations. Using such an approach, the British theoretical astrophysicist Arthur Eddington proposed in 1917 that Cepheids were pulsating variables whose pulsation was caused by periodic changes in the flow of energy from deep inside the star.[15] The luminosity of all normal stars is powered by fusion of lighter chemical elements into heavier chemical elements in the core region. In order for a star to shine steadily, with little or no variability, the energy produced in the core region must flow smoothly to the surface where it can escape into space. How effectively this occurs depends on how opaque the gases surrounding the core are to light. If something causes the opaqueness of the material to get too high, the energy flowing from the center may get "bottled" up, meaning temporarily stalled. A star cannot bottle up its energy without consequence, so if the usual mechanisms of energy flow fail, the star reacts by pulsating in size.

Throughout most of their evolution, stars behave like perfect gases, meaning that when they expand they cool off and when they contract, they warm up. This temperature change, together

---

[14]H. Shapley, *Astrophysical Journal*, **40**, 448 (1914).

[15]A. S. Eddington, *The Observatory*, **40**, 290 (1917); see also, A. S. Eddington, *The Internal Constitution of the Stars* (Cambridge, Cambridge University Press, 1926).

with the size change, is what causes the light variability. When the gases around the core are compressed, which occurs approximately when the star is faintest, the higher temperature causes them to become more opaque to light. This warms the gases further, making the outward thermal gas pressure higher than the inward pull of gravity. In response, the gases expand and become cooler and more transparent. After expanding for a while, the outward thermal gas pressure drops and gravity draws the gases back inward. The star's layers then compress again and the cycle repeats itself. The Cepheid P–L relation results because more luminous Cepheids are generally larger than lower luminosity Cepheids, and therefore have a lower surface gravity[16] than do smaller Cepheids. In pulsation, the timescale of the imbalance between gravity and outward thermal gas pressure depends on the surface gravity, such that the lower the surface gravity, the longer the period of the oscillation.

4.4.   Hubble and the Great Spirals in Triangulum and Andromeda

At the time of Leavitt's discovery of the Cepheid P–L relation, people knew that the spectrum of the Andromeda Nebula, the brightest "white nebula" in the northern sky, indicated it was a system made of stars.[17] If it were mostly a gaseous nebula, it would have had a spectrum dominated by emission lines. Instead, the spectrum showed mostly absorption lines, like the spectra of stars. Even the early photographs of Isaac Roberts indicated a degree of resolution of the Andromeda Nebula into extremely faint stars. M33 in the constellation Triangulum also showed evidence of resolution in early photographs, but into what some thought were "nebulous stars." That is, the arms appeared to have swarms of faint star-like objects that didn't seem as sharp as they should have been if they were ordinary stars. They looked fuzzier, and implied to some astronomers that they could be forming stars.

---

[16]The acceleration of gravity at the surface of a star or planet; determines the force on an object at that surface.

[17]J. Scheiner, *Astrophysical Journal*, **9**, 149 (1899).

Fig. 8.   Modern color image of M33, the Great Spiral in Triangulum, showing resolved star clouds everywhere across its visible surface area. The picture shows a complex dust pattern and that the spiral arms are not well organized. Image source: Adam Block/Mount Lemmon SkyCenter/University of Arizona, with permission.

Mount Wilson Observatory astronomer Edwin P. Hubble (1889–1953) looked into the resolution issue in M33, where the swarms were especially conspicuous, in more detail and argued that the "nebulous" nature of the "condensations" in the arms was illusory, being due to the faintness of the stars and the way a telescope combined with a long exposure of a plate spreads out the light of a star (what astronomers call the "point spread function"). He concluded there was no reason to believe the condensations were anything but ordinary stars, not "stars in the process of formation."

The modern color image of M33 in Figure 8, obtained with the Mount Lemmon Observatory 32-inch telescope, shows resolved stars in extensive star clouds. These stars are mostly bluish in color and there is no ambiguity that they are real stars.

Just as for globular clusters, the key to determining the distance to the Triangulum and Andromeda Nebulae was to identify the types of stars resolved on photographs taken with a large reflector. In 1926 and 1929, Hubble described in two monumental papers[18] his detailed studies of resolved stars in these two objects, based on inspection and analysis of hundreds of plates taken with the Mount Wilson 60- and 100-inch telescopes over a period of nearly two decades. The most critical discovery Hubble made was that he identified Cepheid variable stars in both spiral nebulae. Not only this, but he noticed that these Cepheids were much fainter than those in the LMC and the SMC, implying the Andromeda and Triangulum spirals are much farther away than those objects. He found the first two Andromeda Cepheids in 1923, and even with only these, he was able to conclude that Andromeda was clearly a star system far beyond the Milky Way. He announced his discovery in 1924.

By the time of his detailed studies, Hubble had discovered 50 variable stars in Andromeda, including 40 Cepheids with periods ranging from 10 to 50 days, and 42 variables in Triangulum, of which 35 were Cepheids having periods of 13–70 days. The maxima ranged from magnitude 18.0–19.1 for Triangulum to 18.0–19.3 for Andromeda, implying that the two objects were nearly at the same distance. In spite of the extreme faintness of these stars he was able to see the Cepheid P–L relation clearly in both systems, and by comparing this relation with Shapley's composite globular cluster/SMC P–L relation (Figure 7) he deduced that Andromeda was 275,000 parsecs, or nearly 900,000 light years away, and that Triangulum was 263,000 parsecs, or 850,000 light years away. This proved, once and for all, that the Andromeda and Triangulum spirals were external galaxies, not parts of the Milky Way. When it is

---

[18]E. P. Hubble, *Astrophysical Journal*, **63**, 236 (1926) (M33); E. P. Hubble, *Astrophysical Journal*, **69**, 103 (1929) (M31).

considered that M31 and M33 are among the brightest spiral nebulae in the sky, then the faintest ones must be hundreds of millions to billions of light years away. The concept of a vaster universe of galaxies was born.

As impressive as Hubble's numbers were for the distances to M31 and M33, they were significant underestimates. Hubble used Shapley's calibration of the Cepheid P–L relation to get his distances, but the calibration was flawed. Shapley used globular cluster variables as the "anchor" for the Cepheid P–L relation (open circles in Figure 7). The problem with this is that most globular cluster variables are of the short period type; few clusters have Cepheids with periods longer than a day. In order to better define the relation for $P > 1$ day, Shapley matched the observed relation for the SMC (defined by the 25 variables studied by Henrietta Leavitt; Figure 4) to the globular cluster relation to get the composite relation shown in Figure 7 (where the filled circles are Leavitt's SMC variables). This failed to take into account that globular cluster Cepheids belong to a different stellar population than do SMC Cepheids. This was brilliantly determined by Mount Wilson and Palomar astronomer Walter Baade (1893–1960) and described in a 1956 paper.[19] Twelve years earlier, Baade had discovered that stars in the disk of the Andromeda Galaxy were fundamentally different from those in the bulge and halo. The former were called Population I stars while the latter were called Population II. One basic difference is that Population II stars are deficient in heavy chemical elements ("metals"[20]) compared to Population I stars. Population I Cepheids are about 1.5 magnitudes brighter in photographic luminosity than globular cluster Cepheids. This is approximately shown by the dotted line in Figure 7. SMC Cepheids are Population I stars, but Shapley's composite P–L relation makes them Population II stars snd fainter than they actually are. Because they are more luminous than Shapley assumed, SMC-type Cepheids are the type most likely to be detected in galaxies like the Andromeda and Triangulum spirals, not globular

---

[19]W. Baade, *Publications of the Astronomical Society of the Pacific*, **68**, 5 (1956).
[20]In astronomy, a metal is any chemical element heavier than helium.

cluster Cepheids. It became essential to distinguish Population II Cepheids from Population I Cepheids. Shapley's calibration caused Hubble's distances to be underestimated by at least a factor of two. Modern distance estimates to these objects are 2.5 million light years for M31 and 2.8 million light years for M33. More on the two stellar populations is described in Chapter 10.

The discovery that spiral nebulae are galaxies verified the long-unproven "island universe" hypothesis, and also showed that the Milky Way is simply one of an untold number of galaxies that populate the farthest regions of space detectable with modern telescopes. The Universe leapt in size from a few hundreds of thousands of light years to millions and even hundreds of millions of light years. It was the biggest development in astronomy since the invention of the telescope.

Chapter 5

An Expanding Universe

By the time of Hubble's Cepheid work on M31 and M33, other aspects of the nebulae were being explored, some by Hubble himself. Everyone knew by then that white nebulae were external galaxies. An important question was: what would be the next steps in the study of these objects?

One type of observation that would prove to be critical to the study of external galaxies was the measurement of radial velocities, a step pioneered by Vesto M. Slipher (1875–1969), a research astronomer at the Lowell Observatory in Arizona. Radial velocities are line of sight speeds obtained using the Doppler effect. Slipher was impressed with the discoveries of James Keeler who, as noted previously, was the first to show that the spiral form of nebulae was widespread, populating the sky in much larger numbers than visual observing had led astronomers to believe. Slipher noted that the abundance of spiral nebulae meant they were important objects for further study, and that in addition to the direct photography that had been accumulated in recent years, it was essential to measure radial velocities of these objects. Measuring such velocities accurately meant detecting absorption or emission lines in the spectra of nebulae. Slipher acknowledged that this would be difficult in general owing to the faintness of spirals and the fact that their typical spectra were absorption-line rather than emission-line. The exposure times required with the spectrographs and plates available in his day were generally many hours, sometimes spread over several days, to get a reasonably good estimate of the radial velocity.

Slipher's first target for radial velocity measurement was the Andromeda Nebula.[1] From four plates he found that Andromeda is approaching the Sun at 300 km/s, the highest radial velocity measured for anything at the time. He then made an odd speculation: that the high speed could mean that Andromeda encountered a "dark star," and that this could have given "rise to the peculiar nova that appeared in the nebula in 1885." In a 1915 paper, Slipher presented the radial velocities of 14 other spiral nebulae and found values ranging from −300 km/s to +1100 km/s.[2] He noticed that the average radial velocity of four edge-on nebulae (which he assumed to be spirals) was significantly higher than for the other nebulae he observed, and made the peculiar suggestion that "the spirals move edge-forward." In a 1917 paper,[3] Slipher increases the sample to 25 nebulae and notes that exposure times of 20–40 h are generally needed for this kind of work. He notes that the sample is not spread well enough across the sky to say what the systematics of the radial velocities might be. He further notes that "the average velocity, 570 km/s, is about 30 times the average velocity of the stars. And it is so much greater than that known for any other class of celestial objects as to set the spiral nebulae aside in a class to themselves. Their distribution over the sky likewise shows them to be unique — they shun the Milky Way and cluster around its poles."

The high average positive velocity of the spiral nebulae led Slipher to suggest that perhaps they were "scattering," that is, flying apart, but he noted that the tendency for spiral nebulae to cluster would likely mitigate against such an interpretation. He promised that "A little later a tentative explanation of the preponderance of positive velocities will be suggested."

Slipher also noted that "The form of the spiral nebulae strongly suggests rotational motion" and indeed he found several cases where spectrograms [photographic spectra] showed rotation or internal motion." In some cases, as in NGC 4594 (M104), the lines were

---

[1]V. M. Slipher, *Popular Astronomy*, **22**, 19 (1914).
[2]V. M. Slipher, *Popular Astronomy*, **23**, 21 (1915).
[3]V. M. Slipher, *Proceedings of the American Philosophical Society*, **56**, 403 (1917).

slanted in the manner of those seen in the spectrum of Jupiter, as if the rotation were solid body (like the rotation of the Earth). In other cases, such as the Andromeda Nebula and M65, the rotation was solid body in the inner regions and non-solid-body farther out.

In this same paper, Slipher states that "Because of its bearing on the evolution of spiral nebulae it is desirable to know the direction of rotation relative to the arms of spirals." This was an interesting question, and he knew that the only way to answer it would be to know which side of a tilted spiral is "the nearer us." Spectroscopy can only tell us which half of the major axis of a tilted spiral is approaching us, and which half is receding from us. Spectroscopy alone cannot tell us the sense of winding of the spiral arms: opening outward into the direction of rotation ("leading"), or opening outward opposite the direction of rotation ("trailing"). This ambiguity can be resolved only if we knew which side of the minor axis of a spiral is the near side, because half of the object is tipped towards us and half is tipped away from us. He had already noted in a previous paper how spirals often show lanes of absorbing material, especially in the edge-on view. He states: "If we now imagine we view such a nebula from a point somewhat outside its plane the dark band would shift to the side and render the nebula unsymmetrical — the deficient edge being of course the nearer to us." He further notes that "the inclined ones commonly show this dissymmetry" and that "we may infer their deficient side to be the one towards us." He concluded that "The central part — which is all of the nebulae the spectrograms record — turns into spiral arms as a spring turns in winding up." Finding that "all spiral nebulae rotate in the same direction with reference to the spiral arms," he believed this was a "favorable check on the conclusion as to the nearer edge of the nebulae." In a 1922 paper, Slipher discusses this issue again, and states "as I have previously pointed out I found some years ago an indirect means which is apparently dependable in the general case for deciding which is the nearer edge of an inclined spiral."[4]

---

[4]V. M. Slipher, *Publications of the American Astronomical Society*, **4**, 232 (1922).

Slipher[5] brought up another issue: that the radial velocities of nebulae prominent in the spring sky tend to be positive, while in the opposite part of the sky, some negative velocities are found (as, for example, for the Andromeda Nebula). He believed that if the nebulae with known radial velocities were more uniformly distributed around the sky, one might detect the motion of the solar system relative to the spiral nebulae. Based on the meager dataset he had available he deduced that the solar system is moving towards right ascension 22 h, declination $-22°$, at a speed of 700 km/s. He believed that this apparent "drift through space," which is not detected when the radial velocities of stars are used for reference, implies that our whole "stellar system moves and carries us with it." He was aware of the "island universe" view of the nebulae, and concluded, "This theory, it seems to me, gains favor in the present observations."

Slipher published many short notes about his spectroscopic work on the nebulae frequently in the years following the initiation of his program in 1912. In one such communication,[6] he noted that observations of NGC 4449 and 4214, two similar looking objects with a great deal of structure, showed strong emission lines against a mostly continuous background (with some absorption lines), and that the emission lines were typical of gaseous nebulae and had relative brightnesses that were different in different locations. He concluded from these objects that you could deduce in advance what the spectrum of a nebula would be just from its appearance.

In another short note,[7] Slipher was amazed at the high radial velocities he found for two bright non-spiral nebulae: NGC 584 (1,800 km/s) and NGC 936 (1,300 kms). He concluded that both objects nevertheless belong with the spiral nebulae.

Hubble received his PhD in 1917 from the University of Chicago, where he was a student at the Yerkes Observatory. The title of

---

[5]V. M. Slipher, *Proceedings of the American Philosophical Society*, **56**, 403 (1917).
[6]V. M. Slipher, *Publications of the Astronomical Society of the Pacific*, **30**, 346 (1918).
[7]V. M. Slipher, *Popular Astronomy*, **29**, 128 (1921).

his PhD thesis was "Photographic Investigations of Faint Nebulae," the main goal of which was a photographic investigation of the clustering of extragalactic nebulae, that is, of the nebulae seen away from the band of the Milky Way. This thesis led to Hubble's three signature achievements in astronomy: proof that the "island universe" hypothesis of these nebulae was correct, development of an astrophysically meaningful classification of these objects, and the realization that there is a strong correlation between the distances to these nebulae and their radial velocities.

As with Vesto Slipher, Hubble carried out his thesis work at a time when the field of nebular research was wide open for discovery. Up until about 1880, nebular studies were largely visual in nature. Hubble[8] summarizes well the change in nebular studies by the time of his work: "The study of nebulae is essentially a photographic problem for cameras of wide angle and reflectors of large focal ratio.[9] The photographic plate presents a definite and permanent record beside which visual observations lose most of their significance." Thus, at the time of Hubble's thesis work, nebular research was very active and Hubble was aware of all the types of photographic research that were being carried out, including the radial velocities of Slipher and the rotation studies of F. G. Pease.

The previous chapter described how Hubble deduced that the major bright spirals, M31 in Andromeda and M33 in Triangulum, lay outside our stellar system and were galaxies in their own right. But this research on distances did not end with M31 and M33. Hubble also wanted to get distances to fainter nebulae than these. The reason for this: he had noticed that fainter nebulae tended to have larger radial velocities than did brighter ones. That is, there seemed to be a velocity–distance correlation. In the 1920s, Milton Humason (1891–1972) began using the Mount Wilson 100-inch reflector to measure radial velocities of fainter nebulae than had been observed by Slipher. Hubble's discovery of the velocity–distance correlation

---

[8]E. P. Hubble, *Publications of the Yerkes Observatory*, **4**, 2 (1920).

[9]The ratio of the focal length to the diameter of the telescope's primary mirror.

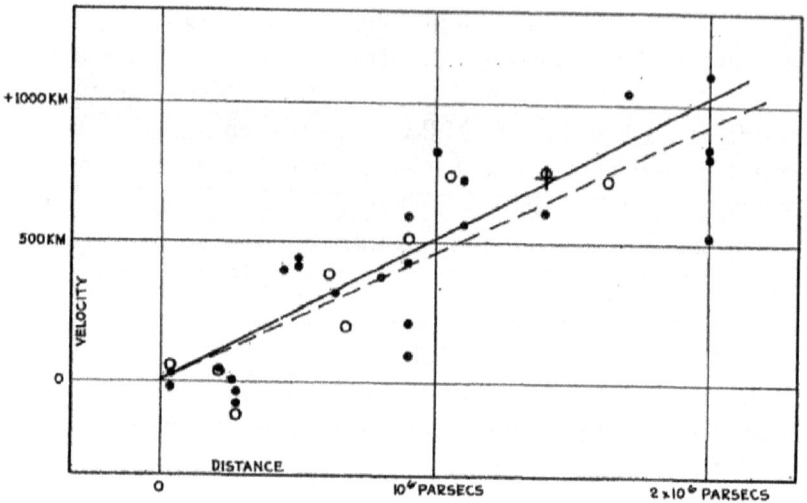

Fig. 1.   Hubble's original velocity–distance graph, published in 1929. The lines are based on different ways to represent the data. © AAS. Reproduced with permission.

was published in 1929.[10] His first major graph of the correlation is reproduced in Figure 1.

Like others at the time, Hubble was interested in using radial velocities of nebulae to detect the motion of the Sun relative to the nebulae. If the nebulae themselves had no intrinsic systematic motions, then their radial velocities would simply reflect the motion of the Sun relative to the reference frame they define. This motion would be reliably determinable only for a sample of nebulae spread around a wide swath of sky. In such a circumstance the radial velocity of any nebula is related to the speed of the Sun projected into the direction of the nebula. Hubble noticed that this simple model did not work. In order to account for the high radial velocities being found for some nebulae, the solar motion had to include a term proportional to the distance to the nebula. The proportionality constant of that term, the units of which were kilometers per second per Megaparsec (km/s/Mpc), later became known as the "Hubble constant," the

---

[10]E. Hubble, *Communications of the Mount Wilson Observatory*, **3**, 23 (1929).

reliable measurement of which became one of the major goals of extragalactic astronomy in the second half of the 20th century.

The importance of the velocity–distance relation, later to be known as the "Hubble Law,"[11] to astronomy cannot be overestimated. First, the law had a huge implication: we live in an expanding universe, an observation that led to the Big Bang Theory, one of the principal ideas in the field of cosmology, the study of the structure, origin, and evolution of the universe. Second, for nearly 150 years, the main problem with understanding the nature of the extragalactic nebulae lay in the uncertainty of their distances. The velocity–distance relation opened up to astronomy a way of getting distances to the faintest and obviously remotest nebulae using only a measurement of radial velocity. Although measuring a nebular radial velocity in Slipher's day involved a major effort of exposure time and telescopic prowess, eventually such measurements would become routine, such that by the beginning of the 21st century, radial velocities are available for about a *million* galaxies.

The measurement of the Hubble constant has never been easy, because in order to derive it reliably, one needs to know the distances to at least some galaxies independent of their radial velocities. This can be done reliably only for the nearest galaxies, because only for the nearest galaxies is it possible for the best distance indicators to be detected. Cepheid variable stars are detectable in other galaxies, but have a limited range. For example, the next nearby spiral beyond the Andromeda and Triangulum spirals is M81, the Great Spiral in Ursa Major, but at 5 times greater distance, Cepheids were not detectable in M81 with the equipment available to Hubble. The same was true for other bright spirals like M94, M101, and M51, ranging from 6 to 9 times the distance of the Andromeda Galaxy. Only the brightest blue stars were detectable for these more distant spirals, and although such stars are good indicators of distance, they do not

---

[11]Recently renamed the Hubble–Lemaitre law to acknowledge Belgian priest Georges Lemaitre's independent analysis (*Annals of the Scientific Society of Brussels*, **47**, 49 (1927)) which led to the same idea.

reach the "standard candle" level of Cepheids. Like Cepheids, the brightest blue stars also have a limited range.

Hubble was aware that the velocity–distance relation has "noise" in the sense that an individual galaxy's radial velocity includes not only a contribution from the "flow" part, but also a random part due to the gravity of other galaxies. The Hubble constant only describes the flow part, or the part of the line of sight speed associated with the general universal expansion. To extend the range of distances and minimize the effects of the random part of the radial velocities, Hubble used not only the brightnesses of individual stars in a galaxy as a distance indicator, but also the brightness of *whole galaxies*. He used 24 nearby galaxies to get the average absolute magnitude of typical galaxies, and compared this to the average apparent magnitudes of galaxies in distant groups or clusters. This is not expected to give distances as accurate as those from resolved stellar populations, since the spread in absolute magnitudes of individual galaxies is large.

Hubble's first estimate of his constant was 513 km/s/Mpc, 7 times higher than the modern value of 73 ± 5 km/s/Mpc. With this finding, Hubble increased the scale of the Universe dramatically. Nebulae were not merely galaxies a few million light years away, but could be seen to distances of hundreds of millions of light years. The constant also had something else to offer: an estimate of the age of the Universe. An expanding universe points to a "beginning" (commonly known as the "Big Bang"), and if the universe has expanded uniformly since this beginning, then the inverse of the Hubble constant gives an estimate of the time since the universal expansion began. With Hubble's original value of 513 km/s/Mpc, this age (called the "Hubble time") was 1.9 billion years, but with the modern value of 73 km/s/Mpc, the age is 13.4 billion years. The latter value is much more consistent with the ages of globular clusters than was Hubble's original value.

The derivation of the Hubble constant is one of the dramatic stories of 20th century astronomy, and is described in more detail in Chapter 23.

Chapter 6

The Hubble "Filing" System

Hubble's interest in nebulae was not confined to figuring out how far away they are. He was also interested in finding an astrophysically useful way of looking at them. Even a century ago, and based on a limited number of plates, it was clear that galaxy morphology is complex and bewildering. In a landmark 1926 paper,[1] Hubble developed a classification system for nebulae that has stood the test of time. There were several realizations that led to this.

First, by 1926 Hubble already knew that the "white nebulae" were galaxies. Extragalactic nebular classification was a classification of stellar systems, not peculiar clouds of glowing interstellar gas. Second, Hubble disagreed with the conclusion of Heber Curtis that all nebulae are spirals, and that any that do not appear to be spiral would appear so when photographed with better equipment. Hubble recognized that there are genuine non-spiral (and therefore non-disk-shaped) nebulae. Third, Hubble believed that the existing classification scheme of nebulae, one which had been proposed by University of Heidelberg astronomer Max Wolf (1863–1932) in 1908 and which was in use from 1908 to 1926, was a "good temporary filing system until something better comes along."[2] He said this in

---

[1] E. P. Hubble, *Astrophysical Journal*, **64**, 321 (1926).
[2] The Wolf classification scheme (*Publications of the Astrophysical Institute of Koenigstuhl-Heidelberg*, **3**, (5) (1908)) was proposed before the true nature of the nebulae had been established, and thus also includes some Galactic nebulae such as planetary nebulae. Also, the Wolf system gave edge-on and face-on galaxies their own filing letters, while Hubble felt he could still type highly-inclined

1917, and by then was probably already thinking about a new "filing system." Finally, Hubble noticed that some spirals had significant central concentration and tightly-wrapped, smooth spiral structure, while others had low central concentration, and open, knotty spiral structure. He recognized this as astrophysically important.

Hubble viewed galaxy morphology in terms of a left-to-right sequence of increasing flattening and increasing degree of structure. In the left part of the sequence, he placed galaxies that were essentially featureless, elliptical-shaped objects. These "elliptical nebulae" had no other characteristics for classification other than the degree of ellipticity,[3] which was specified by a number following the letter E, as in En. The number after the "E" refers to the apparent ellipticity multiplied by a factor of 10 and rounded to the nearest whole integer, that is, $n = 10(1 - \frac{b}{a})$, where a is the long (or major) axis length and b is the short (or minor) axis length. For example, an E5 galaxy is twice as long as it is wide, so $n = 10(1 - \frac{1}{2}) = 5$. Note that Hubble intended for n to be an eyeball-estimated value, not necessarily based on detailed measurement.

The elliptical galaxy sequence (Figure 1) begins with the roundest galaxies in the E category, called E0, and ends with the flattest E galaxies, called E7,[4] where the major axis length is about 3 times the minor axis length. Between these were intermediate cases E1, E2, E3, E4, E5, and E6. The E7 galaxies generally had a "lenticular" shape (like a side-on view of a double convex lens), which led Hubble to believe they blended smoothly with the more highly flattened

---

galaxies within his classification system without inventing a new type class for them.

[3]Later studies (for example, J. Kormendy and R. Bender, *Astrophysical Journal*, **464**, L119 (1996)) were able to find other characteristics that could be used for classifying elliptical galaxies, based on the deviations of the isophotes (contours of constant surface brightness) from a perfectly elliptical shape. More on this is described in Chapter 21.

[4]S. van den Bergh, *Astrophysical Journal*, **694**, L120 (2009) found that very few elliptical galaxies are typed as E7; instead, any ellipticals this flattened are more likely to be recognized as disk-shaped galaxies.

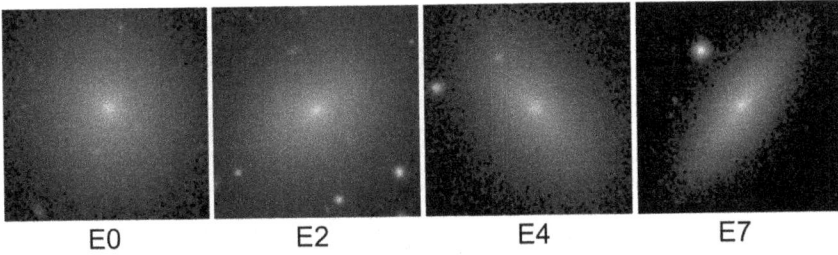

E0	E2	E4	E7

Fig. 1. Hubble's elliptical galaxy sequence based on shapes. The number after the letter "E" is the apparent flattening multiplied by a factor of 10. From left to right, the galaxies are NGC 4434, 1052, 6030, and 4259. Images source: EFIGI/SDSS (NGC 4434, 4259, 6030; Sloan Digital Sky Survey Collaboration, http://www.sdss.org); de Vaucouleurs Atlas of Galaxies published by Cambridge University Press, 2007; reprinted with permission (NGC 1052).

spirals. In Figure 1, the images of all but NGC 1052 are taken from the EFIGI survey.[5]

The next step in the sequence of classification was the spirals. In the edge-on view, disks were more flattened than E7 galaxies. Had there been no barred spirals, the entire Hubble classification system would have been one linear sequence ranging from ellipticals to the best-defined spirals. Instead, Hubble viewed spirals as falling on two parallel sequences (Figure 2), one for non-barred (or "normal") spirals, and one for barred spirals, the objects that Heber Curtis once called "ϕ-type spirals." Along each sequence, the spirals are subdivided according to three characteristics: the degree of central concentration, the degree of openness of the arms, and the degree of resolution of the arms into "knots" or complexes of stars. The sequences begin as tightly-wrapped, smooth-armed spirals with significant central concentration. These were denoted as Sa and SBa galaxies and were considered "early type" spirals. The sequences ended with loosely wrapped spirals where the arms are highly

---

[5]EFIGI is a French acronym for "Extraction of the Forms of Galaxies from Images," a major classification project of galaxies based on Sloan Digital Sky Survey (SDSS) images and carried out by A. Baillard *et al.*, *Astronomy and Astrophysics*, **532**, 74 (2011). EFIGI images are SDSS images that have been homogenized for galaxy classification studies.

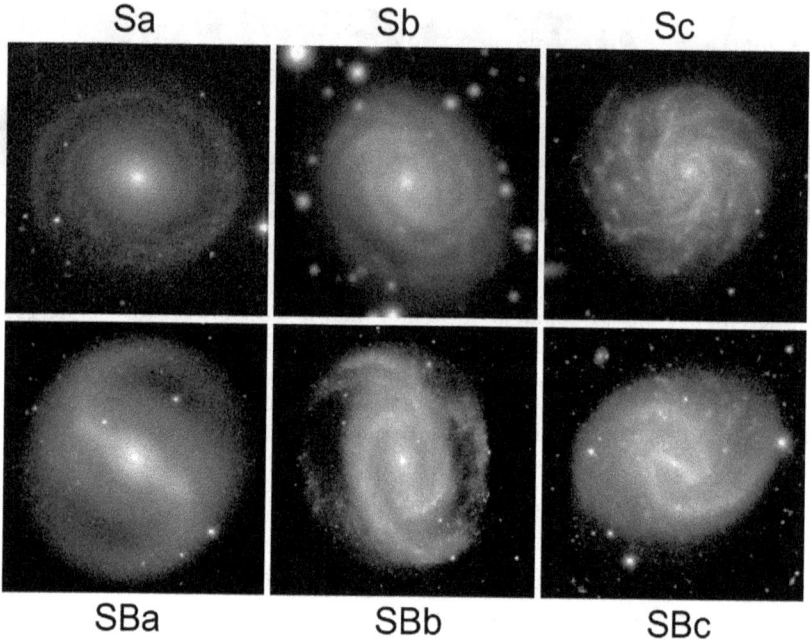

Fig. 2. Hubble's spiral galaxy sequences based on the degree of central concentration, the openness of the arms, and the degree of resolution of the arms into stars or "knots". Top row: normal spirals, left to right: NGC 4378, 7042, and 1376. Bottom row: barred spirals, left to right: NGC 4314, 1300, and 3513. Image sources: NGC 1376 is from the EFIGI/SDSS (credit: Sloan Digital Sky Survey Collaboration, http://www.sdss.org). The rest are from the de Vaucouleurs Atlas of Galaxies published by Cambridge University Press, 2007; reprinted with permission.

resolved into "knots" or clumps of young stars, and the central concentration is barely significant. These were denoted as Sc and SBc galaxies and were considered "late type" spirals. Between these extremes he placed Sb and SBb galaxies which were considered to be "intermediate type" spirals.

The "early" and "late" connotations were not meant to imply anything temporal about the galaxies, just a descriptive term to indicate position in the sequence. The terminology mirrored what was often used to describe stars: stellar astronomers would call stars of spectral classes O and B "early-type stars" while those of spectral classes K and M were called "late-type stars." Here a temporal connotation was implied because O and B stars tend to be in an earlier stage of evolution (for example, the main sequence stage) than K and M stars (the most conspicuous of which tend to be giants or supergiants in later stages of evolution). The irony in this is that, after it became clear what kinds of stars populate the disks and bulges (central regions) of galaxies of different types (Chapter 11), it was realized that the light of early-type galaxies is dominated by late-type stars, while the light of late-type galaxies is dominated by early-type stars, with intermediate-type galaxies having a more balanced mix.

The same six type bins in Figure 2 can be recognized even in highly-inclined and edge-on galaxies (Figure 3). Two of the three

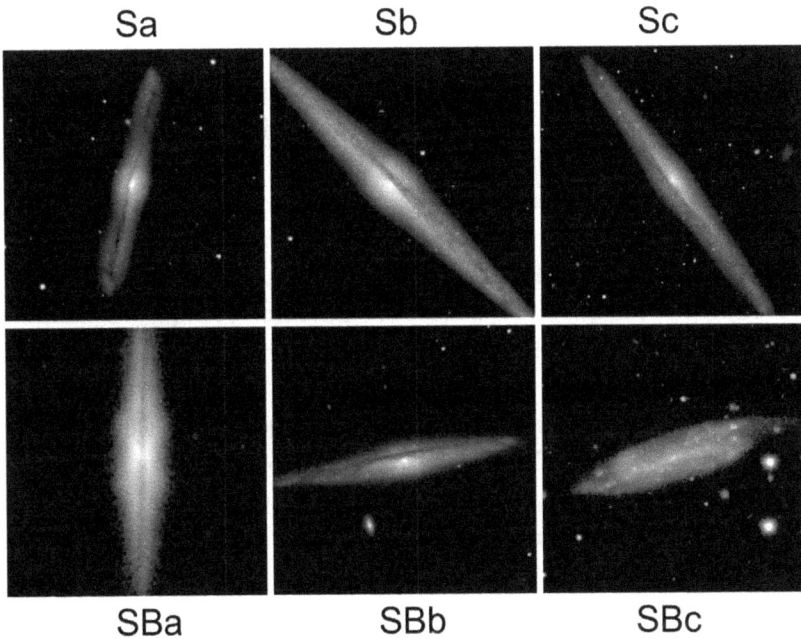

Fig. 3.   The same two sequences can be identified even in nearly edge-on galaxies. Top row: normal spirals, left to right: NGC 678, 4565, and 5170. Bottom row: barred spirals, left to right: UGC 9690, NGC 5965, and UGC 5249. Image sources: NGC 678 and 4565 are from the de Vaucouleurs Atlas of Galaxies published by Cambridge University Press, 2007; reprinted with permission. The rest are from the EFIGI/SDSS sample (credit: Sloan Digital Sky Survey Collaboration, http://www.sdss.org).

criteria for classifying the a, b, and c spiral types: the relative size of the unresolved nuclear region and the degree of condensations in the arm regions, are still distinguishable in such views. What becomes harder to detect in the edge-on view, of course, is the actual spiral arms themselves. Although Hubble did not know it, bars can be detected in edge-on galaxies. The reason is that a bar is not necessarily a flat structure like the disk, but can be much more three-dimensional. Strong bars can manifest themselves as a "boxy/peanut" bulge in the edge-on view. (That is, they are broad, boxy elongated features pinched in the middle, like a peanut.) UGC 9690[6] and NGC 5965 in Figure 3 show this character, the nature of which is discussed in more detail in Chapter 15.

The four galaxies with a prominent bulge in Figure 3 also show an important characteristic: a dark, dividing lane of obviously obscuring matter. It was eventually determined that the material making up such dividing lanes is interstellar dust. How the nature of the material was determined, and how it affects our view of the rest of the universe, is described in Chapters 8 and 9.

Hubble claimed that 97% of extragalactic nebulae fit neatly into his system, which accomodated mainly nebulae having "rotational symmetry around dominant nuclei." The remaining 3% included "irregular nebulae," which lacked such symmetry.

Although barred spirals figured prominently in Hubble's classification, the nature of the bar and the reasons why some galaxies are barred and others are not was not understood at all in his day. But, even as late as 1939, Hubble believed that barred spirals accounted for no more than a quarter of all spirals.[7] The "bar fraction," as it is now called, has in recent years been shown to be close to 70% when examined in infrared light.[8] Many bars stand out better in infrared light because the features tend to be made of old stars, especially the bars in galaxies of types SBa and SBb. In contrast, the bars in

---

[6]UGC stands for Uppsala General Catalogue and is the name assigned in P. Nilson, *The Uppsala General Catalogue of Galaxies* (Uppsala, 1973).
[7]E. P. Hubble, *Publications of the American Astronomical Society*, **9**, 249 (1939).
[8]For example, P. Eskridge *et al.*, *Astronomical Journal*, **119**, 536 (2000).

some of Hubble's SBc galaxies include young stars. The nature of bars and how they are thought to affect galaxy evolution is discussed in Chapter 15. The evidence for a bar in the Milky Way is discussed in Chapter 8.

Hubble in 1926 imagined that the E galaxy sequence and the two spiral sequences joined at the E7 and Sa, SBa junctures, and while he did not provide an illustration of what he viewed, contemporary British astrophysicist Sir James Jeans (1877–1946) provided one for him. In Jeans' 1929 book,[9] he comments that "Hubble finds that it is not possible to place all observed nebular configurations in one continuous sequence; their proper representation demands a Y-shaped diagram such as is shown in Figure 53" (Figure 4). This diagram is thought to have inspired Hubble's later (unacknowledged) depiction of his classification system as a "tuning fork,"[10] although this representation (shown in Figure 5) had something the Jeans representation did not. Hubble had noticed that there was a problem in joining the elliptical and spiral sequences in this manner: Sa galaxies generally had fully-formed spirals, while some apparent SBa galaxies did not. Hubble hypothesized that to account for this, there

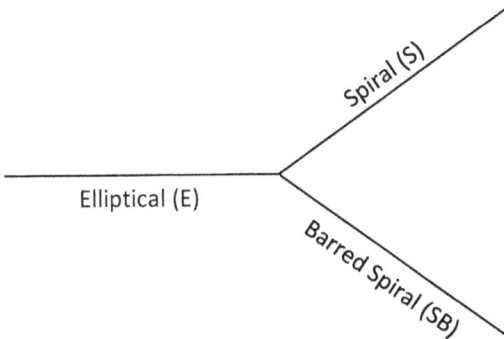

Fig. 4.   Sir James Jeans's 1929 depiction of Hubble's 1926 classification system.

---

[9] J. Jeans, *Astronomy and Cosmogony* (Cambridge, Cambridge University Press, 1929).
[10] See A. Sandage, in *Penetrating Bars Through Masks of Cosmic Dust, The Hubble Tuning Fork Strikes a New Note*, D. L. Block and others, eds. (Dordrecht, Springer, 2004), p. 39.

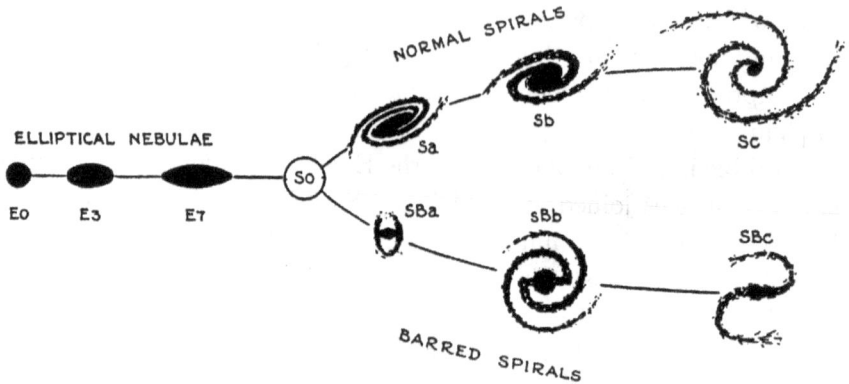

Fig. 5.   The Hubble "tuning fork" of galaxy morphology (see footnote 11).

had to exist non-spiral, non-barred galaxies flatter than type E7. These he called S0 galaxies, where the "0" is meant to imply 0 spiral arms. An S0 galaxy is an armless disk-shaped galaxy, while an elliptical galaxy is essentially a non-disk-shaped galaxy (and by default such galaxies also do not have any spiral arms). Hubble placed the hypothesized S0 class at the juncture between ellipticals and spirals in his 1936 book,[11] and it is in this book that he schematically illustrated his system for the first time. Hubble's "tuning fork" representation of his classification system is one of the most famous renditions in the history of galaxy research.[12]

Hubble's tuning fork connected the non-spiral property of the galaxies in the left half of his sequence to the highly-flattened disk shape of those in the right half of his sequence. The hypothetical intermediate (S0) class was most evident in the edge-on view. Both barred and non-barred S0 galaxies could be recognized. Figure 6

---

[11]E. P. Hubble, *The Realm of the Nebulae* (New Haven, Yale University Press, 1936).

[12]The fact that Jeans depicted Hubble's 1926 classification as a "Y" 7 years before *The Realm of the Nebulae* was published has led to the suggestion that the Hubble tuning fork be renamed the "Jeans-Hubble" tuning fork; see D. L. Block *et al.*, in *Penetrating Bars Through Masks of Cosmic Dust, The Hubble Tuning Fork Strikes a New Note*, D. L. Block and others, eds. (Dordrecht, Springer, 2004), p. 15.

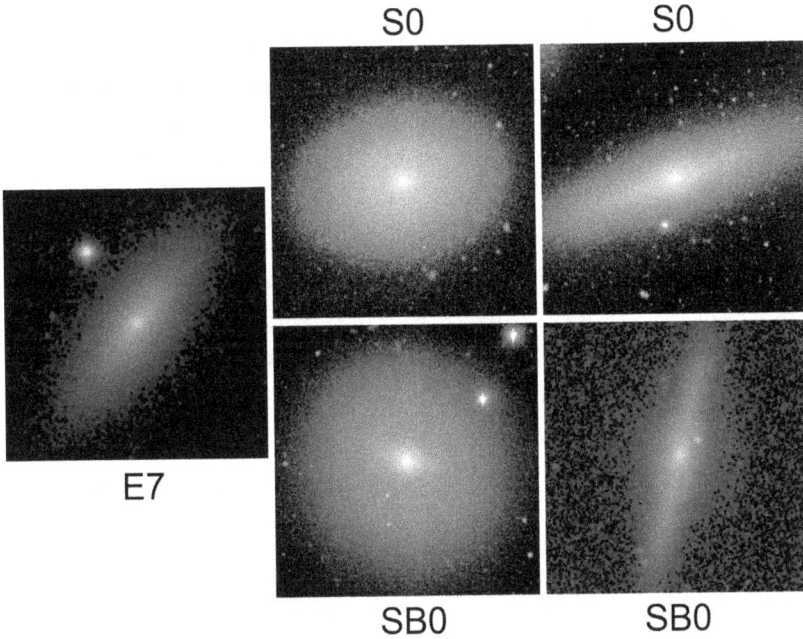

SO   SO

E7

SB0   SB0

Fig. 6.   Four S0 galaxies which go into the juncture between elliptical and spiral galaxies. The face-on examples are NGC 1411 (top) and NGC 1533 (bottom). The edge-on examples are NGC 1596 (top) and NGC 3762 (bottom). The E7 example is NGC 4259. Image sources: NGC 1411, 1533, and 1596 are from the de Vaucouleurs Atlas of Galaxies published by Cambridge University Press, 2007; reprinted with permission. The others are from the EFIGI/SDSS database (Credit: Sloan Digital Sky Survey Collaboration, http://www.sdss.org).

shows two nearly face-on S0 galaxies and two edge-on S0 galaxies. Comparison with edge-on spirals classified in the same Sa–Sc, SBa–SBc type bins (Figure 3) as for the more face-on spirals in Figure 2, we see a distinguishable difference between spirals and S0s: in the edge-on view, many spirals show a dark lane of obscuring matter running parallel to the long axis of the galaxy. If a spiral is viewed exactly edge-on, the nucleus of the galaxy can be completely obscured by this dark lane. In a typical edge-on S0, however, there is no dust lane and the nucleus shines brightly.

Figure 6 also shows how subtle the structure of S0 galaxies can be. The two nearly face-on examples resemble elliptical galaxies, which can lead to misclassification if the type is based on inadequate image material. In some cases, only detailed analysis of the light distribution can give the correct interpretation.

While the tuning fork was meant to be a purely empirical device for looking at the structure of galaxies, it can be tempting to imagine it in terms of an evolutionary sequence. The apparent continuity from the handle to the prongs of the tuning fork suggests that spirals may begin as open, knotty features in the presence of small central concentrations and evolve into tighter spirals with greater central concentration. As intriguing as this may sound, Hubble emphasized that his sequence was purely empirical and not necessarily based on any theory.

6.1. Galaxy Morphology and Classification Circa 1960

In Hubble's day and even decades later, our view of galaxy morphology was limited by the number of images available for classification. Although photography advanced from the multi-hour exposures of Hubble's time to faster plate/filter/telescope combinations, building a sizeable library of images for classification was still a slow process even up until the 1980s. Major observatories would have vaults of plates, and research on galaxy morphology would necessitate visiting these vaults.

After his big discoveries in the late 1920s, Hubble continued his extragalactic research, focussing on a variety of topics. In particular, he continued his work on galaxy morphology by obtaining more plates, and was preparing a revision to his 1936 tuning fork classification at the time of his death in 1953. Although Hubble died before completing his revision, which included real examples of the originally hypothetical class of S0 galaxies, he left behind many notes on specific cases. Beginning in 1957, student assistant Allan R. Sandage used Hubble's notes to prepare the monumental *Hubble Atlas of Galaxies*, which was published in 1961. The atlas used high quality reproductions of photographic plates of 176 mostly normal galaxies to show as much of galaxy morphology as the available plate collection allowed.

Sandage not only followed Hubble's notes on the classification of S0 galaxies, he also added some new features to the classification. In most spiral galaxies, the arms wind directly from near the center in a pattern that resembles a forward or backward s. In other galaxies, particularly barred galaxies, the arms break from a feature called an inner ring. Sandage used the notation (r) if an inner ring was present and (s) if no ring was present. In cases where a partial inner ring (in the form of tightly-wrapped spiral arms) was present, Sandage used the combined notation (rs). These became known as inner "pseudorings" and were one of the earliest recognitions of an apparent *continuity* in spiral galaxy morphology. In the revised classification, there were galaxies of types Sa(r), Sa(rs), and Sa(s), or SBb(r), SBb(rs), and SBb(s), etc. Examples of an Sb(s) and an SBb(r) galaxy are shown in Figure 7.

Around the time Sandage was preparing the Hubble Atlas of Galaxies, another astronomer was showing an intense interest in galaxy morphology and in using the Hubble classification system. This was Gérard Henri de Vaucouleurs, a 1949 graduate of the

Fig. 7.　Two spiral galaxies, one with an inner ring [NGC 2523, right panel, type (r)], and one lacking such a feature [NGC 1566, left panel, type (s)]. Images source: the de Vaucouleurs Atlas of Galaxies published by Cambridge University Press, 2007; reprinted with permission.

University of Paris, Sorbonne Physics Research Laboratory and Institute of Astrophysics. From 1952 to 1955, de Vaucouleurs used the 30-inch Reynolds reflecting telescope atop Mount Stromlo, near Canberra, Australia, to take photographs of galaxies too far south to be accessed from the Mount Wilson and Palomar Observatories in California. Over the nearly 3-year period of his program, de Vaucouleurs obtained classifiable images of 460 southern galaxies. His goal was to judge the Hubble types of the galaxies and to measure from the plates the angular dimensions of each galaxy. Angular sizes were one of the first astrophysical measurements that could be made during the height of the photographic era. In the 1950s, such sizes were not very precise because galaxies do not necessarily have sharp edges. Instead, they fade gradually with increasing distance from the center (see Figure 13).

The de Vaucouleurs revision to the Hubble tuning fork is the most used classification system for galaxies today. In the 1956 publication of his Reynolds telescope survey,[13] de Vaucouleurs describes the basis for his views: "The system of classification used in the present survey represents an elaboration and extension of the scheme originally introduced by Hubble (1926) as modified and improved first by Hubble himself and lately further developed by Sandage. I was very fortunate to become acquainted with this revised scheme during a visit to Pasadena in 1955 and I wish to record my indebtedness to Dr. A. R. Sandage for the generous communication of much unpublished information on the present status of nebular classification work at Mt. Wilson and Palomar; a confrontation of opinions and experiences in this difficult and unsettled field has proved of considerable value to clarify and fit together many parts of the classification puzzle. The present classification scheme and system of notification have resulted from these discussions."

---

[13] G. de Vaucouleurs, *Survey of Bright Galaxies south of −35 degrees Declination with the 30-inch Reynolds Reflector (1952–1955)*, Memoirs of the Commonwealth Observatory, No. 13 (1956).

In 1959, de Vaucouleurs proposed a revision[14] to the Hubble-Sandage classification (outlined in the *Hubble Atlas of Galaxies*) that he believed provided a better description of what a galaxy looked like without being too unwieldy. de Vaucouleurs viewed galaxy morphology as a continuum of forms spread across three major dimensions: the stage or position in the sequence; the family or apparent bar strength; and the variety, referring to the presence or absence of an inner ring or pseudoring. The original Hubble stages a, b, and c were fairly broad categories, which led de Vaucouleurs to use a combined notation such as Sab, Sbc (or SBab, SBbc) for types that looked intermediate between Hubble's main stages.

On the issue of bar classification, both Sandage and de Vaucouleurs recognized that some galaxies have weak bars. Sandage used a combined notation such as "Sb/SBb" for some of these cases, but de Vaucouleurs used a more effective approach. Instead of simply referring to nonbarred spirals as "S", he used "SA," which allowed the recognition of weak bars using the combined notation SAB. This is similar to using (rs) to recognize a partial inner ring. The same varieties as Sandage used were also recognized, except that de Vaucouleurs positioned them differently in his classifications. In de Vaucouleurs' notation, a nonbarred galaxy with an inner ring and a stage between Sa and Sb would be classified as SA(r)ab. A weakly-barred galaxy with an s-shape and a stage Sc would be classified as SAB(s)c. The most "mixed" type of all would be a classification like SAB(rs)bc.

---

[14]G. de Vaucouleurs, *Handbuch der Physik*, **53**, 275 (1959). In a letter dated April 17, 2007, Allan Sandage recounted to the author: "What I never told you, and I suspect that de Vaucouleurs himself did not know, is that I had been asked in 1957 by the editor of the Handbuch der Physik to write the article on galaxy classification for the astronomy volume of the Handbuch, but declined because of the pressure to complete the Hubble Atlas. I recommended de Vaucouleurs for the article, the recommendation that was accepted by [the editor] Flugge. Perhaps if de Vaucouleurs had known he would have modified his later attitude about Pasadena and the astronomers here." (This is likely referring to their disagreements on the extragalactic distance scale, which is described in Chapter 23.)

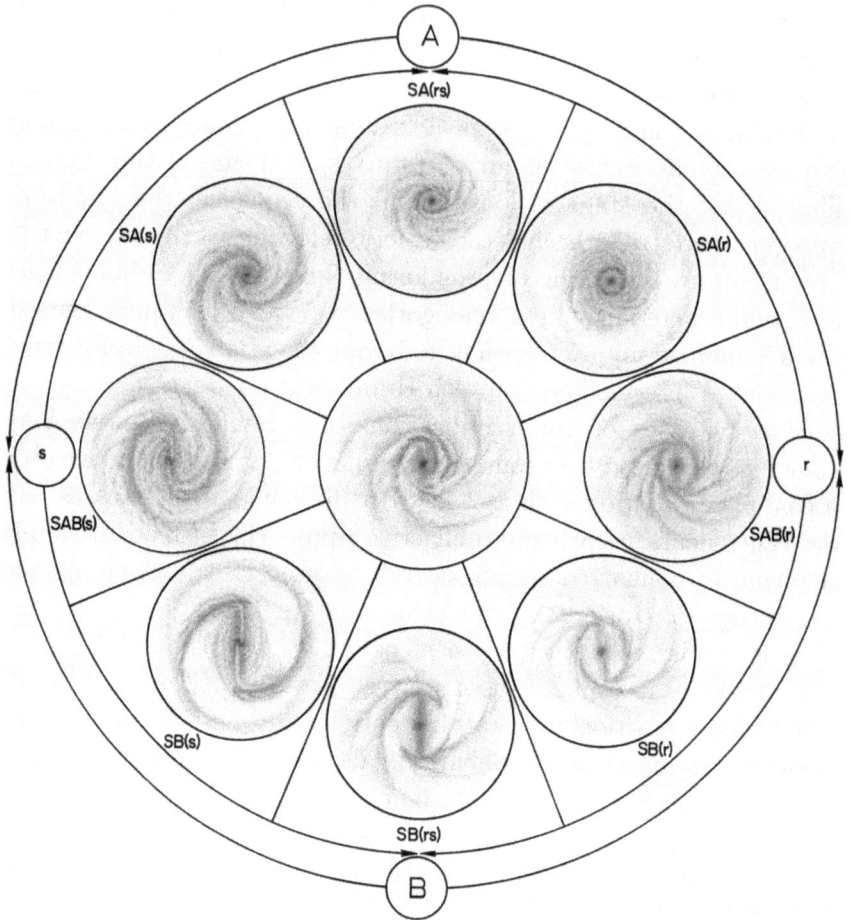

Fig. 8. An illustration of "families" (SA, SAB, SB) and "varieties" [(r), (rs), (s)] in the de Vaucouleurs revised Hubble-Sandage galaxy classification system. All of the types depicted are near the de Vaucouleurs revised Hubble stage Sbc. This figure first appeared in the Reference Catalogue of Bright Galaxies, the first of three major catalogues conceived and prepared by Gérard and Antoinette de Vaucouleurs, and coworkers. Scanned from original art.

Figure 8 shows de Vaucouleurs' famous sketch of the different families and varieties of galaxy morphology near the stage Sbc. The individual drawings highlight a cross-section through the classification volume defined by stage, family, and variety. The beautiful artistic rendition was made on a cloudy night at McDonald

Observatory circa 1962, and first appeared in print as the frontispiece of the Reference Catalogue of Bright Galaxies, de Vaucouleurs' first major galaxy catalogue that was published in 1964 (see Chapter 7).

Although the de Vaucouleurs revision is more complicated than merely a revision of the tuning fork, it can be depicted in part by adding a few extra galaxy types to the right end. Following a suggestion by Harlow Shapley and J. Paraskevopoulos (1889–1951),[15] de Vaucouleurs recognized spirals with very little or no central concentration and very open, complex spiral arms as types Sd and SBd. He also recognized the Magellanic Clouds as galaxies similar to Sd types but showing a characteristic asymmetry in the form of a single-armed morphology.[16] Magellanic-type spiral galaxies became known as Sm and SBm types.[17] Finally, de Vaucouleurs ended the classification with what were known as Magellanic irregular galaxies, or Im and IBm types. Such systems have complex, irregular shapes and often host a great deal of ongoing star formation. Figure 9 shows examples of Sd, Sm, and Im galaxies, and their barred counterparts.

Figure 10 shows all the parts of the revised tuning fork put together. A version of this tuning fork is the basis for the classifications provided in the *Carnegie Atlas of Galaxies*,[18] the most comprehensive photographic galaxy atlas ever made and based on the entire Carnegie Observatories plate collection. As effective as the Figure 10 version of the tuning fork may seem, it has been challenged in recent years. To see this, the nature of both S0 and elliptical galaxies has to be examined (Chapters 21 and 22).

How frequently do the different galaxy types occur when morphology is systematically and carefully observed? In a 1963 study, de Vaucouleurs examined the morphology of 1528 bright galaxies and

---

[15]H. Shapley and J. S. Paraskevopoulos, *Proceedings of the National Academy of Sciences of the United States of America*, **26**, 31 (1940).

[16]G. de Vaucouleurs and K. C. Freeman, *Vistas in Astronomy*, **14**, 163 (1972).

[17]Swedish astronomer Knut Lundmark also recognized the Magellanic Clouds as a distinct type of galaxy.

[18]A. Sandage and J. Bedke, the *Carnegie Atlas of Galaxies* (Carnegie Institution of Washington Publication No. 638).

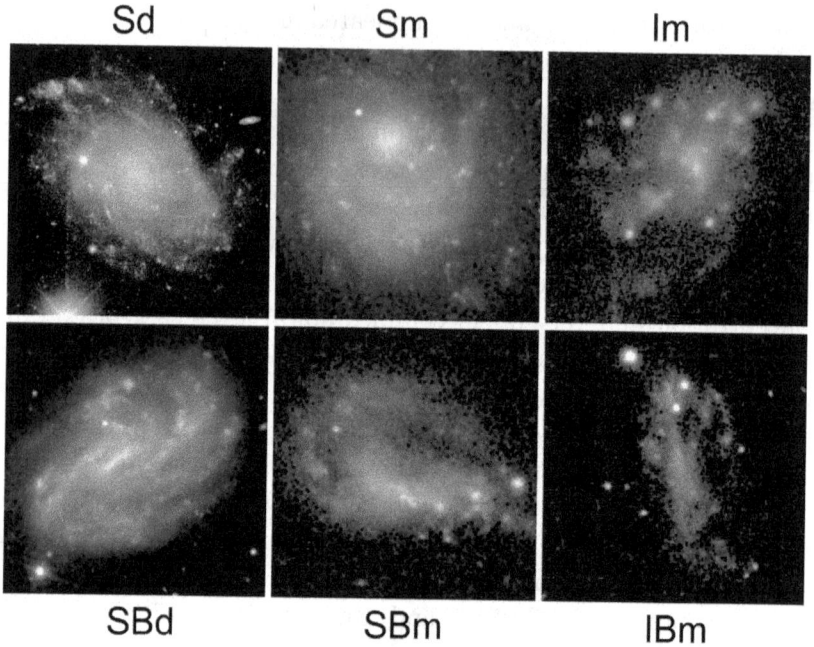

Fig. 9.   de Vaucouleurs's extension of the Hubble tuning fork from Sd, SBd to irregular galaxies. Top row: left to right: NGC 45 (Sd), NGC 5474 (Sm), and UGC 10549 (Im). Bottom row: left to right: NGC 5112 (SBd), UGC 12578 (SBm), and UGC 10795 (IBm). Sources of images: The de Vaucouleurs Atlas of Galaxies published by Cambridge University Press, 2007; reprinted with permission (NGC 45); EFIGI/SDSS (all the others; Credit: Sloan Digital Sky Survey Collaboration, http://www.sdss.org).

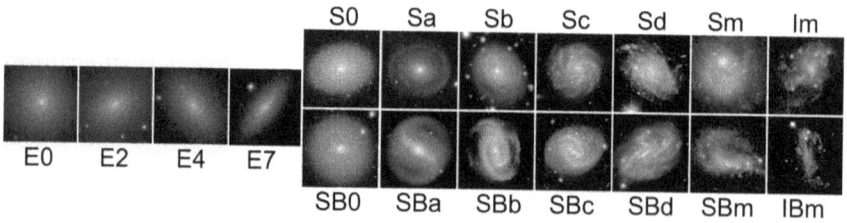

Fig. 10.   The revised Hubble tuning fork with input from Sandage and de Vaucouleurs.

found that 13% were ellipticals, 21% were S0s, 61% were spirals, 3% were irregulars, and 2% were peculiar or unusual.[19] This study confirmed that among luminous galaxies, spirals are the dominant type. de Vaucouleurs also was able to show that (r)-variety spirals are significantly less frequently seen than are (s)-variety spirals, especially among what Hubble called "late-type" spirals.

6.2. "Filing System" Controversies

Given the dynamism of nebular research a century ago, and the speed with which our understanding of these objects progressed, it is perhaps inevitable that there would be controversy. Hubble was not the only one to propose a new "filing system" to replace the old Wolf system. In 1920, 6 years before Hubble's monumental paper on galaxy classification and morphology, British amateur astronomer John H. Reynolds (1874–1949) proposed a simple classification of spiral nebulae[20] that is remarkably similar to that of Hubble, but was never referenced by Hubble. Reynolds envisioned seven categories of spirals, labeled with Roman numerals, that ranged from those having smooth, "amorphous" (or without a clear form) nebulosity (Type I); to those having significant inner amorphous structure but outer "condensations" (Type II); to those having even less amorphous inner structure (Type III); to those where only a small central region is amorphous, the rest of the structure being covered with condensations (Type IV, which was regarded by Reynolds as the category which "includes the great majority of spirals"); to those cases where the knots and condensations dominate the morphology completely (Type V); to cases where the star formation occurs in intermediate regions (Type VI); and finally to cases where the spiral structure lacks regularity (Type VII). Reynolds gave examples of

---

[19] G. de Vaucouleurs, *Astrophysical Journal Supplement Series*, **8**, 31 (1963).

[20] J. H. Reynolds, *Monthly Notices of the Royal Astronomical Society*, **80**, 746 (1920).

each type, and from these we can see that Type I is similar to Hubble type Sa, Types II and III are similar to Hubble type Sb, Type IV is similar to Hubble type Sc, Type V is similar to Sc, but with even less central concentration than Type IV (de Vaucouleurs would say the Hubble type of these is Scd), Type VI could include star-forming rings since these often are of intermediate scale, and finally Type VII includes some extremely dusty, irregular spirals. The system did not recognize barred spirals in any explicit way.

David L. Block and Kenneth Freeman, in their book *Shrouds of the Night*,[21] carried out extensive historical research on the origins of the Hubble classification system, and concluded that there is good reason to believe that Hubble would have been aware of Reynolds' paper, and therefore should have referenced it in his 1926 classification paper.

Ironically, in Hubble's 1926 paper, he accuses another person, Swedish astronomer Knut Lundmark, of publishing in 1926[22] a classification system that, except for the letter notation, was practically identical to his system. Lundmark used the term "Anagalactic nebulae" for the white nebulae, and divided them into three main classes[23]: spirals (As), "globular, elliptical, ovate, and lenticular nebulae" (Ae), and "Magellanic nebulae" (Am). The spirals were placed into subclasses As1c-As5c and As1b-As5b according to the appearance of the arms and the "degree of compressibility towards the center." Ringed spirals (Asr), Curtis ϕ-type spirals (Asp), Saturn-shaped spirals (Ass), and spirals with a companion or "appendix" nebula (Asa) were also recognized. The Ae's and Am's were subdivided into further categories depending on degree of central concentration.

Lundmark's classification system does have elements in common with Hubble's system and, by the same token, Hubble's system had elements in common with Reynolds. In the field of nebular

---

[21]D. L. Block and K. Freeman, *Shrouds of the Night* (New York, Springer, 2008), p. 205.

[22]K. Lundmark, *Archive for Mathematics and Physics*, **19**, 8 (1926).

[23]K. Lundmark, *Studies of Anagalactic Nebulae* (Royal Society of Science of Upsala, Almqvist and Wiksells, Uppsala, Sweden, 1927).

morphology, different observers would have emphasized details differently, but basic classes such as non-spirals, spirals, and irregulars would likely have been evident to anyone who carefully examined many plates of nebulae. Only Curtis challenged the existence of non-spiral nebulae. The question was, which of these classification systems provided the best astrophysical description of the white nebulae?

In 1958, University of Chicago astronomer William W. Morgan (1906–1994) outlined a galaxy classification system that built on an apparent correlation between the form and central concentration of a galaxy and the spectroscopic character of its nuclear light.[24] As had already been noticed by Vesto Slipher (Chapter 5), it was possible to tell from the appearance of a nebula what its nuclear spectrum would be like. This turned out to be an effective classification but, as noted by Morgan, it was still just a modification of Hubble's system.

In the end, all of these other classification systems besides Hubble's were abandoned, except for elements of Morgan's system, one of which is his recognition of supergiant amorphous galaxies called cD galaxies. These are frequently found at the centers of rich galaxy clusters and are believed to have formed by the "cannibalism" (merger) of smaller galaxies over billions of years. Thus, cD galaxies are a genuinely distinct class of objects. Figure 11 shows two examples of a Morgan cD galaxy.

Why has the Hubble classification survived in some form to the modern era, while the other classifications did not? I believe the reasons are many as follows:

(1) First and foremost, the original Hubble system was fairly simple to apply in practice. Based on the plate material available in his day, Hubble found that only a few percent of nebulae could not fit easily into one of his categories.

(2) The Hubble "tuning fork," also frequently called the "Hubble sequence," provided an organized way of looking at galaxies. Physical parameters of galaxies, such as their stellar content and

---

NGC 6166 cD A2199   **IC 1101 cD A2029**

Fig. 11.   Two examples of a supergiant amorphous galaxy type called cD from William W. Morgan's 1958 galaxy classification system. The "c" borrows from an obsolete way of denoting supergiant stars in spectral classification. For example, a supergiant star of spectral type F5 would be denoted cF5. Both of the galaxies illustrated are located in the centers of rich galaxy clusters, numbers 2199 and 2029 in George Abell's famous galaxy cluster catalogue.[25] The deep orange-reddish colors are due to the extremely old stellar populations that make up these galaxies. Images credit: Sloan Digital Sky Survey Collaboration, http://www.sdss.org.

amount of interstellar gas, were found to correlate with position on the fork (for example, see Figure 1 of Chapter 11).

(3) The effort of Allan Sandage to prepare the *Hubble Atlas of Galaxies* firmly cemented Hubble's ideas into galaxy research. Neither the Reynolds nor the Lundmark classifications were ever illustrated in such a manner, and indeed no later astronomers ever actually applied those systems to long lists of galaxies. The only other classification system to be so illustrated is the de Vaucouleurs revision to Hubble's system, described in the *de Vaucouleurs Atlas of Galaxies*.[26]

(4) The way edge-on galaxies can be classified within Hubble's system (as shown in Figures 3 and 6) without inventing new

---

[26]R. Buta, H. Corwin, and S. Odewahn, 2007, *The de Vaucouleurs Atlas of Galaxies* (Cambridge, Cambridge University Press, 2007).

subcategories for them is a definite advantage of Hubble's classification over the others (including Wolf's 1908 system). A view expressed by Hubble in a 1930 paper is that "The ideal classification should indicate the position of individual objects in the sequence of actual forms, independent of orientation."[27]

(5) Although the placement of S0 galaxies at the juncture of the tuning fork between elliptical galaxies and spirals has been challenged in recent years (Chapter 21), the fact that Hubble's classification gives prominent recognition to these objects inspired a great deal of informative research that led to investigations into the role of nature and nurture on galaxy evolution and morphology.

(6) We live in an era where there are publicly available, online databases that provide high-quality digital images of large numbers of galaxies. For example, the Sloan Digital Sky Survey (SDSS)[28] provides multi-filter, classifiable images of several hundred thousand galaxies. In the face of such numbers, we need a classification system that can accomodate most of what we see.

(7) The launching of the *Hubble Space Telescope* in the 1990s opened up research into the morphology of very distant galaxies, known as high redshift galaxies (Chapter 20). This renewed interest in galaxy morphology and how it might have changed over time kept the Hubble classification system itself from becoming as obsolete as the other systems.

(8) The way the original Hubble system could be revised and updated to account for new or other morphological subtypes that were not fully appreciated or recognized in Hubble's time is perhaps the system's most remarkable strength of all. The Comprehensive de Vaucouleurs revised Hubble-Sandage (CVRHS) system, illustrated in the *de Vaucouleurs Atlas of Galaxies*, represents one of the latest of these revisions.[29]

---

[27] E. Hubble, *Astrophysical Journal*, **71**, 231 (1930).

[28] J. E. Gunn and others, *Astronomical Journal*, **116**, 3040 (1998); D. G. York *et al.*, *Astronomical Journal*, **120**, 1579 (2000).

[29] The CVRHS system is described in more detail in R. Buta *et al.*, *Astrophysical Journal Supplement Series*, **217**, 32 (2015).

The extension of the Hubble system to higher degrees of complexity (as in the Hubble, Carnegie, and de Vaucouleurs atlases of galaxies), may be considered a regressive step. This was the view of Walter Baade, who felt there was no need to try and have the system recognize finer details than were needed to place a given galaxy into the system. He argued that "if you want to study the variations on the theme Sc, you simply have to take the plates and examine them — only then do you get the full story."[30] In our time, the "plates" have all been taken, by the millions in fact, and the variations on a theme such as "Sc" have never been better delineated than in the modern extension of the Hubble system.

6.3. A "Zoo" of Galaxy Morphologies

One of the most interesting developments in galaxy research the past 15 years has been the Galaxy Zoo project, a crowd-sourced astronomy effort to provide classifications for large numbers of galaxies from inspection of images provided by the SDSS. The authors of the project, led by Chris Lintott and Kevin Schawinski (both at the University of Oxford at the time), wanted to do a census of galaxy types in the SDSS database, which provided classifiable images of the largest number of unexamined galaxies ever available, on the order of a million. The problem was that galaxy classification can be a slow, difficult process, especially Hubble–Sandage–de Vaucouleurs classification, which requires paying attention to fine details. Even the most expert galaxy morphologist, working 8 h a day only classifying galaxies, would take decades to classify a million galaxies. It was clear that providing the coveted census that Lintott and Schawinski wanted required more than what the limited number of galaxy morphology experts worldwide could provide.

Because distinguishing between the very basic classes of galaxies (for example, disk-shaped versus non-disk-shaped, spiral versus non-spiral) is something that does not require a PhD in astrophysics,

---

[30] W. Baade, *Evolution of Stars and Galaxies*, C. Payne-Gaposchkin, ed. (Harvard University Press, 1963), p. 19.

Lintott and Schawinski decided to take the task of classifying a million galaxies to the general public. Classifications were performed using a web-based interface which allowed "citizen galaxy morphologists" to select choices of morphological characteristics using a small number of "buttons." More than 100,000 people worldwide participated in the first Galaxy Zoo,[31] which was followed up with a greater selection of buttons in Galaxy Zoo 2.[32] This latter project led to more sophisticated classifications for 304,122 galaxies. If a galaxy classification expert were to judge Hubble–Sandage–de Vaucouleurs types for a sample this large, at a rate of 1,200 galaxies per month, it would take nearly 20 years for that expert to get through the whole Galaxy Zoo 2 sample. By crowd-sourcing the project, the Galaxy Zoo 2 leaders reduced the classification time to two years, and even though the Zoo classifications were less comprehensive than, say, CVRHS types, the information was still extremely valuable.

The SDSS database included so many previously unexamined galaxies that Zoo classifiers were bound to see something new. One of these discoveries was a peculiar ionized, galaxy-sized intergalactic cloud dubbed "Hanny's Voorwerp" (Hanny's Object), which lies near the spiral IC 2497 (Figure 12). Hanny's Voorwerp was discovered by Dutch schoolteacher Hanny van Arkel, who noticed it while participating as a citizen morphologist in the Galaxy Zoo classifications. It has been proposed that the peculiar cloud is being photoionized by radiation redirected (light-echoed) from a now defunct quasar.[33] It is thought that IC 2497 suffered a major gravitational interaction with another galaxy in its past that distorted its disk, triggered a period of intense nuclear activity, and flung out a tail of material, some of which became the Voorwerp. The green color of the Voorwerp in the *Hubble Space Telescope* image in Figure 12 is due

---

[31] C. J. Lintott *et al.*, *Monthly Notices of the Royal Astronomical Society*, **389**, 1179 (2008).

[32] K. Willett *et al.*, *Monthly Notices of the Royal Astronomical Society*, **435**, 2835 (2013).

[33] C. J. Lintott *et al.*, *Monthly Notices of the Royal Astronomical Society*, **399**, 129 (2009).

Fig. 12.   The green object at right is the peculiar, galaxy-sized cloud known as Hanny's Voorwerp. The spiral at left is IC 2497. The two objects are 670 million light years away in the constellation Leo Minor. Image credit: NASA/ESA/ HST/W. C. Keel, with permission.

to doubly-ionized oxygen atoms.[34] Quasars and nuclear activity in galaxies are described further in Chapter 19.

6.4. In What Manner Do They Fade?

The era of photography in astronomy not only allowed astronomers to measure angular sizes of galaxies, but also to examine how galaxies actually fade with distance from the center, what Hubble called the "distribution of luminosity." That is, in what manner does the surface brightness of galaxies decline with increasing galactocentric distance?[35] This was for a while a complex problem, on one hand because galactic structure is so varied, and on another hand because

---

[34]W. C. Keel *et al.*, *Astronomical Journal*, **144**, 66 (2012).

[35]Distance from the center of a galaxy. Can be expressed in angular units such as degrees, arcminutes, and arcseconds, or in physical linear units such as light years or parsecs.

surface photometry[36] was so difficult to do reliably. Nevertheless, several important deductions were made that are still relevant to the study of galaxies today.

Looking for a way of characterizing the luminosity distribution in galaxies began with the work of English astronomer John H. Reynolds who, in a 1913 paper,[37] examined the photographic brightness distribution in the Andromeda Galaxy using plates taken with a 28-inch reflector he had built himself. Reynolds found that a simple formula like $(x + 1)^2 y = $ constant could represent fairly well the decline in brightness of the central region of Andromeda with increasing angular distance from the center. In Reynolds' little formula, y stands for surface brightness in linear units, and x is related to the angular radius of a point from the center. The constant is a number that would be different for different galaxies.

Hubble used a slightly modified version of the same formula for his study of the distribution of luminosity in elliptical galaxies in a 1930 paper.[38] Hubble found that what seemed to work for the bulge of the Andromeda Galaxy also worked for the amorphous elliptical galaxies, but a question was, over what range of angular distance would this work, and would the formula break down at some point? Hubble found reasonably good but not perfect agreement between the formula and the observed brightness distributions in a sample of elliptical galaxies. The Reynolds formula could not represent the brightness distribution in elliptical galaxies from their centers to their outer parts equally well. Beginning in 1946, Gérard de Vaucouleurs used his extensive knowledge of photography and photographic techniques (a topic he had written books about) to explore anew the distribution of luminosity in E galaxies, and found that if one graphs the surface brightness in magnitudes per square arcsecond versus the galactocentric angular radius r raised to the

---

[36]The measurement of the distribution of apparent brightness per unit area in a galaxy.

[37]J. H. Reynolds, *Monthly Notices of the Royal Astronomical Society*, **74**, 132 (1913).

[38]E. P. Hubble, *Astrophysical Journal*, **71**, 231 (1930).

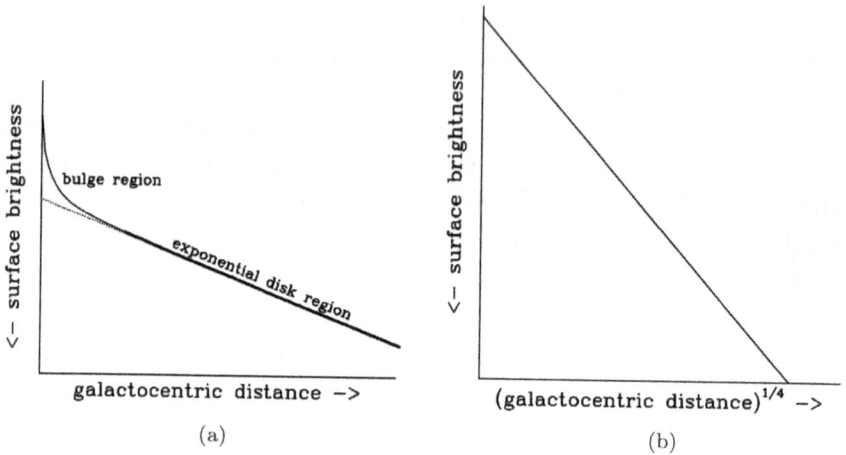

Fig. 13.   (a) A schematic of surface brightness in magnitudes per square arcsecond versus galactocentric distance in arcseconds or kiloparsecs for a typical disk-shaped galaxy. When plotted in this manner, an exponential becomes a straight line. The regions dominated by the bulge light and the disk light are indicated. (b) A schematic of surface brightness in magnitudes per square arcsecond versus galactocentric distance raised to the one-quarter power for a typical massive elliptical galaxy. When plotted in this manner, the de Vaucouleurs $r^{\frac{1}{4}}$ law becomes a straight line. The arrows on the vertical axes point downwards because larger magnitudes correspond to fainter light.

$\frac{1}{4}$ power, the points would follow a straight line over a wide range of surface brightness (Figure 13(b)).[39] This representation, which eventually became known as the "de Vaucouleurs $r^{\frac{1}{4}}$ law," seemed to work well for elliptical galaxies in general.

The distribution of luminosity in spiral galaxies was found to be different. Florence Shirley Patterson-Jones (1913–2000) was a student of Harlow Shapley, and one of the first to do a dissertation on galaxy surface photometry. In her 1940 study, she found that the surface brightness distribution in the Triangulum Spiral M33 declined in an *exponential* manner from the inner to the outer parts of the galaxy. That is, the surface brightness distribution could be written as $I = I_0 e^{-\frac{r}{a}}$, where r is the angular distance from the

---

[39] G. de Vaucouleurs and M. Capaccioli, *Astronomical Journal Supplement Series*, **40**, 699 (1979).

galaxy's center and I_0 and a are constants that would be different for different galaxies. The exponential nature of galaxy disks stands out best when one graphs the surface brightness in magnitudes per square arcsecond versus galactocentric radius r. In this kind of representation, an exponential is a straight line (Figure 13(a)).

De Vaucouleurs and others had also noticed that in galaxies consisting of a bulge and a disk, the bulge could sometimes be modeled using the $r^{\frac{1}{4}}$ law while the disk could be modeled as an exponential. Deriving the parameters of each component, known as *photometric decomposition*, became the staple of extragalactic astronomy for many years. With knowledge of fitted parameters, it became possible to search for *scaling relations* between parameters that could shed light on how galaxies formed and evolved. Examples of scaling relations would be how the sizes and luminosities of galaxies scale with galaxy mass.

In 1968, José Luis Sérsic (1933–1993) of the Cordoba Observatory in Argentina proposed a more general way of expressing the known luminosity laws of disk galaxies.[40] Noting that an exponential disk declines with increasing radius as $r^{\frac{1}{1}} = r$, he proposed describing disk galaxies using a general $r^{\frac{1}{n}}$ law, where $n = 4$ for spiral galaxy bulges and ellipticals, and $n = 1$ for exponential disks. Photometric decomposition today has almost entirely reverted to the Sérsic approach, but without forcing n to be 4 for bulges.

An interesting question is: if an exponential disk could be followed to infinity (or an approximation thereof), would the slope be the same? Some disks show this kind of behavior. Others, however, change slope at large distances from the center, either fading more rapidly than in the inner regions (a "truncated" disk) or less rapidly (an "anti-truncated" disk).[41] Determining the causes of these changes has been an important recent topic of research.

---

[40] J. L. Sérsic, *Atlas de Galaxias Australes* (Cordoba, Argentina, Observatorio Astronomico, 1968).

[41] P. Erwin, J. L. Tonry, and M. Pohlen, *Astrophysical Journal*, **626**, L81 (2005).

Chapter 7

From RSA to NED

The New General and Index Catalogues were valuable compilations of nebulae and clusters, but were lacking in astrophysical value because no physical measurements were attached to specific visual descriptions. For example, in these catalogues one could ask: how faint is "very faint" and how bright is "pretty bright?" How large is "considerably large," and how extended is "very much extended?" An astrophysically-oriented catalogue is one that contains measured physical quantities that can be used for studying the astrophysics of galaxies.

Photography opened the door to the production of such catalogues. The NGC/IC descriptions of the 19th century were no longer needed, because with photography, it was possible to measure the things that could only be described before. By the early 20th century, photography made it possible to:

(1) reliably judge the morphology of nebulae (that is, to classify them as elliptical, spiral, or irregular, for example);
(2) estimate real apparent brightnesses (magnitudes);
(3) quantify color through the use of light filters;
(4) measure apparent angular sizes in units like degrees, arcminutes, or arcseconds;
(5) measure bodily movement (called radial velocities or line of sight speeds);
(6) measure other parameters that became practical as time went on.

The first list of nebulae presented as an astrophysically-oriented catalogue was titled "A Survey of External Galaxies brighter than the 13th Magnitude," prepared by Harlow Shapley and Adelaide Ames, and published in 1932 in the journal *Harvard Annals*, Volume 88, Number 2. The catalogue included 1249 objects selected on the basis of apparent brightness, the first major catalogue so compiled. The title of the list is noteworthy. Recall that Shapley had been the proponent of the "one-universe" viewpoint at the Great Debate in 1920, favoring the idea that the universe was one galaxy, the Milky Way, and that the spiral nebulae were nearby objects and not external galaxies (Chapter 4). By 1932, however, Hubble, Curtis, and others had laid out the evidence for the "island universe" idea, and Shapley converted by now referring to the nebulae as external galaxies. In spite of Shapley's modern terminology, astronomers still persisted in using the term "nebulae" for galaxies even up to the 1960s.

One of the most important astrophysical quantites that could be measured for any galaxy is its apparent brightness. The measurement of apparent brightness in astronomy, called photometry, has a long history.[1] As already noted in Chapter 1, astronomers use the magnitude system for quantifying apparent brightness.

A perhaps unappealing aspect of the magnitude system is that smaller values of magnitudes correspond to higher apparent brightnesses. The magnitude system is logarithmic, meaning it is based on powers of 10. A first magnitude star is brighter than a second magnitude star, a second magnitude star is brighter than a third magnitude star, etc. With the Pogson scale, some former "first magnitude" stars came out to be magnitude 0, meaning brighter than magnitude 1. In fact, the brightest star in the night sky, Sirius, comes in at magnitude -1.5, a *negative* magnitude. This means the more negative the magnitude, the brighter the star (example: the Sun has a magnitude of -26.7). By the same token, the larger the magnitude, the fainter the object. Telescopes allow us to see stars fainter than the sixth magnitude, with bigger telescopes revealing fainter stars.

---

[1]For example, R. Miles, *Journal of the British Astronomical Association*, **117**, 172 (2007).

For example, a 2-inch telescope can reveal stars as faint as 11th magnitude, while a 200-inch telescope can reveal stars as faint as magnitude 21. These are visual limits with ground-based telescopes. With the *Hubble Space Telescope*, objects as faint as magnitude 31 have been detected in very deep digital images.

A critical task in photometry is defining a *photometric system.* Usually this involves choosing which stars will be defined to have magnitude "0" (or some other magnitude) in the system. Up until about the 1950s, photographic magnitudes m_{pg} based on blue-sensitive photographic plates were the basic type of magnitude astronomers derived. However, such magnitudes could not really stand alone; it was essential to define a second magnitude system at a different effective wavelength so that *color* could be judged from the difference between two magnitudes. The one chosen was a more yellow-sensitive filter-plate combination that sampled the wavelength domain of the human eye. This became known as a photovisual magnitude (m_{pv}). The difference $m_{pg} - m_{pv}$ is sensitive to color and became known as a "color index," being larger for redder stars and smaller for bluer stars. The "zero points" of the photographic and photovisual systems used stars having surface temperatures[2] of close to 10,000 K to define the color "0," or $m_{pg} - m_{pv} = 0$. For the m_{pg}, m_{pv} systems, the "standard stars" used would come from locations on the sky known as the Kapteyn selected areas.[3]

The magnitudes given in the Shapley–Ames catalogue are the most important part of Shapley and Ames's contribution to astronomy. Not only did they need magnitudes in order to decide which objects to include in the catalogue, but also they needed an effective method to judge those magnitudes, because galaxies are extended sources of light and not point sources like stars. The key to their success was to use telescopes having a very small image scale, such that galaxies looked like stars on plates taken with the telescopes. Shapley and Ames state: "The images of the nebulae were compared directly with the images of the stars of the magnitude sequences [that is, the "standard stars], and the magnitudes estimated to tenths."

---

[2] The Kelvin temperature scale is described in more detail in Chapter 10.

[3] These are described in Chapter 8.

As for which galaxies to actually observe this way, Shapley and Ames assumed that Dreyer's New General Catalogue was complete to the 13th magnitude, allowing them to focus their work on already known objects. They say that "The magnitude estimates were made almost entirely on plates of 90 min exposure ... taken with the 2-inch Ross–Tessar and the 2-inch Zeiss–Tessar at centers regularly spaced to cover the sky."[4]

In addition to magnitudes, the Shapley–Ames catalogue compiled available published angular dimensions of each galaxy in the form of a major axis dimension and a minor axis dimension, each in units of arcminutes.[5] Although useful for estimating physical sizes when distances are known, Shapley showed in a later study[6] that visual estimates of angular diameters from available plate material at the time were significant underestimates of the actual sizes.

7.1. Gérard de Vaucouleurs and the "Reference Catalogues"

The Shapley–Ames Catalogue was published at a time when detailed astrophysical extragalactic research was still in its infancy. As time went on, both telescopes and the quality of photographs improved. Although highly regarded, the catalogue inevitably would need to be revised as new information was gathered on its many entries. An ongoing program to obtain high-quality photographs of as many entries as possible was carried out by Hubble and others using the sophisticated facilities of the Mount Wilson and Palomar Observatories in the north and the Las Campanas Observatory in the south. This program eventually led to the Revised Shapley–Ames Catalogue (RSA) that was prepared by Allan Sandage and

---

[4]The Tessar was a special wide-field lens combination used in regular cameras in the early 20th century and manufactured by the Zeiss and Ross companies, among others.

[5]K. Reinmuth, *Publications of Heidelberg Observatory*, **9**, 1 (1926).

[6]H. Shapley, *Harvard Annals*, **88**, 4 (1934).

Gustav A. Tammann, and published in 1981.[7] The main justification for the effort was summarized by Sandage and Tammann, who argued that "surveys of bright galaxies have provided the foundation upon which much of observational cosmology rests." The extensive new photographic material was needed to "classify the galaxies for morphological studies, a process which ... leads directly to the central problem of galaxy formation and evolution."

Beginning in 1949, Gérard de Vaucouleurs took on the task of not only revising the Shapley–Ames Catalogue, but also of expanding its scope. His approach to making an astrophysically useful catalogue of galaxies was to gather data from multiple sources, compare observations in common between sources, and then *homogenize* the data by correcting for any systematic differences between sources and combining sources where possible using a weighting scheme. At the same time, Gérard and his wife, Antoinette, collected much new basic data on galaxies themselves, especially at McDonald Observatory of the University of Texas at Austin, whose faculty Gérard joined in 1960. Since data collection was an ongoing process, and new sources of galaxy data appeared regularly in the journals, Gérard began his expansion of the Shapley–Ames Catalogue by using a card index where he would add information on notecards, restricting at first to Shapley–Ames galaxies, and then expanding the sample to include all cases having at least one piece of information. Galaxies were excluded from the final catalogue only if they were too small (<1 arcmin) in angular size, fainter than magnitude 15, or had a line of sight velocity greater than 15,000 km/s.

The result of Gérard and Antoinette's efforts was a series of remarkable "Reference Catalogues" of bright galaxies, published in 1964, 1976, and 1991. The hallmark of the catalogues was homogenized data accompanied by lists or notes with references of data sources and other information on specific galaxies.

---

[7]A. Sandage, G. A. Tammann, *A Revised Shapley–Ames Catalogue of Bright Galaxies* (Carnegie Institution of Washington Publication No. 635, 1981).

RC1, titled the *Reference Catalogue of Bright Galaxies*[8] was prepared by Gérard and Antoinette and included 2599 galaxies, more than twice the original Shapley–Ames catalogue. In making RC1, Gérard noted that he and Antoinette tried to include any large but low surface brightness cases that were found on photographs but were missed by the visual observers, and which as a result did not have NGC or IC numbers. In so doing, he found that the original Shapley–Ames catalogue was only 50% complete to magnitude 12.5 (when it was supposed to be complete to magnitude 13.0). Not only were large, low surface brightnesses galaxies missed, but very small, high surface brightness galaxies had also gone unnoticed by visual observers, because these look too much like stars. Gérard concluded that "it is impossible in practice to produce a galaxy catalogue complete to any specified limit of magnitude or diameter." Even in the modern age this may still be true, because digital imaging is not yet available for the whole sky.

RC2, or the *Second Reference Catalogue of Bright Galaxies*, was coauthored by the de Vaucouleurs' and Harold G. Corwin, Jr., and published in 1976.[9] An explosion in data for individual galaxies, and many new references between 1964 and 1975, was what necessitated the new catalogue. Although RC1 had much useful data, there were nevertheless many blank entries because so many bright galaxies still did not have much basic data in 1964. As data accumulated, more galaxies started to have at least one piece of information, and the number of entries in the catalogue rose to 4,364.

RC3, or the *Third Reference Catalogue of Bright Galaxies*, was coauthored by the de Vaucouleurs' and Harold G. Corwin, Jr., Ronald J. Buta, Georges Paturel, and Pascal Fouqué and published in 1991.[10] Again, a continued explosion in basic data necessitated this next iteration in the catalogue series. The amount of information

---

[8]G. de Vaucouleurs and A. de Vaucouleurs, *University of Texas Monographs in Astronomy* No. 1 (Austin, University of Texas Press, 1964).
[9]G. de Vaucouleurs and A. de Vaucouleurs, *University of Texas Monographs in Astronomy No. 2* (Austin, University of Texas Press, 1976).
[10]G. de Vaucouleurs *et al.*, *Third Reference Catalogue of Bright Galaxies* (New York, Springer-Verlag, 1991).

available, the number of galaxies having any information at all, and the number of references was so large that the catalogue had to be printed in three volumes, with one volume devoted to references alone. The final catalogue, spread in two parts over the remaining two volumes, had 23,024 entries.

The value of Gérard's work can be seen in the way he derived magnitudes for galaxies. Prior to the mid-1930s, the only way to get the magnitude of a galaxy was photographically. Such a magnitude could be derived in a number of ways, as in the Shapley–Ames method of using plates of very small image scale so that galaxies look more like stars. A direct plate with a large image scale resolving details of morphology could also be used for this purpose, provided the relationship between photographic image density and actual intensity of the light is known. This is the basis of the technique known as *surface photometry*, and it can yield magnitudes by direct integration of the light. However, prior to 1970, surface photometry was a difficult process prone to significant systematic errors, and it was not practical to rely on such photometry for an astrophysically-useful galaxy catalogue.

In the mid-1930s a new method of doing astronomical photometry began to be applied, called *photoelectric photometry*. This technique, which was pioneered in astronomy by Joel Stebbins (1878–1966) and Albert E. Whitford (1905–2002)[11] measured apparent brightness using a piece of metal, known as a photocathode, that emits electrons when light shines on it (the "photoelectric effect").[12] An electric field accelerates the dislodged electrons to a series of detectors which amplify the signal. The final counts outputted by the photometer are in direct proportion to the apparent brightness of the object. Although photoelectric photometry did not provide images, it was nevertheless a more effective and efficient tool for deriving galaxy magnitudes than was surface photometry. As noted in Chapter 1, during the period 1977–1983, I used the McDonald Observatory 30-,

---

[11] J. Stebbins, A. E. Whitford, *Astrophysical Journal*, **86**, 247 (1937).
[12] In 1921, Albert Einstein was awarded the Nobel Prize in physics for explaining the photoelectric effect.

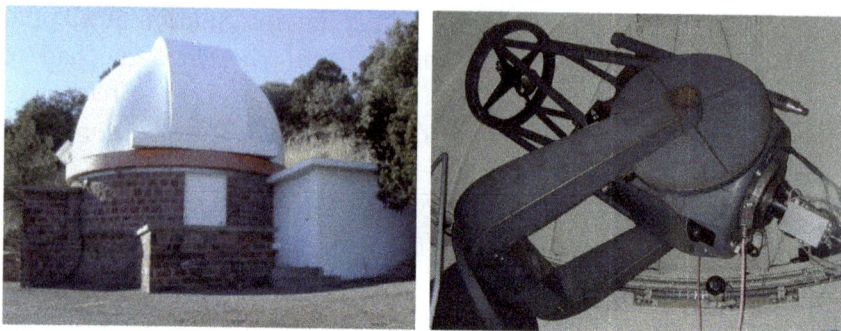

Fig. 1.   The McDonald Observatory 36-inch telescope. Images credit: mcdonaldobservatory.org.

36- and 82-inch telescopes on about 80 nights to do photoelectric photometry for a variety of projects, mostly involving galaxies. The 36-inch telescope and dome are shown in Figure 1. These observations were later augmented by similar photoelectric measurements at Mount Stromlo and Siding Spring Observatories in Australia from 1984 to 1986 and then again at McDonald Observatory from 1986 to 1988. All of my photoelectric work at McDonald and other observatories were used to derive parameters for RC3.

The main kind of photoelectric photometry that people did from about 1955 to 1980 was UBV photometry. This was a standard system of broadband photometry developed by Harold L. Johnson (1921–1980) and William W. Morgan, and involved measuring apparent brightnesses through what were known as standard UBV filters. The U-band filter has an effective wavelength of 360 nm[13] and is most sensitive to the near-ultraviolet part of the spectrum.[14] The B filter measures blue light magnitude similar to photographic m_{pg} magnitudes, and has an effective wavelength of 440 nm. The V filter has an effective wavelength of 540 nm and is designed to approximate the human eye response, just as photovisual magnitudes once did.

The UBV system was supported by an extensive set of standard stars accessible to both celestial hemispheres and had a zero point

---

[13]A nanometer (nm) is one billionth of a meter.

[14]The near ultraviolet is the longer wavelength part of the ultraviolet spectrum.

set by main sequence stars having a spectral type of A0 (meaning a surface temperature of about 10,000 K).[15] For A0 main sequence stars not significantly reddened by interstellar dust, the color indices in the UBV system were set to $U - B = B - V = 0$, and each filter corresponded to a specific flux above the atmosphere at these magnitudes. The basic zero point of the system ($V = 0$) was defined by the bright A0 star Vega. Doing UBV photometry involved bracketing "program" objects with periodic observations of UBV standard stars to correct for atmospheric extinction and any effects due to the use of filters or a photometer slightly different from the ones used to define the system.

For getting the magnitudes of galaxies, photoelectric photometry was superior to photographic plates. Photocathodes are more sensitive to light than are photographic plates, and the use of three filters (UBV) versus only two (m_{pg}, m_{pv}), provided more color information in the form of the color indices $U - B$ and $B - V$. The redder the star or galaxy, the larger the values of $U - B$ and $B - V$. The UBV system could classify stars on the basis of color.

The general procedure for doing UBV photometry in the late-1970s was to point the telescope at an object, view it in a finding eyepiece with illuminated cross-hairs, center the object on the cross-hairs, flip a mirror to pass the light into a measuring aperture (circular opening), center the object as well as possible in the aperture, and then flip another mirror to allow the light to fall on the photocathode. Most photoelectric photometers had a filter wheel that could be rotated automatically. At McDonald Observatory in the late-1970s, a flick of a switch would start an integration sequence, 10 s in each filter in the automatic order UBV, the numbers being recorded on paper tape. This 30-s period was the astronomers resting time. After the integrations were finished, one would manually move the telescope off of the object and sample the sky background brightness, choosing a location with no conspicuous objects based on a finding chart. The background was not very significant when

---

[15] The "0" in A0 refers to the first of 10 temperature subdivisions of the A spectral class.

observing bright stars but was usually very significant for galaxies, to the point that this kind of photometry could only be done well for galaxies when the Moon was close to a new phase (that is, during "dark time"). To get a reasonable signal-to-noise ratio, our general procedure at McDonald Observatory was to observe a program object five times, interspersed with four pointings on the blank night sky. The integrations were generally short enough that no specific guiding was needed.

When an observer's program mainly involved stars, the measuring aperture used had to be fairly small because stars are point sources and small apertures were generally used to define the photometric system. Comparison with standard stars would directly give the UBV magnitudes of any program stars. Galaxies, however, are extended sources of light having different angular sizes and shapes. A typical McDonald Observatory photoelectric photometer would have up to five measuring apertures: 1, 2, 4, 8, and 12 mm in diameter. Individually, none of these alone could be used to get the *total magnitude* of a galaxy, because galaxies do not have sharply defined edges. The method de Vaucouleurs used to get total magnitudes of galaxies was the technique of *multi-aperture* photometry: a galaxy is observed with all or most of the apertures available in a given photoelectric photometer (Figure 2(a)). The magnitudes are graphed versus the size of the aperture, and a curve is fitted to the points that can be extrapolated to give the total magnitude (Figure 2(b)), which is equivalent to what would be derived with an infinite-sized aperture. This was usually done for the blue light filter, and a similar analysis was made for the total $U - B$ and $B - V$ color indices.

7.2. The World of Dwarf Galaxies

Surface brightness is an important characteristic of galaxies that has an impact on their visibility. Surface brightness is brightness per unit area, which we have seen (Chapter 6) declines from the center to the outer parts of a galaxy in a relatively smooth manner. Surface brightness is not the same as total brightness. The total apparent

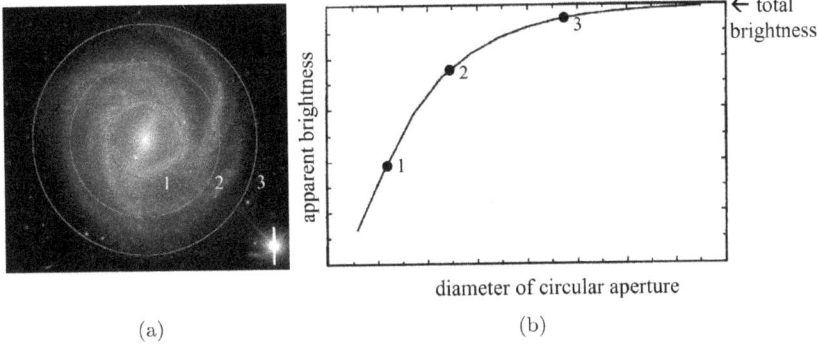

Fig. 2. Illustration of multi-aperture photometry for NGC 4902. The three circles represent three apertures in which the apparent brightness of the galaxy is measured using a photoelectric photometer. The brightnesses are plotted as magnitude versus the size of the measuring aperture. The total magnitude is derived by extrapolating a curve that represents the average shape of a magnitude versus aperture curve based on other well-observed galaxies. The shapes of these curves are sensitive to galaxy type (elliptical, spiral, irregular). Galaxy image source: Sloan Digital Sky Survey Collaboration, http://www.sdss.org.

brightness of a galaxy is its brightness integrated over its whole angular extent on the sky. Knowledge of the total apparent brightness as well as the distance to a galaxy allows us to determine the absolute total magnitude and therefore the luminosity of the galaxy. Surface brightness, on the other hand, is independent of the distance to a galaxy. Since galaxies appear as two-dimensional objects against the background brightness of the night sky, it is surface brightness that ultimately determines the visibility of an object, not total apparent brightness. If a galaxy is easy to see against the sky background, we say it has high surface brightness, while if it is hard to see against the night sky, we say it has low surface brightness. In galaxies like the Milky Way, the surface brightness is generally highest in the center and lowest in the outer regions far from the center.

During the late 1970s and early 1980s, I "learned the ropes" of *UBV* photoelectric photometry through an observing project that was the ultimate challenge when it came to measuring galaxy magnitudes. This was my work with Gérard and Antoinette de Vaucouleurs on photometry of dwarf galaxies. These galaxies are characterized by very low average surface brightnesses, smaller physical sizes, and

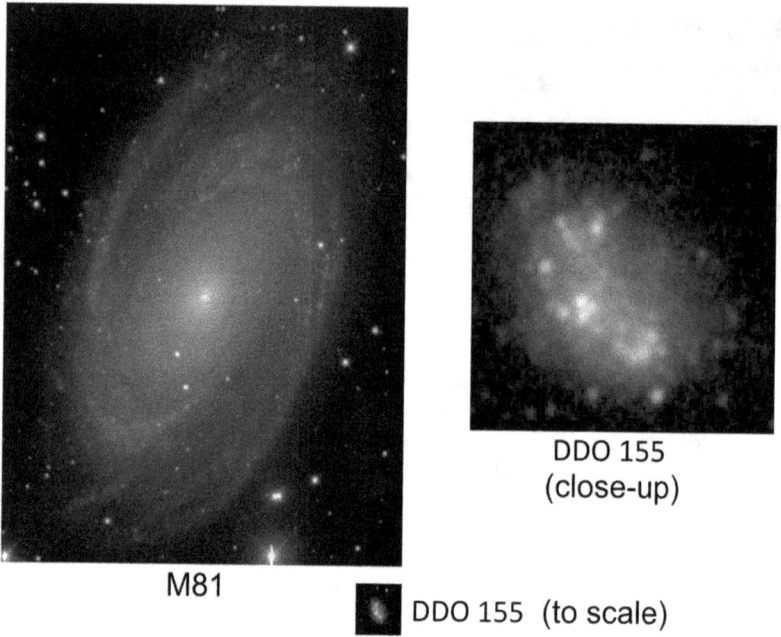

M81

DDO 155
(close-up)

DDO 155 (to scale)

Fig. 3.   Comparison between the giant spiral galaxy M81 and the dwarf irregular galaxy DDO 155. In the left panel, the two galaxies are compared on the same scale. In the right panel is a close-up of DDO 155, which is 4,000 times less luminous than M81. Adapted from R. Buta, *Secular Evolution of Galaxies*, J. Falcon-Baroso and J. Knapen, eds. (Cambridge, Cambridge University Press, 2013), p. 155. Reproduced with permission.

lower luminosities than typical spiral galaxies (Figure 3). Even under the darkest skies possible, dwarf galaxies were difficult to observe photoelectrically.

Dwarf galaxies are relevant to the study of spirals because large spirals like the Milky Way and the Andromeda Galaxy have many dwarf companions. The Milky Way, for example, has nearly 60 small companions. Undoubtedly, some of the dwarfs these big galaxies once had were absorbed by them, causing the dwarfs to lose their identities. Thus, dwarf galaxies can affect the evolution of spirals. Also, when proper samples are examined, it is found that dwarf galaxies are the most abundant type of galaxy in the Universe.

The dwarf observing project was part of Gérard's comprehensive program of photoelectric photometry of late-type spirals and Magellanic irregulars at McDonald Observatory. All of the observations made for the project became a part of RC3, and gave the catalogue a unique flavor because prior to the start of the project, dwarfs were sorely neglected for photometric observations.

The dwarf galaxies we observed were found on Palomar Sky Survey[16] prints by Canadian astronomer Sidney van den Bergh, who catalogued 243 examples larger than 1 arcminute in angular diameter in two papers published in 1959 and 1966.[17] At the time, van den Bergh worked at the David Dunlap Observatory in Toronto, and so the galaxies in the list became known as "DDO Dwarfs." Mainly on the basis of low apparent surface brightness, it was presumed that these were candidates for low luminosity dwarf systems, but apart from this, little was known about these rather insignificant-looking objects.

In the 1970s, radio astronomers R. Brent Tully and J. Richard Fisher were surveying the DDO galaxies in the 21-cm line of neutral atomic hydrogen, and found the objects generally rich in neutral interstellar gas.[18] For detailed analysis, and especially to determine how the dwarfs "fit in" with more luminous galaxies, Tully and Fisher needed measures of the total apparent magnitudes of the dwarfs, but could only get rough estimates from Palomar Sky Survey prints. On learning of this, Gérard realized that the dwarfs needed "proper photometry" based on photoelectric measurements, not photography. Undaunted by the dwarfs' low surface brightness, Gérard and Antoinette began observing the DDO objects photoelectrically at McDonald Observatory, using the multi-aperture approach. The program began in earnest in 1975 and involved not only both de Vaucouleurs, but Harold G. Corwin, Jr., Jean-Luc Nieto, William D. Pence, and myself.

---

[16]The Palomar Sky Survey was a photographic survey of most of the sky visible from Mount Palomar in California. The survey was conducted in the 1950s using a wide field Schmidt telescope.

[17]S. van den Bergh, *Publications of the David Dunlap Observatory*, **2**, 147 (1959); S. van den Bergh, *Astronomical Journal*, **71**, 922 (1966).

[18]R. B. Tully and J. R. Fisher, *Astronomy and Astrophysics*, **44**, 151 (1975).

At the time I started to work on the DDO program, I was amazed that anyone would try and observe such low surface brightness galaxies photoelectrically. This was because doing photoelectric multi-aperture photometry required visual centering of an object within a measuring aperture, and I could not imagine how one would do this if you could barely see what you were measuring. Because of this challenge, I have always looked on the project as one of de Vaucouleurs's boldest observational studies. He was the first to take on detailed photoelectric photometry of a large sample of low surface brightness galaxies.

My work on the DDO project began in April, 1978, when I joined Gérard and Antoinette for a three night observing run on the McDonald 36-inch telescope (Figure 1). The observations were made with what was known as the "Galaxymeter," a multi-purpose galaxy observing instrument that likely was designed by Gérard himself. Multi-aperture photoelectric photometry was only one of the capabilities of the Galaxymeter. It could also be used for spectroscopy. The dwarfs we observed one night were DDO 72, 112, and 120. Gérard did all the photometry for these and showed me the various steps. I looked through the main eyepiece in each case and could barely see the galaxies. The finding chart for DDO 72 showed it to have a small edge-on companion galaxy, which I later determined had the name UGC 5461. After observing DDO 72, Gérard moved the telescope to UGC 5461 and said: "You do this one." UGC 5461 thus became the first galaxy I observed photoelectrically.

Gérard later sent me out on other observing runs by myself. For these runs, I did not use the Galaxymeter, but instead used one of the generally available "people's photometers." A wide field offset guider eyepiece was a standard part of the setup of these photometers, which I found to be a little easier to use than the Galaxymeter whose finding eyepiece had a much more limited field of view. To reduce the noise in the integrated brightnesses, it was necessary to cool the instrument using dry ice. It was never hard to reach the offset guider eyepiece. The dome of the 36-inch telescope had a moveable floor that could be adjusted for any observing orientation.

I viewed the approach of a photometric (i.e., perfectly clear) night with great anticipation. As the Sun set, I eyed any residual daytime clouds with suspicion. If the night was my first night on the telescope for a given observing run, I would go through a checklist of things that had to be done once it was dark enough. The most important test was of the flatness of the field, that is, of the uniformity of the photocathode response across the measuring apertures. Another issue was the alignment of the crosshairs in the main offset guider eyepiece with the center of the measuring apertures. This was important because the dimmest, most invisible dwarfs could only be centered using the crosshairs and reference to a finding chart. By this I mean that if two foreground stars visible in the offset guider field formed a triangle with a dwarf as seen on the chart, I would try and make the same triangle using the center of the cross hairs. By centering locations on the crosshairs, flipping a mirror would place a galaxy as close to the center of the apertures as possible, such that even if I could not see some of the dwarfs, the photometer always detected them. In the end, I observed 143 DDO dwarfs at McDonald Observatory, of which I was able to see only half. My observing runs were recorded by Gérard and Antoinette as N01B–N19B, based on 19 separate runs of 3–7 nights each. By run N19B, I had made 488 photoelectric measurements of those 143 dwarfs.

In 1981, Gérard, Antoinette, and I prepared a detailed paper[19] summarizing all of the observations of DDO galaxies made up to May 1980 (corresponding to run N13B). We listed all measurements and derived the coveted "proper magnitudes" and other parameters needed by the radio astronomers. The main result of the 1981 paper was that virtually all of the DDO objects we observed are fainter than the dark (moonless and non-light polluted) night sky brightness. In visual light, the darkest sky corresponds to a surface brightness of 12.6 magnitudes per square arcminute. For 132 of

---

[19] G. de Vaucouleurs, A. de Vaucouleurs, R. Buta, *Astronomical Journal*, **86**, 1429 (1981).

the dwarfs I measured, the average visual surface brightness was 14.8 magnitudes per square arcminute, or only 14% of the dark night sky brightness. This is also only 1/3 the mean surface brightness of the large but faint Triangulum Spiral M33.

In a subsequent 1983 paper,[20] we analyzed the derived apparent brightnesses to get some idea of the average properties of the DDO galaxies. Were they all really dwarfs, as their low surface brightnesses seemed to imply? To assess this, we needed distance estimates to the galaxies, and Gérard and Antoinette collected all the information they could find, often relying on group memberships. The results showed that while the more nearby DDO objects are in fact real dwarfs (like DDO 155 shown in Figure 3), the more distant ones are likely to be intermediate-sized systems like the Triangulum spiral or the Large Magellanic Cloud.

Although the DDO objects were faint and showed little detail, they were still interesting objects to see. These were the main objects I got to observe under the darkest and clearest skies possible at McDonald Observatory. Even on photometric nights, the dwarfs' visibility was very sensitive to any moonlight, solar activity, atmospheric dust, humidity, zenith distance (angular offset from the overhead point), even a single car passing the mountain at night. Compared to ordinary bright galaxies, the main characteristic lacking in DDO dwarfs was a noticeable central concentration. This is what made them hard to observe using the photoelectric technique.

Surprisingly, not all dwarf galaxies have low surface brightness. Some high surface brightness galaxies are also dwarfs. An example is NGC 3928, a small, E0-like galaxy lying 47 million light years away in the constellation Ursa Major. NGC 3928 is unusual in that it has an embedded mini-spiral, an exceedingly rare example of a dwarf Sa-Sb spiral.[21] It is shown in Figure 4 as compared with UGC 6614, an exceptionally large spiral (more than 250,000 light years in diameter) lying 281 million light years away in the constellation Leo.

---

[20] G. de Vaucouleurs, A. de Vaucouleurs, R. Buta, *Astronomical Journal*, **88**, 764 (1983).

[21] S. van den Bergh, *Publications of the Astronomical Society of the Pacific*, **92**, 409 (1980).

Dwarf spiral NGC 3928
versus giant low surface
brightness spiral galaxy
UGC 6614

NGC 3928 B

|............... 4.5 kpc .... |

→ NGC 3928 at same distance

|........................... 81.6 kpc .... |

UGC 6614 B

Inner ring: 29 kpc

Inner ring: 260 pc

Fig. 4.   Comparison between blue-light (*B*-band) images of the giant ringed spiral galaxy UGC 6614 and the dwarf Sab spiral galaxy NGC 3928. Both images are in blue light, and the diameters of the rings in parsecs (pc) or kiloparsecs (kpc) are indicated. Note that UGC 6614 overflows the field shown, which corresponds to a diameter of more than 250,000 light years. The *B* after each name refers to the *B* filter. From R. Buta, *Secular Evolution of Galaxies.* J. Falcon-Barroso and J. Knapen, eds. (Cambridge, Cambridge University Press, 2013), p. 155. Reproduced with permission.

Both galaxies have an inner ring, but while the ring of NGC 3928 is only 260 pc (850 light years) in diameter, that in UGC 6614 is 29 kpc (95,000 light years) in diameter. This means the inner ring of UGC 6614 is as large as the Milky Way. UGC 6614 has been classified as a giant low surface brightness galaxy because the spiral arms and disk light outside the bright ring are so dim.[22] These two galaxies illustrate the extremes which are possible for spirals.

---

[22]S. S. McGaugh *et al.*, *Astronomical Journal*, **109**, 2019 (1995).

7.3. Magnitudes Today: The CCD Revolution

Photoelectric multi-aperture photometry became obsolete as a means of getting total galaxy magnitudes in the 1990s. Photography also became obsolete. At that time, digital detectors using charge-coupled devices (CCDs) took over as the tool of detection in astronomy. CCDs have several major advantages over photographic plates and photoelectric photometers:

(1) CCDs have a very high "quantum efficiency," or sensitivity to light. Compared to photographic images, CCD images are faster to produce, do not require development, are already digital, do not have any special storage requirements, and do not degrade significantly over time.

(2) A CCD is a more linear detector than is a photographic plate. This means that the numbers in a CCD image depend directly on the number of photons[23] that strike a given part of the detector. CCDs do become nonlinear above certain brightness thresholds, but these can be avoided by combining multiple frames of shorter exposure time.

(3) The impact of foreground or background objects on photometry with CCD images can be dealt with much more effectively than these could with a photoelectric photometer. Many galaxies have superposed field stars that, if left unremoved, would perturb the magnitude-aperture graph and lead to large errors in the total magnitudes. The largest apertures, especially, would often have such objects within their area. With CCD images (and with calibrated photographic plates), the removal of foreground stars and background galaxies can be done largely automatically, and total magnitudes free of foreground or background object contamination can be determined more accurately than they could from multi-aperture photometry.

(4) With CCD images, one can do standard star photometry in the exact same manner as one does photoelectric photometry, that is, program object observations interspersed with standard star

---

[23] A photon is a particle of light that has no mass but has an energy that depends inversely on the wavelength. The shorter the wavelength, the higher the energy.

observations. Unlike in photoelectric photometry, where only a single standard star can be observed at a time, CCD images often have more than one useful standard in a given image field.

(5) CCDs have considerable sensitivity to red and infrared light. Harold Johnson added longer wavelength filters to his standard UBV system to broaden the base of the system and open the sky to these important wavelengths. The new filters were labeled R (red, 700 nm) and I (infrared, 900 nm). However, the Johnson filters were not adopted by the astronomical community who preferred slightly different R and I filters proposed by A. W. J. Cousins. In Cousins' system, R is at 650 nm and I is at 790 nm. The combination of the Johnson UBV and Cousins RI systems is called the $UBVRI$ standard system. CCDs can be effectively used for $UBVRI$ photometry of galaxies.

7.4. Luminosities and Angular Dimensions

Having well-defined total magnitudes for galaxies allows us to estimate how luminous these systems actually are. As already noted in Chapter 4, the luminosity of any object, star or galaxy, is the amount of radiant energy it emits every second. A total blue magnitude, like that provided in the de Vaucouleurs Reference Catalogues, can be used to determine the luminosity, at least in blue light. After corrections for estimates of foreground and internal dust, the Andromeda Galaxy M31 has a blue-light luminosity of 40 billion suns while the Triangulum Spiral M33 has a blue-light luminosity of 5 billion suns. The Small and Large Magellanic Clouds have blue-light luminosities of 600 million and 2 billion suns, respectively, and are thus much dimmer systems than are M31 and M33. The SMC is close to what is considered a dwarf galaxy, but it is much more luminous (by a factor of 60) than the dwarf DDO 155 (Figure 3). One of the most massive spirals known, NGC 1961, has a blue-light luminosity of nearly 180 billion suns.

Shapley[24] showed that different plate collections and observers can yield very different estimates of the angular sizes of galaxies.

---

[24] H. Shapley, *Annals of Harvard College Observatory*, **88**, 91 (1934).

Galaxies tended to look larger on longer exposure plates than they did on shorter exposure plates. This occurred because galaxies generally do not have sharp edges, but instead fade gradually until their light is so faint that it blends into the noise of the sky background (Chapter 6).

The best way to define the size of a galaxy is to choose a specific level of surface brightness as a standard. Because galaxies are extended sources of light, the way they look defines their surface brightness, or "luminosity," distribution. A particular galaxy may have a bright center of high surface brightness, or may be characterized by low surface brightness everywhere. Spiral arms are generally low surface brightness features easily lost to light pollution by eye. In terms of magnitudes, surface brightness has units of magnitudes per square arcsecond. To give angular sizes of galaxies more physical meaning, de Vaucouleurs chose 25.00 magnitudes per square arcsecond in blue light as the reference surface brightness level for angular sizes in the reference catalogues.

Figure 5 shows an ellipse that corresponds to the 25.00 magnitudes per square arcsecond dimensions of the spiral NGC 5364 in the constellation Virgo. The ellipse characterizes not only the size of the galaxy, but also its tilt to the line of sight, much like a circle appears elongated if you view it from an inclined perspective.

The angle of the long axis of the 25.00 magnitudes per square arcsecond ellipse also carries information. How the rotation of a disk-shaped galaxy projects onto the sky depends not only on the tilt of the galaxy, but also the line around which it is tilted. The long axis of the ellipse in Figure 5 is where the galaxy plane and the sky plane intersect, and is called the photometric *line of nodes*.[25]

The measured angular diameter (length of the long axis of the ellipse) is 6.6 arcmin. With an estimate of the distance to NGC 5364, we can convert this angular size to a linear (physical) diameter. Modern estimates place NGC 5364 at a distance of 54.5 million

---

[25]In celestial mechanics, a node is a point where an object moving in a particular orbital plane crosses a reference plane. In the case of a galaxy, the orbital plane is the plane of the galactic disk, while the reference plane is the plane of the sky.

Fig. 5. The angular dimensions of NGC 5364 are represented by the red ellipse where the surface brightness in blue light is 25.00 magnitudes per square arcsecond. Image source: Adapted from the de Vaucouleurs Atlas of Galaxies published by Cambridge University Press, 2007; reprinted with permission.

light years which, together with the measured angular size, tells us that NGC 5364 has a linear diameter of 105,000 light years. This is very similar to the number that is often used to describe the size of the Milky Way: 100,000 light years. Both the Milky Way and NGC 5364 are typical spirals, the kinds that are found in abundance in catalogues. For comparison, the dwarf DDO 155 has a diameter of only 1,600 light years, while the supergiant cD galaxy IC 1101 (Figure 11(right) of Chapter 6) has a diameter of nearly 350,000 light years, at the same surface brightness level.

Being rotating stellar systems, the tilt of a spiral galaxy to the line of sight is an important parameter that determines how much of the rotation speed of a galaxy at any given galactocentric distance projects to the line of sight. If we assume that the shape of the ellipse in Figure 5 gives an indication of tilt, then a measurement of the projected line of sight speed at the radius of the ellipse, together with the tilt, would tell us the amount of mass enclosed by the ellipse. In the case of NGC 5364, available data give a mass of 100 billion suns, which implies the presence of 100-billion stars if all of the galaxy's stars were the same mass as the Sun. This is unlikely because most stars in the Milky Way are less massive than the Sun. The actual value could be as high as 200 billion stars.

De Vaucouleurs' approach to measuring angular diameters of galaxies was largely successful. Any visual surveys of angular sizes of galaxies could be compared with his 25.00 magnitudes per square arcsecond dimensions and reduced to a standard system, thereby increasing the astrophysical value of his catalogues.

7.5. The Last of Its Kind

RC3 was the last galaxy catalogue of its type published. The National Aeronautics and Space Administration/Infrared Processing and Analysis Center (NASA/IPAC) Extragalactic Database (NED)[26] was conceived by Caltech astronomers George Helou and Barry F. Madore in the late 1980s and is the modern approximation to a de Vauouleurs-style reference catalogue in that references, positions, magnitudes, line of sight speeds, and other information are gathered for convenient use by astronomers. There are, however, several major differences: NED is strictly online and therefore can be kept up to date; NED includes more than a million galaxies, making it on a scale unprecedented in extragalactic astronomy; and NED has many items that were not practical or timely to include in RC3, such as distance estimates, images, space-based observations, a more extensive cross-index of identifications, and estimates of Galactic extinction[27] in many different filters. NED also includes a knowledge database of review articles and other publications in extragalactic astronomy.

The one major distinction between RC3 and NED is that the latter mostly collects data, but, except for standardization of units,

---

[26] See ned.ipac.caltech.edu.

[27] Galactic extinction (with a capital G) refers to extinction due to interstellar dust in the Milky Way.

neither homogenizes it nor necessarily derives any additional parameters from the collected data. This is left to individual astronomers to extract what they might need from the database. In spite of what NED has to offer, RC3 has not become obsolete since it was published 30 years ago. From 2015–2019, RC3 has averaged nearly 80 citations per year and continues to be useful to the astronomical community.

Chapter 8

Our Home Spiral Galaxy

In Chapter 1, I described the spectacular appearance of the Milky Way as seen from a high mountain top in Chile on a moonless, crystal clear night in July, 1981. At midnight, the Milky Way in Sagittarius stood overhead, and looked like a bird with enormous wings. This view of the Milky Way is special because Sagittarius is in the direction of the Galactic center. Based on extensive photographic surveys of nebulae, many astronomers noticed that the appearance of spiral galaxies depends in part on how they are viewed. The spirals appear roundish when oriented face-on, and more ray-like when oriented edge-on. Also, the more inclined the galaxy, the more apparent lanes of obscuring material become, until in the exactly edge-on view, the ray shape is divided almost evenly by a single, narrow dark band. An example of a nearby edge-on spiral showing this is NGC 891 in Andromeda (Figure 1 of Chapter 2).

The dark material seen in highly-inclined spirals like NGC 891 (and others shown in Figure 3 of Chapter 6) is interstellar dust. Although manifested most clearly in photographs, such lanes were actually detected visually well before the photographic era of astronomy. For example, in Figure 4 of Chapter 2, nebula 12 shows the detection by William Herschel of a clear planar dust lane in an edge-on, ray-shaped object he called V. 19, which turns out to be NGC 891. In his 1811 paper, Herschel describes V. 19 as a "considerably bright nebula about 15′ long and 3′ broad; its length is divided in the middle by a black division at least three or four minutes long."[1] The most famous edge-on spiral galaxy is the Sombrero Galaxy M104 in Virgo, shown previously in Figure 1 of Chapter 4. In this spectacular object, the disk of dust and spiral structure is tipped about 6° from edge-on and is silhouetted against a massive, bright bulge. Another remarkable case is NGC 5866, where the dust lane appears to have a very limited length (Figure 1).

Figures 1 of Chapter 2 and 1 of this chapter show what planar disks of dust look like from a great distance, well outside the galaxies themselves. In this view, the lanes are well defined and relatively thin. But viewed from within, the perspective would be different. For example, if one were to "fly into" the dust lane of NGC 891, the dust would obscure large regions and give an irregular appearance to the galaxy's star clouds. In addition, the bright center of the galaxy, known as the nucleus, would be completely obscured, and the spiral structure in the disk would not be evident at all. This is exactly how we view the Milky Way. Our perspective is an edge-on view as in NGC 891. We know it is not like our view of the Sombrero, because we can see the bright center of the Sombrero.

The existence of obscuring material in the disk planes of other galaxies seemed clear even before Hubble established the extragalactic nature of nebulae, and one could also make the case for some in our Galaxy. For example, it had been known since the 1780s that the common nebulae with fairly regular shapes (that is, the ones later proven to be other galaxies) largely avoided the band of the Milky Way, while in the directions of the Galactic poles, perpendicular to the plane of the Milky Way, rich fields of nebulae were found. So few nebulae were found in the direction of the Milky Way that Hubble named the region the "Zone of Avoidance." Even globular clusters to a significant extent also avoided the Milky Way. Once it was established that most nebulae are galaxies, it became imperative to determine how the obscuring material causing the Zone of Avoidance is distributed and what it might be made of.

---

[1]W. Herschel, *Philosophical Transactions of the Royal Society of London*, **101**, 300 (1811).

Fig. 1. A *Hubble Space Telescope* image of the edge-on disk galaxy NGC 5866, showing a truncated and slightly warped dust disk and an extended bluish-tinted stellar disk. The truncation of the dust disk could signify a past gravitational interaction with another galaxy. Image credit: NASA, ESA, and the Hubble Heritage Team (STScI/AURA).

The key to proving the existence of obscuring interstellar material in our Galaxy is to focus on a class of objects located in the Galactic plane, and for which the distance can be determined using at least two techniques: one based on measures of star brightnesses, and one based on some geometric property, such as the size. In 1930, Robert J. Trumpler (1886–1956) of Mount Hamilton Observatory in California

Fig. 2.   The Hyades star cluster in Taurus. The bright reddish-colored star is Aldebaran, a red giant lying 85 light years in the foreground of the cluster. Image copyright: Alan Dyer; used with permission.

used open star clusters for this purpose.[2] Open clusters are moderate to loosely bound groupings of a few hundred stars that are found throughout the Galactic plane. A few examples of open clusters are shown in Figures 2 and 3.

The nearest open cluster is the Hyades in Taurus which, at its distance of only 46 pc (150 light years), is close enough to us for its distance to be determined using trigonometric parallax. This is the most accurate way of getting the distance to any star. Most other open clusters are, however, too distant for this method to be used. Instead, a technique known as "main sequence fitting" is used. All open clusters have main sequence stars (Chapter 10), and by comparing the brightnesses of main sequence stars in distant clusters with the same stars in the Hyades, one gets an estimate

---

[2]R. J. Trumpler, *Publications of the Astronomical Society of the Pacific*, **42**, 214 (1930).

(a)   (b)

Fig. 3.   (a) The open star cluster M25. Image credit: CFHT/Coelum — J.-C. Cuillandre/G. Anselmi; with permission. (b) The open star cluster M39. Image credit: H. Schweiker, WIYN, NOAO, AURA, NSF.

of the distance to the cluster. One can also define a physical size for a cluster. Individual clusters tend to have a bright core region a few light years across, and are generally no bigger than 20–30 light years in diameter. The distance to a cluster can be determined by assuming all have the same physical size.

Comparison of the two distance estimates is an effective way of detecting the presence of interstellar absorbing material. This is because the first method is sensitive to any obscuration, while the second method, being purely geometric, is not. Extinction can make a nearby star appear farther away than it actually is. If absorbing material is present and spread widely across the Galactic disk, then we would notice that the difference between the main sequence star distance and the linear diameter distance for clusters would systematically increase with increasing distance. This is exactly what R. J. Trumpler found in his detailed study.

The nature of the absorbing material is no longer a mystery. The obscuring material selectively scatters and absorbs light in the sense that bluer light is more extinguished than redder light. Thus, interstellar extinction is accompanied by interstellar reddening. This tells us that the obscuring material is in the form of dust grains having a size comparable to the wavelength of visible light, that is, on the order of one 10-millionth of a meter. Such grains are mostly built around heavy elements like carbon and silicon. The more dust starlight must pass through to reach us, the greater the amount of extinction and reddening of that starlight. There is so much dust between us and the Galactic center that we cannot see the galactic nucleus at all in visible light.

Where does the dust come from? It is believed that in the normal course of star evolution, dust condenses in the atmospheres of highly evolved stars, especially low mass stars which are in a second red giant phase. Such stars are so cool and so bloated that dust grains can form around elements like carbon that are created in the core region and carried into the star's atmosphere by hot bubbles of rising gas (an energy transport mechanism known as *convection*). The dust can be carried out into interstellar space by the pressure of the star's light or by a "wind" of particles, like a scaled-up version of the solar wind. Also, these second stage red giants are the progenitors[3] of planetary nebulae, the colorful expanding gases that are silently ejected into space at the end of a low mass star's life cycle. Any dust created by the star and still in its vicinity would also be carried into space. Although most of the dust in a spiral likely comes from low mass stars, which are much more numerous than high mass stars (Chapter 13), dust can also form in the atmospheres of red supergiants and in the violence of a supernova explosion.[4]

After a long period of time, the expelled dust mixes with interstellar gas and collects in the Galactic plane into a layer that is generally thinner than the stellar distribution perpendicular to the plane. After many generations of stars have come and gone, the dust fills much of the disk plane and its distribution can react to the presence of noncircular structures, such as bars and spiral arms. Dust is virtually never found in the halo regions of spirals because particle collisions would not allow it to easily pass through the Galactic plane.

The dustiness in the Galactic plane has always made determining the structure of the Milky Way challenging. By "structure," I mean its physical shape, how the matter is distributed in the Galactic

---

[3]The objects that expelled their outer layers of gas.
[4]For example, A. Sarangi *et al.*, *Astrophysical Journal*, **859**, 66 (2018).

disk, the properties of the Galactic bulge, and the nature of any fine structure in the Galactic disk, such as spiral arms, a bar, and a ring. Early attempts to determine the structure of the Milky Way were based on star counts in different directions. The most celebrated attempts were made by William Herschel and Jacobus C. Kapteyn (pronounced Cap-tine; 1851–1922). Herschel used his 18.7-inch telescopes to visually count stars in 600 different directions and, in 1785, deduced that the Galaxy was a flattened system centered not far from the Sun. Kapteyn, of the University of Groningen, approached the problem a century later using photography. Because it was not practical to count stars in all directions from plates, Kapteyn chose 206 directions uniformly spread over the whole sky to act as sampling tools, each about 1 square degree in area.[5] In each of these directions, which he called "selected areas," he compiled basic star data, such as positions, apparent brightnesses, spectral classes, trigonometric parallaxes, and sky-plane (proper) motions, to try and improve on Herschel's work. Because his institution lacked an observatory at the time, Kapteyn made no observations himself. Instead, he made arrangements with astronomers from other institutions that had observatories to collect the data needed, which he would process with the help of assistants. The result was a model of the Milky Way consistent not only with star counts, but also with what he thought were the systematic motions in the Galactic system.[6]

Although Kapteyn's approach to the structure of our Galaxy was far more comprehensive and sophisticated than what Herschel was able to do, he left out one effect: the possibility of obscuring material in the plane of the Galactic disk. As a result, the location of the Sun in his system was much closer to the center than it was in Shapley's model based on the distribution of globular star clusters. Because globular clusters are not confined to the Galactic plane, but instead

---

[5] *Astronomical Laboratory at Groningen*, 1906; see also F. H. Seares, *Proceedings of the National Academy of Sciences*, **3**, 188 (1917); B. T. Lynds, *Astronomical Society of the Pacific Leaflets*, **9**, 89 (1963).
[6] P. C. van der Kruit (2014), arXiv 1407.2632.

follow a roughly spherical distribution around the galactic center, these objects give a more reliable picture of the Galactic system than do general star counts. The problem of Kapteyn's model was not so much the assumption of no absorbing material, but the general uncertainty at the time of the reality of such material. Even though astronomers had seen clear evidence of the existence of absorbing material in photographs of edge-on spirals, this did not necessarily translate into definitive evidence for such matter in the Milky Way.

Our perspective on the Milky Way does not allow us to see its spiral structure directly. Our view is too edge-on, which allows the dust clouds to affect everything we see. The discovery of spiral structure in nebulae nevertheless fueled speculation that our Galaxy was also a spiral. In a remarkable 1852 paper,[7] Princeton professor of mathematics and astronomy Stephen Alexander (1806–1883) took seriously descriptive comments on the appearance of the Milky Way made by the Herschels, as well as the observations by Lord Rosse, to deduce that the Milky Way was a multi-branched spiral. In order to make such a judgment, Alexander had to first believe that spiral nebulae were, in fact, systems of stars like the Milky Way. In a 1958 paper,[8] radio astronomers Jan Oort (1900–1992), Frank Kerr (1918–2000), and Gart Westerhout (1927–2012) argued that the "the extreme flatness of the Galactic System, as well as the frequency of large groupings of O and B stars, had for a long time been convincing evidence that [the Milky Way] belonged to the class of spiral galaxies."

We would not have to go too far out of the plane of the Milky Way to see the arms. As already shown in Figure 7 of Chapter 3, the spiral structure of the galaxy NGC 3877 is still detectable even at an inclination of 80°. But to view our Galaxy from such a vantage point would require traveling nearly 5,000 light years out of the Galactic plane. One of the fastest moving spacecraft ever launched by NASA, Voyager 1, travels at only 17 km/s (compared to 300,000 km/s for the

---

[7]S. Alexander, *Astronomical Journal*, **2**, 95 (1852).

[8]J. Oort, F. Kerr and G. Westerhout, *Monthly Notices of the Royal Astronomical Society*, **118**, 379 (1958).

speed of light). At 17 km/s, which corresponds to 38,250 miles/h, it would take 85 million years to travel 5,000 light years.

It would thus be fairly hopeless for us to detect and map the spiral arms of the Milky Way were it not for the fact that certain types of objects and certain types of interstellar material tend to *trace* the spiral structure we see in other galaxies. These tracers mostly involve star-forming regions and interstellar clouds of atomic hydrogen. If we could detect these tracers in the Galactic plane and figure out their distances from us, then we would have a mapping of the Milky Way's spiral arms. Ideally, we would like to have a mapping of the arms across the entire disk of the Milky Way, but in attempting to do so, we run into a major hurdle. There is so much interstellar dust in the disk of the Milky Way that our view into the disk, at least in visible light, is very limited. Nevertheless, through observations at radio, microwave, and infrared wavelengths, it has been possible to get a consistent view of the Milky Way's structure.

We can understand the tracers of spiral arms by examining hydrogen, the most abundant chemical element in the universe and also the simplest element from a structural point of view. As an atom, hydrogen consists of 1 proton in the nucleus and a single electron in an orbital around the proton. Orbits around the Sun can be at any distance from the Sun, that is, there are no natural restrictions on where an object can orbit the Sun. The same is not true for atoms. Every atom has what are known as allowed energy levels, orbital positions which an electron can occupy. The lowest energy level, or the one closest to the nucleus, is called the ground state. All other energy levels, farther from the nucleus, are called excited states. In hydrogen, the energy levels can be characterized by an integer index n, where $n = 1$ is the ground state and $n > 1$ refers to excited states (Figure 4).

The energy level structure of hydrogen leads to a rich spectrum of easily recognizable emission lines. The most important of these lines, H-alpha (or Hα), is part of what is known as the Balmer series, a set of optical emission lines that arise from quantum leaps[9] of electrons

---

[9] Jumps or transitions from one energy level to another.

excited energy levels

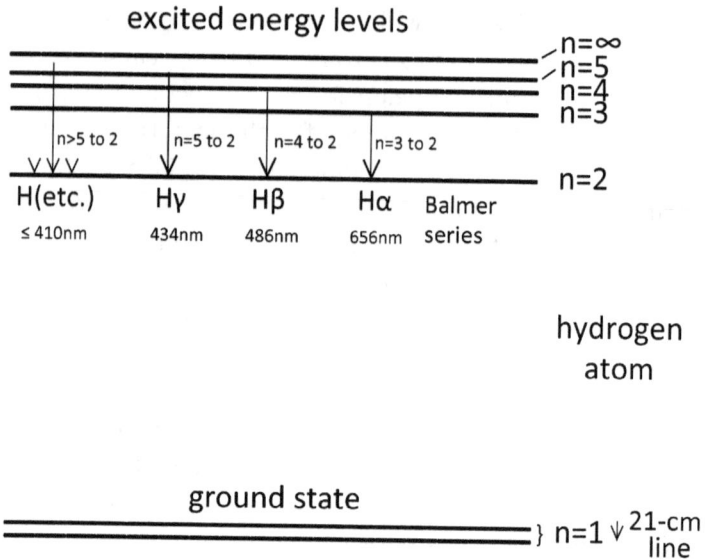

Fig. 4.   Schematic of energy levels in the hydrogen atom. Different levels are labeled by an index n, where $n = 1$ is the ground state (lowest energy level) and $n > 1$ refers to higher, excited energy levels. Quantum leaps of electrons from $n > 2$ excited levels to $n = 2$ defines the Balmer series of hydrogen emission lines as indicated, where Hα refers to the transition $n = 3$ to $n = 2$, Hβ refers to $n = 4$ to 2, Hγ refers to $n = 5$ to 2, and H(etc.) refers to the multitude of leaps from $n > 5$ to 2. The wavelengths of these lines are also indicated. The ground state is shown as two close energy levels (not to scale). A quantum leap of an electron from the upper ground state level to the lower ground state level produces the 21-cm emission line.

momentarily occupying high ($n > 2$) excited energy levels to the first excited energy level ($n = 2$) (long downward arrows in Figure 4). A leap from $n = 3$ to $n = 2$ is what produces Hα in emission, a red-colored photon having a wavelength of 656 nm. The transition from $n = 4$ to $n = 2$ produces an aqua-colored line called H-beta (or Hβ) at a wavelength of 486 nm, while $n = 5$ to 2 produces a bluish-colored line called H-gamma (or Hγ) at a wavelength of 434 nm.[10]

---

[10]In principle, at least 24 Balmer series emission lines can have a Greek letter designation, but in practice we do not see most of these lines because they blend together as the wavelength approaches 365 nm, the shortest wavelength possible in the series.

In order to see the Balmer series in emission, there has to be a mechanism for populating the excited energy levels. This can occur in an HII region, a generally pinkish interstellar cloud where new stars have formed. The notation "HII" refers to ionized hydrogen, a form of hydrogen where the electron has been ejected from an atom, leaving behind basically a free proton. The key to making an HII region is the presence of one or more very hot, massive main sequence stars. Such stars emit most of their light in the ultraviolet part of the spectrum at wavelengths too short for our eyes to see. When hydrogen atoms are bathed in ultraviolet light having a wavelength of 91.2 nm or less, the light photoionizes the atoms, meaning the atoms lose their electron by absorbing the energy of the light. Only main sequence stars hotter than about 30,000 K are capable of photoionizing a significant-sized HII region. Such a star would have a mass of 18 suns and a main sequence lifetime[11] of only about 7 million years. For this reason, an HII region is a star-forming region because the presence of massive stars tells us that the stars involved are not more than a few million years old.

At a distance of 1344 light years, the Orion Nebula, M42, is the nearest major HII region and a stellar nursery that can be studied in great detail. M42 is visible to the naked eye in the Sword of Orion. At the center of the nebula, a compact group of hot stars is found, the brightest of which, θ_1 Orionis C, has a surface temperature of 39,000 K[12] and is the star believed to be most responsible for photoionizing the surrounding gases.

Calculations have shown that as much as 99.9% of the hydrogen in an HII region is ionized at any given instant. However, the separated protons and electrons do not swim freely around each other and stay permanently separated. At any instant, large numbers of free protons and electrons are attracted to each other and will "recombine" into new atoms. It can be shown that a statistical kind of equilibrium is set up, such that there is a balance between

---

[11] The length of time a star shines as a normal star powered by core hydrogen fusion; see Chapter 10 for more about this number.

[12] Temperature on the Kelvin scale which is described further in Chapter 10.

the number of photoionizations per unit time and the number of recombinations to different energy levels, per unit time. It is the process of recombination that populates the excited energy levels of hydrogen and leads to the emission line spectrum. Any transition from a higher n energy level to a lower n energy level is possible, but the one most responsible for the pink color of an HII region is $n = 3$ to $n = 2$, or Hα. The Hα wavelength of 656 nm is in the red part of the visible light spectrum, and is most responsible for the pinkish color of HII regions.

HII regions only involve interstellar hydrogen gas near hot, massive stars. As a result, the clouds themselves are hot (6,000–10,000 K) and extremely low in density, such that virtually all of the energy levels of hydrogen are populated in such a region. In the vast reaches of the Milky Way, however, abundantly spread throughout the Galactic disk, and far from any star-forming region, there are clouds of hydrogen gas that are so cold that all of the atoms in the cloud are in the ground state. It is a remarkable aspect of nature that hydrogen atoms in the ground state can still emit light. This is because the ground state of hydrogen is actually two closely-spaced energy levels (schematically shown in Figure 4) tied to the spins (quantum rotations) of the proton and electron: the lower level corresponds to anti-parallel spins, while the upper level corresponds to parallel spins.[13] The difference in energy between these two configurations corresponds to a photon of wavelength 21 cm, which is in the radio region of the electromagnetic spectrum. The transition leading to 21 cm radiation being observed is highly improbable on both Earth and in space, but there are so many hydrogen atoms in interstellar space that we see 21 cm radiation from most spiral galaxies. In spirals, the arms are usually bright at 21 cm, and therefore 21 cm radiation acts as a tracer of spiral structure (for example, NGC5850; Figure 5).[14]

---

[13] Anti-parallel means that if around a given axis, one particle is seen to be rotating clockwise, the other particle would be seen rotating counter-clockwise around the same axis. Parallel means that each particle is rotating around the same axis in the same sense.

[14] J. L. Higdon *et al.*, *Astronomical Journal*, **115**, 80 (1998).

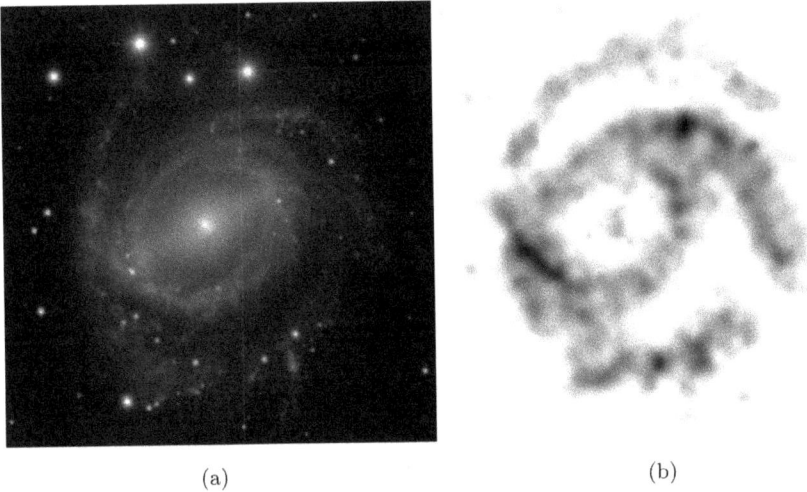

(a)   (b)

Fig. 5. NGC 5850 is a nearby spiral galaxy whose distribution of atomic hydrogen has been mapped at 21 cm. (a): Blue-light (*B*-band) image; image source: the de Vaucouleurs Atlas of Galaxies published by Cambridge University Press, 2007, reprinted with permission; (b) 21 cm map obtained by James L. Higdon and coworkers in 1998 with the Very Large Array radio observatory in New Mexico. The 21-cm map shows that atomic hydrogen in NGC 5850 follows the spiral arms closely, including the inner ring encircling the prominent bar. The bar itself is not seen in the 21-cm map. The asymmetry of the arms is caused by a local gravitational interaction. From J. L. Higdon *et al.*, *Astronomical Journal*, **115**, 80 (1998). © AAS. Reproduced with permission.

The great advantage of 21 cm radiation for exploring the structure of the Galactic disk is that the wavelength is so long as to be completely unaffected by interstellar extinction. In principle, we could detect 21-cm emission from anywhere within the Galaxy. In practice, however, we can reliably determine its location only in parts of the disk. For example, in a swath on the opposite side of the Galaxy, there is plenty of neutral hydrogen, but it is difficult to determine exactly where the gas at a given line of sight speed is located.

Figure 6 shows one of the first maps ever made of the spiral structure in the Galactic disk. It is based on observations of the spectrum of 21-cm radiation in all directions along and near the Galactic plane, using the equations of Galactic rotation to determine where the 21-cm emitting clouds are located. The map is actually

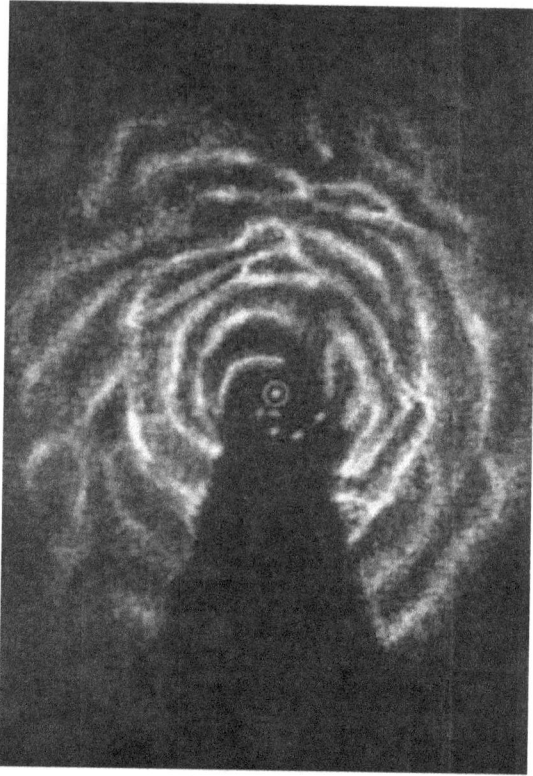

Fig. 6. Structure of the Milky Way based on the 21-cm radiation of cold, neutral atomic hydrogen (HI). The map is an artistic representation of the atomic hydrogen in the Galactic disk prepared by Gart Westerhout, based on 21-cm data obtained at the University of Leiden, the Netherlands and the CSIRO Radiophysics Laboratory, Sydney, Australia in the 1950s. The empty region below center is on the side of the Galaxy opposite the Sun, and is a region where it is difficult to unambiguously map the structure. From G. Westerhout, in *The Galaxy and the Magellanic Clouds*, F. J. Kerr and A. W. Rodgers, eds. (Canberra, Australian Academy of Science, 1964), p. 78.

mostly an artistic impression constructed from 21-cm contour plots by influential radio astronomer Gart Westerhout for a meeting that was held in Canberra, Australia in 1963.[15] Westerhout's map is actually not easy to interpret. The first impression it gives is that the

---

[15]G. Westerhout, in *The Galaxy and the Magellanic Clouds*, F. J. Kerr and A. W. Rodgers, eds. (Canberra, Australian Academy of Science, 1964), p. 78.

Milky Way is a multi-armed spiral, but while this seems a reasonable interpretation, the 21-cm map shows mainly a web of glowing ridges of gas and disjointed spiral-shaped arcs. Although Westerhout urged caution in its interpretation, Gérard de Vaucouleurs interpreted it to indicate that the Milky Way has a "high multiplicity of the spiral pattern."[16]

The most famous of the Milky Way's arms was first detected in 21 cm observations and is called the "3 kpc expanding arm."[17] Radio astronomers had graphed the brightness of Galactic 21 cm radiation towards the central parts of the Galaxy versus the line of sight speed at which the emitting hydrogen clouds were moving. They noticed something in the graph, a distinct "ridge", that they deduced had to be a spiral arm. They located the arm at 5.2 kpc from the Sun, and 3.3 kpc from the Galactic center; the latter distance is how the feature got its name, "3 kpc arm." Remarkably, the gases showed evidence for motion towards us at a speed of 53 km/s. Since this motion is nearly along the line of sight towards the Galactic center, and not in the direction of the Galaxy's rotation, this means there are significant non-circular motions in the inner regions of the Milky Way. In order for the arm to be coming towards us, it would have to be expanding away from the center of the Galaxy. This suggested to some that the arm resulted from an explosion at the center of the Galaxy. The nature of the purported explosion was uncertain.

In 1964, Gérard de Vaucouleurs realized that finding evidence for non-circular motions in the inner regions of the Milky Way meant there was a strong chance that our Galaxy is a barred spiral.[18] Bars are highly-elongated features in disk-shaped galaxies, and interstellar gas moving in the gravitational field of such a feature can follow oval streaming lines.[19] In 1970, de Vaucouleurs took information on all of

---

[16] G. de Vaucouleurs, in *The Galaxy and the Magellanic Clouds*, F. J. Kerr and A. W. Rodgers, eds. (Canberra, Australian Academy of Science, 1964), p. 88.

[17] "kpc" stands for kiloparsec, which means 1,000 parsecs or 3,260 light years.

[18] G. de Vaucouleurs, in *The Galaxy and the Magellanic Clouds, Proceedings of International Astronomical Union Symposium 20*, F. J. Kerr and A. W. Rodgers, eds. (Canberra, Australian Academy of Science, 1964), p. 195.

[19] Like currents in a river; see Figure 4 of Chapter 14 in the discussion of non-circular motions in NGC 6300.

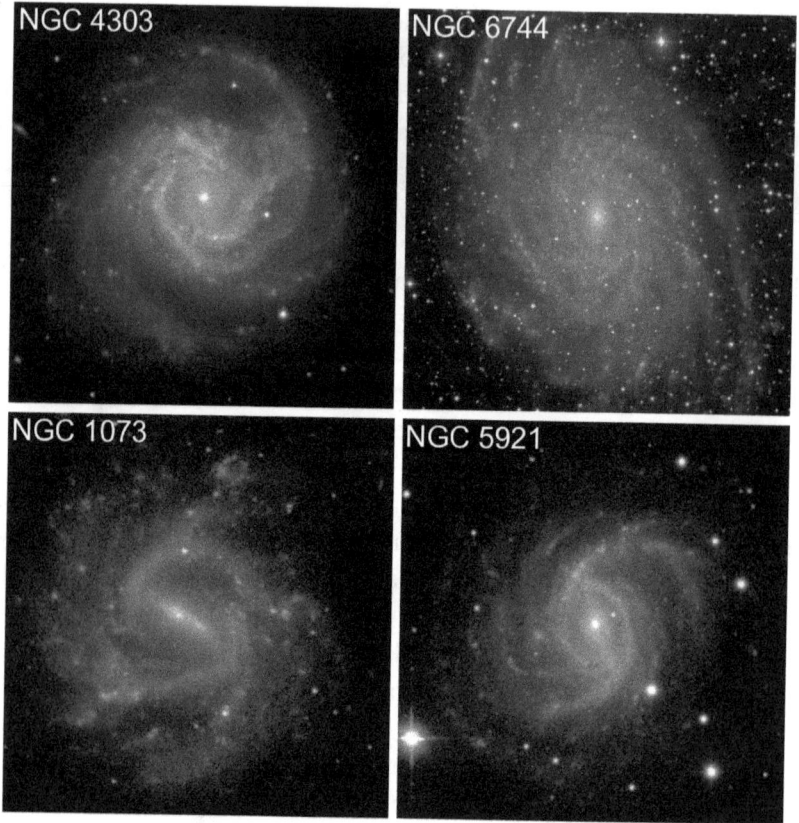

Fig. 7.   Gérard de Vaucouleurs proposed that our Galaxy might look like one of these four galaxies if it could be seen from the outside. Images source: The de Vaucouleurs Atlas of Galaxies published by Cambridge University Press, 2007; reprinted with permission.

the known features of the Galaxy, including its possible multi-armed pattern, the inner non-circular motions, and the 3-kpc arm itself, and narrowed down the face-on morphology of the Milky Way to be similar to the four galaxies in Figure 7.[20] Gérard proposed that our Galaxy is an SAB(rs)bc galaxy in his revised Hubble–Sandage classification system (Figure 8), or Sbc/SBbc(rs) in the system that

---

[20]G. de Vaucouleurs, in *The Spiral Structure of Our Galaxy, Proeedings of International Astronomical Union Symposium 38*, W. Becker and G. Contopoulos, eds. (Dordrecht, Reidel, 1970), p. 18.

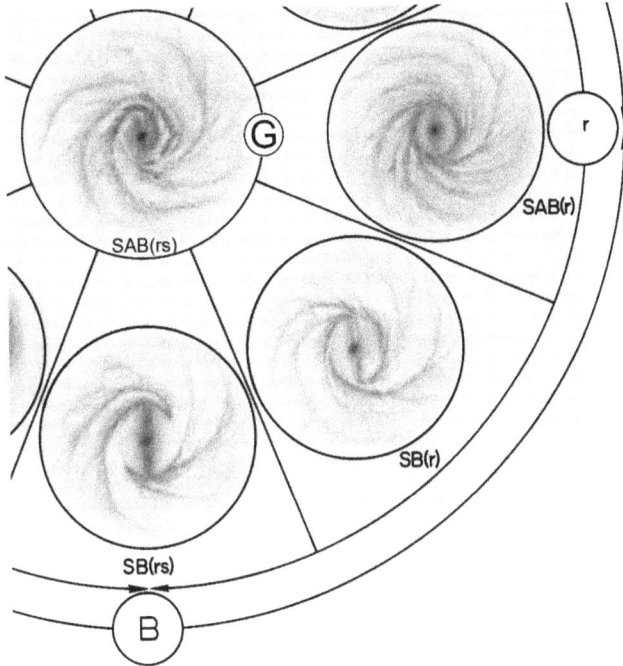

Fig. 8. de Vaucouleurs' interpretation of where the Milky Way "fits in" to the grand scheme of the Hubble classification, as modified by him. The estimated position of the Milky Way in the cross-section that was described in Chapter 6 is indicated by the circled G. This position is mostly based on the high multiplicity of the arm-like features in the early HI map of the Galaxy (Figure 6).

was outlined in the Hubble Atlas of Galaxies.[21] To Gérard, the "3 kpc expanding arm" was likely to be nothing more than an inner pseudoring made of inner spiral structure which each of the galaxies in Figure 7 shows; there was no need to hypothesize a catastrophic event like an explosion to explain the feature.

For a while, the idea that the Milky Way is a barred spiral floundered. The presence of a bar was only inferred and not really seen. A breakthrough occurred in the early 1990s when the Cosmic Background Explorer (COBE) satellite mission was launched with an

---

[21]G. de Vaucouleurs, in *The Galaxy and the Magellanic Clouds, Proceedings of International Astronomical Union Symposium 20*, F. J. Kerr, ed. (Canberra, Australian Academy of Science, 1964), p. 89.

on-board project called the Diffuse Infrared Background Experiment (or DIRBE). The DIRBE project mapped the Milky Way in a set of near-infrared filters, and revealed an asymmetry in the infrared stellar distribution that could be interpreted as perspective on a bar that was tipped at an intermediate angle to our line of sight, that is, was being seen neither end-on nor broadside-on. If the bar instead was being viewed exactly end-on or exactly broadside-on, we might still be able to tell there is a bar, but it would not show the observed asymmetry. The asymmetry appeared in the shape of the Milky Way bulge: half of it was wider than the other half. This is expected because half of the bar is closer to us than the other half.

In spite of the interesting results from the DIRBE, there was some concern about an apparent lack of symmetry in the inner regions of the Galaxy. The originally observed 3 kpc arm is seen in the same quadrant of the Galactic disk as the near end of the purported bar, and is now called the "Near 3 kpc Arm." But if there really is a bar in the central parts of the Milky Way, and if the Near 3 kpc Arm is simply a spiral arm breaking from near one end of this bar, then the symmetry of the bar would demand that there be a far counterpart to the Near 3 kpc Arm. In 2008, astronomers Thomas Dame and Patrick Thaddeus announced that they had found the previously unnoticed "Far 3 kpc Arm."[22] This considerably strengthened the idea that we live in a barred spiral galaxy.

The most recent studies have examined the bulge of the Milky Way in terms of a box/peanut shape and X pattern which have been seen in many edge-on galaxies likely to have bars. Boxy/peanut bulges are described in Chapter 15, and an X pattern is shown for NGC 7020 in Figure 29 of Chapter 15. Some studies have also suggested that in addition to the boxy bulge bar, there is a longer, narrower feature called the "long bar." The nature of this feature is uncertain.

The best tracings of the spiral arms in the Milky Way have come from studies of the locations of star-forming regions (HII regions) and young stars. The most recent of such studies use tracers that

---

[22]T. M. Dame and P. Thaddeus, *Astrophysical Journal*, **683**, L143 (2008).

can be detected at microwave wavelengths.[23] Light in this wavelength domain is also not affected significantly by interstellar dust. The tracers are called "masers," which is an acronym for "microwave amplification by stimulated emission of radiation." A maser is a source of naturally-amplified microwave emission. The source is usually a high mass star-forming region associated with a giant molecular cloud. Interstellar molecules can be excited in these regions, and can be stimulated to de-excite by microwave radiation from the molecular cloud. This amplifies the light into a bright emission line that can be seen through almost any amount of interstellar dust and from great distances. More than a dozen types of molecules have been observed to show maser emission.

Figure 9 shows the spiral arms of the Milky Way as determined from masers by Center for Astrophysics astronomer Mark Reid and coworkers.[24] The interesting feature of this is that the masers were located in the Galactic disk by measuring their *trigonometric parallaxes*. With an ordinary optical telescope, this would have been impossible, not only because of interstellar dust, but also because most of the Galactic masers that could be used for this purpose are much too far away for trigonometric parallaxes to be measured reliably. But in the microwave part of the spectrum, it is possible to do what is known as "very long baseline interferometry," a technique for simulating a very large telescope by pointing a large number of small telescopes at the same object at the same times. VLBI, as the technique is called, simulates a microwave telescope whose resolving power is more than a thousand times better than a single optical telescope would give. At such resolution, much smaller trigonometric parallaxes are measureable, and accurate distances to masers across a wide swath of the Galactic disk allow the spiral structure of the Milky Way to be mapped reliably. Figure 9 shows multiple well-defined arms in our vicinity, some 27,000 light years from the Galactic center.

One of the best-determined arms identified in such studies is the Perseus Arm, whose (non-maser) tracers include the Double

---

[23]The wavelength domain between infrared and radio waves.
[24]M. J. Reid *et al.*, *Astrophysical Journal*, **885**, 131 (2019).

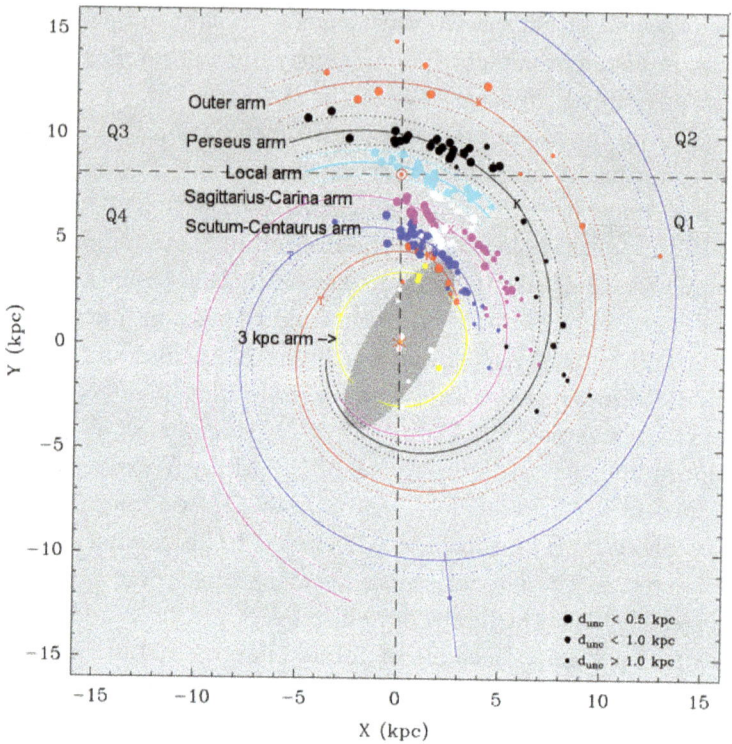

Fig. 9.   Structure of the Milky Way from observations of masers in the Galactic plane. The different arms are labeled with the main constellations in which they have been traced. The Local Arm has in the past been referred to as the Orion Spur. The location of the Sun is indicated by the ⊙ symbol. The Qs refer to quadrants in the Galactic plane relative to the Sun–Galactic center line. The different-sized symbols indicate uncertainties in distances, with the larger symbols having greater precision. Adapted from Reid *et al.*, *Astrophysical Journal*, **885**, 131 (2019). © AAS. Reproduced with permission.

Cluster in Perseus as well as other well-known clusters in Auriga and Cassiopeia. The Perseus arm is considered one of the two most prominent Milky Way arms, the other being the Scutum–Centaurus Arm. The constellation Scutum has one of the brightest star clouds in the Milky Way, and includes the famous Wild Duck Cluster M11. The arm extends southward into Centaurus where it could include the famous Jewel Box star cluster NGC 4755 in Crux. Another weaker arm has been mapped towards the constellation Sagittarius. The maser arms also show that our Sun lies in a distinct feature

Fig. 10.   Artistic rendition of a face-on view of the Milky Way. The grid lines centered on the Sun show directions along the Galactic plane, called Galactic longitude. The Galactic center is at 0° Galactic longitude, while the Galactic anti-center is at 180° Galactic longitude. The circles centered on the Sun give distances in any direction. Image credit: NASA/JPL-Caltech/R. Hurt (SSC-Caltech, 2008); with permission.

that Mark Reid and coworkers call the Local Arm. In the past, this feature was referred to as the "Orion Spur," where spur means a minor arm feature, but Reid and coworkers note that the Local Arm is defined by as much massive star formation as are the Perseus and Sagittarius Arms.

Figure 10 shows the most recent artistic conception of the face-on morphology of the Milky Way. The galaxy is depicted as a mainly two-armed spiral with weaker arms in between these features. The bar is tipped about 45° to the line of sight towards the Galactic center, and a faint inner ring, defined by the Near and Far 3 kpc arms, could be intrinsically elongated along the bar axis as is often observed in other galaxies.

Chapter 9

Galactic Extinction and the Zone of Avoidance

Because angular sizes of galaxies are defined by a specific surface brightness level, they can be affected by Galactic extinction. Foreground interstellar dust can shrink the angular dimensions of a galaxy, and make it appear smaller than it actually is. This is in addition to making an object look dimmer and redder than it actually is. These effects can help us understand why the Milky Way band is the "Zone of Avoidance," the directions in the sky where few or no "white nebulae" were seen by early observers.

As shown in Figure 1 of Chapter 2, seen from outside, the interstellar dust in a disk-shaped galaxy is usually obvious because it manifests itself as dark lanes where light has been made deficient due to extinction. The effects of dust become more noticeable as a galaxy becomes more tilted, until they are most apparent in the edge-on view. However, seen from within the perspective changes, and while some dust clouds would still be obvious, as they are in the Milky Way, the dust will most often manifest itself as extinction and reddening that varies along different lines of sight.

Maffei 1 and Maffei 2 are two heavily obscured nearby galaxies located in the Perseus-Cassiopeia border region that serve as excellent examples of how the brightness, size, and color of galaxies are affected in the Zone of Avoidance. The two galaxies are the main members of a small group consisting of about a dozen objects, and would be among the brightest galaxies in the sky were it not for their location in a direction close to the Galactic plane. The galaxies were

Fig. 1.   The field in the northern Milky Way where the Maffei Galaxies were discovered in 1968 (arrows). The large nebulae are IC 1805 (right) and IC 1848 (left); both are photoionized clouds connected with active star formation. The image is a composite optical, near-infrared, and mid-infrared construction, which is why the Maffei Galaxies are conspicuous. Image credit: NASA/ESA/DSS2/WISE/IRSA; composition by Giuseppe Donatiello, reproduced with permission.

first noticed in 1968 by Italian astronomer Paolo Maffei (pronounced "mahf-ay") during a search for variable stars near the large galactic HII region known as IC 1805 (Figure 1).[1] Probably no one expected to find galaxies in this region, which is not far from the Double Cluster in Perseus, in a particularly conspicuous part of the northern Milky Way.

The key to penetrating the Zone of Avoidance is to observe those directions at wavelengths longer than UBV. Maffei found his two galaxies after obtaining blue, red, and infrared plates of the region; he noticed two nebulous objects on his infrared-sensitive plates which were less visible on his red-sensitive plates and invisible on his blue-sensitive plates. Because longer wavelengths penetrate dust more

---

[1]P. Maffei, *Publications of the Astronomical Society of the Pacific*, **80**, 618 (1968).

easily than do shorter wavelengths, Maffei's infrared plates had the edge over his blue and red plates for detecting galaxies in the Zone of Avoidance.

At first, Maffei was not sure of the nature of his two objects. This wasn't necessarily obvious on his plates, because the objects were barely detected. Nevertheless, his paper caught the attention of observers in California intrigued by the possibility that the objects could be heavily reddened, and possibly nearby, external galaxies. An extensive multi-wavelength observing program was carried out in the early 1970s that led to the possibility that the brighter of the two galaxies, Maffei 1, was a member of the Local Group, the small cluster of galaxies to which our Galaxy belongs. The California observers were able to deduce that Maffei 1 is a large elliptical galaxy whose apparent brightness has been reduced by a factor of 120 by foreground dust, and that Maffei 2 is a spiral, perhaps of Hubble type Sb–Sc, extinguished by a factor of more than 300.

These results were based on photographic plates and were published in *Astrophysical Journal* papers in 1971 and 1973 by Hyron Spinrad and coworkers.[2] Although very useful, there nevertheless was considerable uncertainty remaining in the distances and extinctions of the two galaxies. In 1999, a new study of Maffei's galaxies by myself and Marshall L. McCall (York University) was published in the *Astrophysical Journal Supplement Series*,[3] based on observations using a CCD camera attached to the Kitt Peak 0.6/0.9-m Burrell–Schmidt Telescope, a type of telescope designed to have a wide field of view. The goal of our study was to obtain wide-field images in blue, visual, and near-infrared (BVI) filters to see if we could get more accurate information on the morphologies, total brightnesses, angular sizes, extinctions, and reddenings not only for the Maffei galaxies, but also for several other galaxies in the same general area. At the time, we thought that Maffei 1 and 2 were part of a single group

---

[2]H. Spinrad *et al.*, *Astrophysical Journal*, **163**, L25 (1971); H. Spinrad *et al.*, *Astrophysical Journal*, **180**, 351 (1973).
[3]R. J. Buta and M. L. McCall, *Astrophysical Journal Supplement Series*, **124**, 33 (1999).

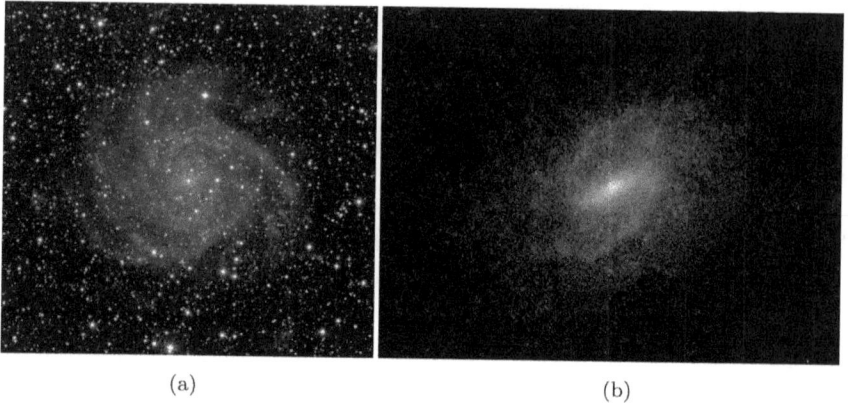

(a)   (b)

Fig. 2. (a) Blue-light (*B*-band) image of the nearby spiral galaxy IC 342;
(b) near-infrared (*I*-band) image of the nearby, heavily obscured spiral galaxy
Dwingeloo 1. The image of Dwingeloo 1 is shown with foreground stars removed in
order to highlight the galaxy's strong resemblance to the Triangulum Spiral, M33
(Figure 5 of Chapter 3). Images source: R. Buta and M. L. McCall, *Astrophysical
Journal Supplement Series*, **124**, 33 (1999). © AAS. Reproduced with permission.

including the large face-on spiral IC 342 and a number of smaller
galaxies. More recent studies have suggested that IC 342 is part of a
separate group. The Buta and McCall image of IC 342 is shown in
Figure 2(a). Note that although IC 342 suffers high extinction and
reddening compared to galaxies outside the Zone of Avoidance, the
Maffei galaxies are much more severely affected.

Figures 3 and 4 show the *B* and *I* images Marshall and I obtained
for Maffei 1 and 2, side-by-side. For both Maffei galaxies, compared
to the *I*-band, the *B*-band brightness distributions are much smaller
and dimmer. Given what we see for these galaxies, it is not hard to
imagine the extinction being so high that a galaxy would be rendered
completely invisible in filters like *B* or *V*. It turns out the Maffei
Galaxies are projected against the Perseus spiral arm of the Galaxy,
which happens to be in the direction of the Galactic anti-center. Had
they instead been in the direction of the Galactic center, they would
likely not have been found.

The images also highlight how, even in the *I*-band, it is difficult
to study the galaxies because of the myriads of foreground stars
covering the fields. Figures 5(a) and 6(a) show what Maffei 1 and 2

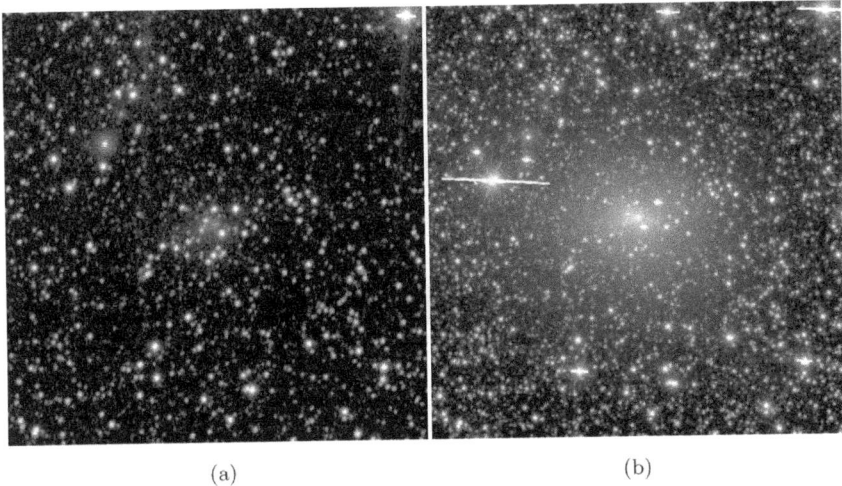

(a)   (b)

Fig. 3. The effect of extinction on the angular size and brightness of the nearby elliptical galaxy Maffei 1, based on blue (a) and near-infrared (b) images. Images source: R. Buta and M. L. McCall, *Astrophysical Journal Supplement Series*, **124**, 33 (1999). © AAS. Reproduced with permission.

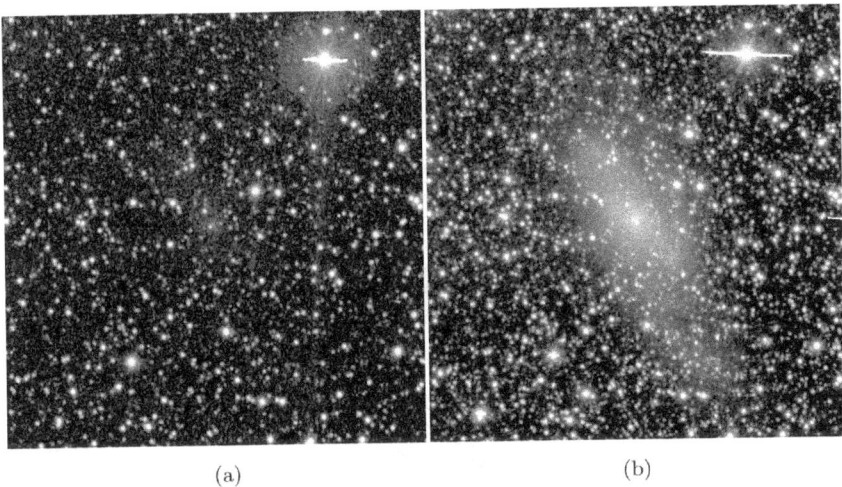

(a)   (b)

Fig. 4. The effect of extinction on the angular size and brightness of the nearby spiral galaxy Maffei 2, based on blue (a) and near-infrared (b) images. Images source: R. Buta and M. L. McCall, *Astrophysical Journal Supplement Series*, **124**, 33 (1999). © AAS. Reproduced with permission.

(a)   (b)

Fig. 5.   (a) Infrared image of Maffei 1 after removal of foreground stars; (b) color image of Maffei 1 on same scale; the white arrow at upper left points to a Galactic reflection nebula that happens to be nearly in the same direction as Maffei 1. Images source: R. Buta and M. L. McCall, *Astrophysical Journal Supplement Series*, **124**, 33 (1999). © AAS. Reproduced with permission.

(a)   (b)

Fig. 6.   (a) Infrared image of Maffei 2 after removal of foreground stars; (b) color image of Maffei 2 on same scale. Images source: R. Buta and M. L. McCall, *Astrophysical Journal Supplement Series*, **124**, 33 (1999). © AAS. Reproduced with permission.

look like with the foreground stars largely removed. These highlight how Maffei 1 is an elliptical (E3) galaxy while Maffei 2 is an inclined barred spiral galaxy of type SBbc. Figures 5(b) and 6(b) show the two galaxies in color. Both are redder than most of the field stars. It is interesting that the color image of Maffei 1 reveals a bluish-colored nebula surrounding a faint star in nearly the same direction as the galaxy. This object is a reflection nebula, a cloud of dust surrounding a hot, blue-colored star, some 9,800 light years away.[4]

The cleaned *I*-band image in Figure 5(a) shows that Maffei 1 has a somewhat irregular appearance. Using Hubble Space Telescope data, Marshall and I were able to determine that this irregularity is due to foreground dust, not to dust in Maffei 1 itself.[5] Although the effects of dust are considerably reduced in the *I*-band compared to the *B*-band, dust still significantly impacts the *I*-band. The only way to reduce the effects of extinction altogether is to observe at even longer wavelengths.

In 1994, astronomers at the Dwingeloo Radio Observatory, located near the village of Dwingeloo, the Netherlands, searched the area of the Maffei galaxies at 21 cm wavelength for previously undiscovered members of the galaxy group, in a project called the Dwingeloo Obscured Galaxies Survey (DOGS).[6] The only major galaxy the DOGS project found was the barred spiral shown in Figure 2(b), an object that became known as "Dwingeloo 1." The survey also netted a small irregular companion to Dwingeloo 1 that became known as "Dwingeloo 2." Like the Maffei galaxies, both Dwingeloo 1 and Dwingeloo 2 are heavily obscured by foreground interstellar dust. Because the two galaxies were discovered in the 21-cm emission line, the radio observations provided their line of sight speeds, which were consistent with their membership in the Maffei galaxy group.

---

[4]R. Buta, M. McCall, and A. Uomoto, *Publications of the Astronomical Society of the Pacific*, **92**, 715 (1980).

[5]R. J. Buta and M.L. McCall, *Astronomical Journal*, **125**, 1150 (2003).

[6]R. C. Kraan-Korteweg *et al.*, *Nature*, **372**, 77 (1994).

The Buta and McCall I-band imaging survey of the same area also revealed three previously unrecognized possible members of the group.[7] MB1 is a small irregular galaxy that appeared on the images of Maffei 1, while MB3 is a dwarf elliptical that appeared on the images of Dwingeloo 1. There is an "MB2," also found on the I-band images of Maffei 1, but it is uncertain whether the object is a galaxy or a foreground galactic nebulosity of some sort. Unlike the 21-cm observations, the I-band images do not automatically tell us the line of sight speeds of the galaxies.

Even though studies of the IC 342/Maffei galaxy group(s) have yielded a few previously uncatalogued galaxies, the Zone of Avoidance is still relatively unexplored. The Sloan Digital Sky Survey (SDSS) covers a significant fraction of the sky, but is focussed mainly around the north Galactic pole far from the Zone of Avoidance. Extensive coverage of the Zone of Avoidance has not been a major part of the SDSS. The great benefit of SDSS imaging is that it is accompanied by a catalogue of positions, apparent brightnesses, colors, and line of sight speeds of whatever galaxies happen to lie in the survey's sky area. The survey was carried out in five filters, $ugriz$, where the filters u (354 nm) and g (477 nm) transmit light in the ultraviolet and blue-green parts of the electromagnetic spectrum, while r (623 nm), i (763 nm), and z (913 nm) transmit light in the red and near-infrared parts of the spectrum. A century ago, blue-sensitive photographic plates were the staple of astronomy, but as shown with the observations of the Maffei galaxies, the visibility of galaxies is much more reduced by interstellar extinction in the blue B-band than in the near-infrared I-band. The SDSS g-band is close to the B-band, and thus is strongly affected by extinction while the SDSS i-band is close to the I-band and is much less affected. The z-band transmits an even longer wavelength than does the i-band. If and when the survey ever covers the Zone of Avoidance, this means that the SDSS has considerable power to penetrate the dust in those areas.

---

[7]M. L. McCall and R. J. Buta, *Astronomical Journal,* **109**, 2460 (1995); M. L. McCall and R. J. Buta, *Astronomical Journal,* **113**, 981 (1997).

9.1. The Distribution of Nearby Galaxies

The Shapley–Ames catalogue was the first compilation of the brightest galaxies in the sky, and it provided more than just magnitudes and angular sizes. It also told us where the brightest galaxies in the sky are located. Location is an important issue, because it tells us how matter is distributed in the nearby universe.

To examine this question, Shapley and Ames graphed the locations of their galaxies on a homolographic (Mollweide) projection map, a way of showing the three-dimensional sky on a two-dimensional plane. One of their maps is shown in Figure 7. Individual galaxies are shown as the filled and open circles, while the plane of the Milky Way is shown by the solid curve, distorted by the projection. This map clearly shows that bright galaxies avoid the band of the Milky Way (labeled the Galactic equator), but it also shows more: There is a clear unevenness in the distribution of bright galaxies: a large number of galaxies lie near the north Galactic pole (NGP) in the Virgo-Leo-Coma Berenices region of the sky, and there are twice as many bright galaxies in the northern Galactic hemisphere as in the southern Galactic hemisphere. This peculiar distribution of the nebulae has been known since the time of the Herschels and Messier. The excess in the Virgo-Leo-Coma Berenices region was so high that Shapley and Ames referred to the region as the "Virgo Supersystem," since the main part of the area includes the Virgo Cluster of galaxies.

In a 1953 paper,[8] Gérard de Vaucouleurs proposed that the distribution of Shapley–Ames galaxies, as well as many fainter NGC galaxies, can be explained if the Local Group is part of a large, flattened *system of galaxies* known as the Local Supergalaxy (or Local Supercluster). This unusual system is shaped like a disk galaxy, but is made of thousands of individual galaxies, many of which are concentrated in groups. Not only this, but clustering in the southern Galactic hemisphere suggested the presence of a second, smaller supergalaxy called the Southern Supergalaxy, having a different plane from the Local Supergalaxy.

---

[8]G. de Vaucouleurs, *Astronomical Journal*, **58**, 30 (1953).

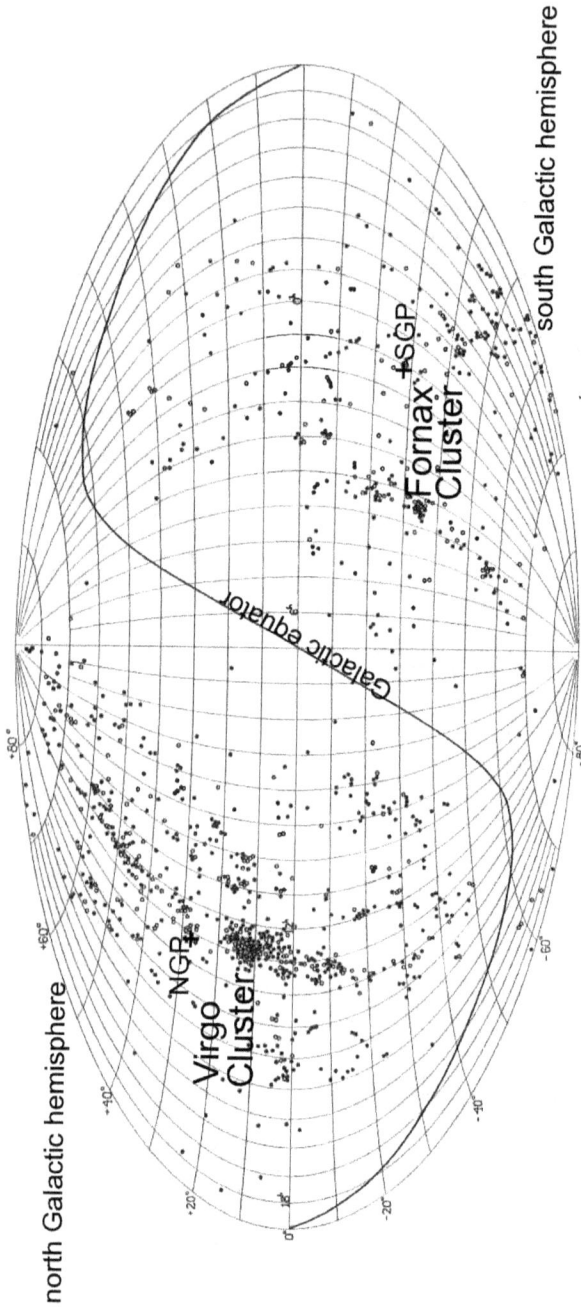

Fig. 7.   The distribution of 1025 galaxies brighter than total photographic magnitude 13.0, based on the Shapley–Ames Catalogue. The north and south Galactic poles (NGP and SGP, respectively) are indicated, as are the locations of the Virgo and Fornax galaxy clusters. The Zone of Avoidance is the blank areas, mostly devoid of nebulae, flanking either side of the Galactic equator, which is distorted into a curve because the graph plots positions in right ascension and declination coordinates. Illustration adapted from H. Shapley and A. Ames, *Harvard Annals*, **88**(2) (1932).

The estimated extent of the Local Supergalaxy was about 100 million light years, making it one of the largest mass concentrations known at the time. In 1958, de Vaucouleurs used available line of sight speeds to deduce that the system might be rotating around a center located in the Virgo Cluster; this rotation could account for the highly flattened shape. He also suggested that the system may be expanding at a faster rate in regions of low galaxy density as opposed to regions of higher density.

Since de Vaucouleurs' work in the 1950s, there has been considerable accumulation of data that has allowed revised interpretations of the distribution of both nearby and more distant galaxies. When positions and line of sight speeds are measured for large numbers of galaxies in a wide swath of sky, we can see how galaxies are distributed on scales ranging from millions to nearly a billion light years. It is found that galaxies congregate in groups, clusters, superclusters, filaments, and walls, all surrounding large, mostly empty regions called voids. A group typically consists of a few large galaxies and numerous dwarfs, an example being the Local Group to which our Galaxy belongs. A cluster can have from several hundred to several thousand members, and may include an even greater population of dwarfs. A supercluster is a type of large-scale structure made of multiple galaxy clusters on a scale much larger than any individual cluster. A filament is a chain-like distribution of galaxies that can be even larger than a typical supercluster. The largest structures are found to be walls, which are filaments having a broader extent in one dimension.

The most remarkable finding is that both the Local Supergalaxy and the Southern Supergalaxy are parts of a larger structure called the Laniakea Supercluster.[9] The structure of this supercluster was determined not only by the apparent distribution of galaxies on the sky, but also using knowledge of actual distances and speeds relative to the cosmic expansion of thousands of individual galaxies. The result is one of the most sophisticated analyses of large-scale cosmic structure ever attempted. The study of such structures also led to the discovery of a mysterious force in the nearby universe

---

[9]R. B. Tully *et al.*, *Nature*, **513**, 71 (2014).

called the "Great Attractor" whose gravity is acting to slow down the participation of our Local Group of galaxies in the universal expansion.[10] The latest results interpret the Great Attractor as being a massive supercluster of galaxies lying 800 million light years away in the constellation Vela, the Sails.[11] This supercluster cuts across the Galactic plane, making our ability to study it complicated by interstellar dust.

9.2. Galactic "Lonely Hearts"

Given the clustering penchant of galaxies, it is interesting to ask if genuine isolated galaxies exist. Isolated galaxies have attracted attention in recent years for what they have to offer about how nature and nurture contribute to the structure, dynamics, and evolution of a galaxy. Because galaxies are large compared to their typical separations, external interactions can play a significant role in galactic evolution, leading to the idea that nurture is important. But if a galaxy is isolated, such that it has had no major gravitational interactions with another galaxy for several billion years, then the structure we see in it must be largely due to nature. In the early 1970s, Russian astronomer V. E. Karachentseva scanned the charts of the Palomar Sky Survey looking for isolated galaxies. The result of her effort was the *Catalogue of Isolated Galaxies* (CIG),[12] a list of about 1050 cases, the bulk of which turned out to be spirals. Andalucia Institute of Astrophysics (Granada, Spain) astronomer Lourdes Verdes-Montenegro and her coworkers have been studying CIG galaxies for many years, and defined an isolated galaxy as one that has not experienced a major interaction with a comparable-sized companion galaxy in at least 3 billion years.[13]

Figure 8 shows six galaxies from the CIG. The six are astonishing in the sense that, in the absence of any significant companions the

---

[10]A. Dressler, *Astrophysical Journal*, **329**, 519 (1988).
[11]R. C. Kraan-Korteweg *et al.*, *Monthly Notices of the Royal Astronomical Society*, **466**, L29 (2017).
[12]V. E. Karachentseva, *Communications of the Special Astrophysical Observatory, USSR*, **8**, 1 (1973).
[13]L. Verdes-Montenegro *et al.*, *Astronomy and Astrophysics*, **436**, 443 (2005).

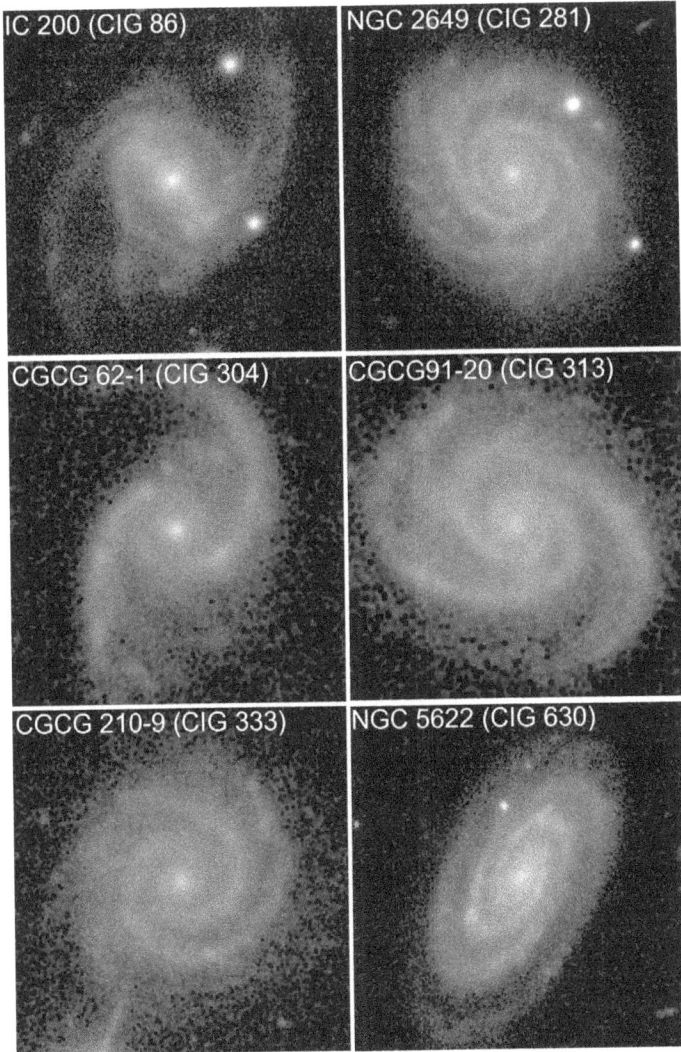

Fig. 8. Six galaxies from the Karachentseva Catalogue of Isolated Galaxies, published in 1973. The images are from the isolated galaxy database described by R. Buta *et al.*, *Monthly Notices of the Royal Astronomical Society*, **488**, 2175 (2019). In addition to NGC or IC names, each galaxy is identified by its CIG number. Galaxies with CGCG numbers are from the Zwicky *Catalogue of Galaxies and of Clusters of Galaxies*.

spiral patterns are well defined. Four (CGCG 62-1, CGCG 91-20, CGCG 210-9, and NGC 5622)[14] are mostly two-armed, non-barred spirals, one (NGC 2649) is a multi-armed non-barred spiral with two strong inner arms, and one (IC 200) is a barred spiral with strong outer arms and a bright inner pseudoring.

More recent studies have shown that the most isolated galaxies are found in voids, extremely large volumes of space including few or no galaxies. A galaxy in a void may have no comparable-sized companion closer than 50–150 million light years away.

The existence of galaxies like those in Figure 8 has important implications for the nature of spiral structure. Do spirals depend on some kind of external influence, or is spiral structure something driven from within a galaxy? This will be discussed further in Chapter 16.

---

[14]CGCG refers to numbers in the six-volume *Catalogue of Galaxies and of Clusters of Galaxies*, F. Zwicky *et al.* (Pasadena, California Institute of Technology, 1961–1968).

Chapter 10

The Stars Within

A galaxy is defined by the kinds of stars it contains. In general, a spiral galaxy is not a uniform mix of star types but instead is a massive gravitational system built of several structural components, each with its own mix of star types. The disk is only one of the major components of a spiral. Many spirals also have a bulge component, a feature that Hubble focussed on for galaxy classification. In the face-on view, the bulge appears as a bright inner section, sometimes including a sharp, stellar nucleus. In many cases, a bulge appears as a more roundish component than the disk, such that in the edge-on view, the bulge sticks out, or "bulges" out of the inner part of the disk component. If the galaxy is not exactly edge-on, the nucleus may still be seen, as in NGC 4565 (Figure 1). In contrast, NGC 891 is exactly edge-on, which is why we do not see its nucleus (Figure 2).

Surrounding the disk component is a spherical volume of space called the "halo" which, from a distance, looks sparse and unpopulated but when examined closely includes globular clusters and some field stars. Both the bulge and the halo together are sometimes combined into a feature called the "spheroidal component," where spheroidal means shaped like a mildly flattened sphere. This is a fair characterization only for a nonbarred spiral. The apparent bulge of a strongly-barred galaxy would be affected by the three dimensional structure of the bar in the edge-on view.

Bulges and halos are interesting because they appear to be made of different stars compared to the disk component. The clue that

Fig. 1.   Close-up of the bulge region of the edge-on spiral galaxy NGC 4565 in Canes Venatici. The galaxy is not exactly edge-on, because the nucleus (bright star-like object in the center) just misses being obscured by the prominent planar dust lane. Image source: The de Vaucouleurs Atlas of Galaxies published by Cambridge University Press, 2007; reprinted with permission.

bulge stars are different from disk stars came from studies of the Andromeda Galaxy, which has both a prominent bulge and a massive, extended disk component (Figure 4 of Chapter 3). Even a century ago, large reflectors like the Mount Wilson 100-inch telescope could resolve the disk of the Andromeda Galaxy into individual stars. These disk stars tended to be massive and luminous compared to the Sun, and are resolvable with much smaller telescopes today (Figure 3).

Given the resolution of the disk of the Andromeda Galaxy into recognizable types of luminous stars, it was a puzzle that the bulge of the galaxy resisted resolution. Presumably, this meant that bulge stars in Andromeda were in general fainter than the resolved disk stars. In the 1940s, Mount Wilson astronomer Walter Baade (1893–1960) was the first to successfully resolve the bulge of Andromeda into stars. An impediment to doing so was the growth

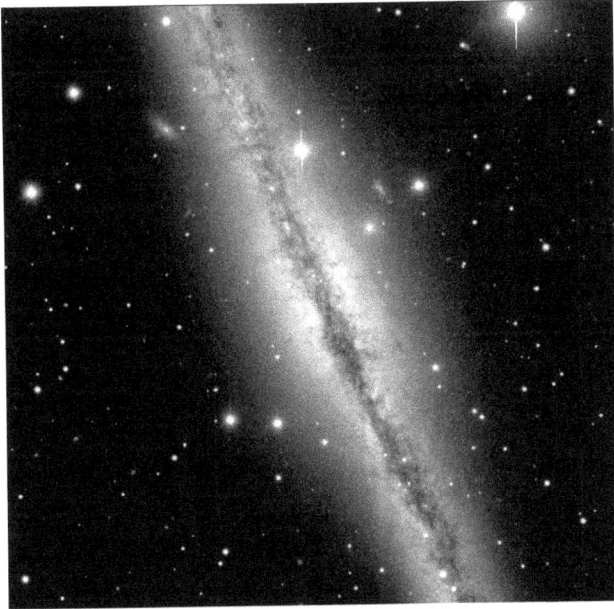

Fig. 2. Close-up of the bulge region of the edge-on spiral galaxy NGC 891 in Andromeda. The galaxy is exactly edge-on, because the nucleus in this case is likely completely obscured by the prominent planar dust lane. Image source: The de Vaucouleurs Atlas of Galaxies published by Cambridge University Press, 2007; reprinted with permission.

of Los Angeles at the time and the light pollution that accompanied this growth. Mount Wilson Observatory was built only 33 miles from LA in the San Gabriel Mountains. Excessive sky brightness would overexpose and darken the background on photographic plates before the bulge stars could resolve. In 1944, however, something happened that allowed the 100-inch telescope to finally resolve the bulge of Andromeda: a blackout over Los Angeles initiated to protect the city from air attack during World War II. During the blackout, the sky was much darker than usual on Mount Wilson, and Baade took plates that successfully resolved the bulge into stars.

Baade discovered that the stars populating the bulge of the Andromeda Galaxy were the same kinds of stars that were found in globular clusters: low mass, giant stars of extremely old age

Fig. 3.   NGC 206 is a star cloud in the Andromeda Galaxy that was resolvable even a century ago with the Mount Wilson 100-inch reflector. In this modern color CCD image, which was obtained with the Kapteyn 40-inch Telescope in the Canary Islands, the object is seen to be made mostly of bluish-colored stars that are likely of spectral classes O and B, making it a large OB stellar association like those seen in M51 (Chapter 12). Image credit: W. C. Keel, with permission.

and deficient in heavy chemical elements.[1] Baade concluded that there were two main populations of stars in spiral galaxies: disk or population I stars, and bulge or population II stars. Population I includes any star belonging to the disk, even very old stars such as the Sun. It is the youngest Population I stars that are detectable from a great distance. Population II includes any star belonging to the bulge or halo of a spiral.

One of the most powerful tools for studying stars and galaxies is a Hertzsprung–Russell (HR) diagram, a graph of the luminosities of stars versus their surface temperatures. This famous diagram, which is named after Danish astronomer Ejnar Hertzsprung (1873–1967) and American astronomer Henry Norris Russell (1877–1957), brings to light different subgroups of stars that can be connected

---

[1]In astronomy, all chemical elements besides hydrogen and helium are referred to as "metals" even if they exist only in gaseous form on Earth, like oxygen and nitrogen.

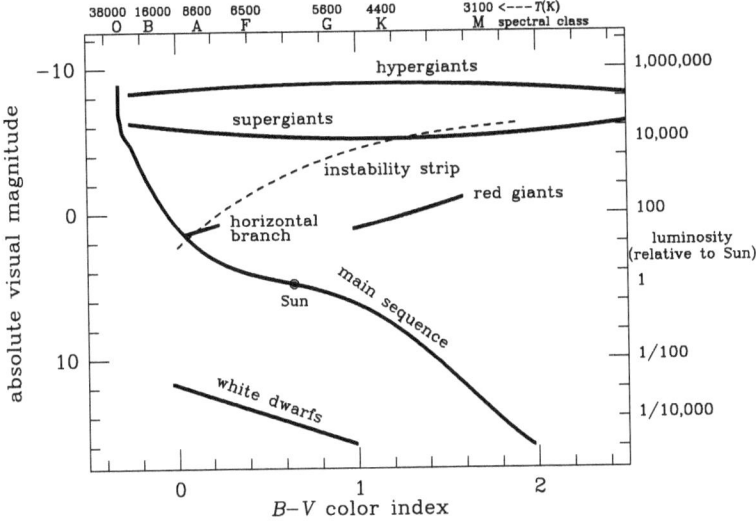

Fig. 4. A general HR diagram for field stars (not necessarily clusters). The main parts of the diagram are labeled. Based on data extracted from wikipedia. The *B–V* color index is used as a proxy for surface temperature. The spectral classes, temperatures, and relative visual luminosities are also indicated.

directly to star evolution. Figure 4 shows the HR diagram of a general population of field stars, including both the brightest stars and the nearest stars.

Luminosity and surface temperature are intrinsic properties of stars, the former being the amount of luminous energy radiated by a star every second, and the latter being a measure of the temperature at the visible apparent surface of a star. In the case of the Sun, the visible surface is called the photosphere. Photospheres are gaseous surfaces, and only for a handful of stars is the photosphere actually resolvable with Earth-based telescopes. Most stars are point sources, meaning we do not see their surfaces directly. The temperature at the surface of any star is easy to measure because it is the main factor that determines the *color* of a star. In fact, astronomers often use the *B–V* color index, corrected for foreground reddening due to dust, as a proxy for surface temperature (as in Figure 4). Also useable as a proxy for surface temperature is the spectral classification of a star. Surface temperature is the main factor that determines the relative strengths of absorption lines in the spectra of stars, a correlation that led to the famous spectral classification sequence OBAFGKM. In this sequence, the hottest atars are spectral class O while the coolest stars are spectral class M.

The units of surface temperature are degrees on the Kelvin scale. Astronomers do not use the Fahrenheit or Celsius scales of temperature because both of these scales allow negative temperatures. The lowest temperature on the Kelvin scale, 0 K, is called "absolute zero" and is defined to be the condition where all random motions of atomic or subatomic particles in a substance cease. On the Kelvin scale, an O star has a surface temperature of about 40,000 K, while an M star has a surface temperature of about 3,000 K. In terms of color index, an O star has $B-V = -0.3$ while an M star has $B-V \approx 2$. The great aspect of surface temperature is that it is an intrinsic property of a star that can be determined *without* knowledge of the distance to the star. This is demonstrated by the picture of the constellation Orion that is shown in Figure 2 of Chapter 2. The contrast in color between Betelgeuse (the red star at upper left in the figure) and Rigel (the bluish-white star at lower right), is very striking. Betelgeuse has a color index of $B-V = 1.85$ and a surface temperature of 3,590 K, while Rigel has a $B-V$ color index of -0.03 and a surface temperature of 12,100 K (wikipedia).

The apparent brightness (magnitude) of any star depends on its luminosity *and* its distance, therefore luminosity is not as easy to determine as is surface temperature. Chapter 7 described how apparent magnitudes have been measured. The challenge in getting the HR diagram of any sample is deriving distances to the stars involved. Trigonometric parallax can be used for the nearest stars. For the vast majority of stars, other ways of getting distances must be used. Once the distance is known, astronomers often specify the luminosity not in its normal physical units, watts, but as absolute visual magnitudes.[2]

The main part of an HR diagram is the main sequence, the band of stars falling on a broad curve from the upper left to the lower right.

---

[2]See Chapter 4 for more discussion of absolute magnitudes.

The main sequence is important for several reasons: approximately 90% of all stars are main sequence stars; the luminosities of all main sequence stars are powered by the fusion of hydrogen into helium in their core or central region; and the more luminous the main sequence star, the higher its mass and surface temperature, and the larger its size. The mass a star is born with tells us something very important: how long it will shine as a normal star. The main sequence is the longest phase of star evolution, because hydrogen is the basic fuel all stars begin with. More massive stars start out with more hydrogen fuel than do lower mass stars, but, perhaps not surprisingly, massive stars use up their fuel at a much faster rate. The reason for this is that stars evolve according to their ability to balance their great mass against gravity. The only way for them to do this is to generate enough internal heat to produce the outward thermal gas pressure needed for the balance. Since high mass stars have more mass to balance, they need higher central temperatures than do lower mass stars, and as a consequence they use up their fuel at a much faster rate than do lower mass stars. In general, high mass stars will be much rarer than low mass stars, even in galaxies with the most ongoing star formation.

The consequence of all this is that normal stars shine steadily for periods that drastically depend on their mass. For example, the main sequence lifetimes of stars having masses of twice, equal to, or half the mass of the Sun are 2 billion, 10 billion, and 60 billion years, respectively. Some stars are so massive that their main sequence lifetimes are only a few *million* years. For example, a star born with 20 times the mass of the Sun will shine as a main sequence star for only about 6 million years. Such stars are so hot that they appear bluish-white in color and often photoionize their surrounding hydrogen gases into a pink HII region.

The remaining 10% of stars in the sky are found in the other parts of Figure 4. Red giants, supergiants, hypergiants, horizontal branch stars, and white dwarfs all represent phases of star evolution that come *after* a star exhausts all of the hydrogen fuel in its core. The luminosities of these kinds of stars are therefore not powered by fusion of hydrogen to helium in the core. The best a star can do when it runs out of hydrogen in its core region is to adjust its structure in order to raise the core temperature high enough to allow it to be powered by other fusions, beginning with helium fusion. This is not easy, because while the temperature required for hydrogen fusion is 15,000,000 K, that required for helium fusion is 100,000,000 K. In order to achieve this high temperature, the core must shrink in size. This converts gravitational potential energy into thermal kinetic energy, and raises the core temperature.

Red giants are evolved low mass stars. Any star having a birth mass between 1/2 and 8 times the mass of the Sun is considered to be low mass. Supergiants are evolved high mass stars, and generally include stars from 8 to 30 times the mass of the Sun. Hypergiants are generally the most massive and most luminous stars in spirals like the Milky Way. Hypergiants usually show evidence of shedding mass at a very high rate in a "superwind" version of the solar wind.[3] Horizontal branch stars are generally whitish to yellowish low mass stars that have evolved past the red giant phase and are powered by core helium fusion. White dwarfs are basically dead low mass stars. The dashed curve in Figure 4 is known as the "instability strip" and is where radially-pulsating stars are found. The most luminous Cepheid variable stars lie among the red supergiants while the least luminous pulsators, the cluster variables, lie on the horizontal branch.

The general implication from Figure 4 is that, from a great distance, the stars we will see in any spiral galaxy will be those found at the top of the HR diagram: hypergiants, supergiants, the most luminous Cepheid variable stars, and the brightest upper main sequence stars. All of these types will involve mainly the most massive stars in the system.

A stellar population can be defined by the HR diagrams of the *star clusters* it contains. Star clusters are extremely useful for understanding stars and stellar evolution because virtually all of the stars in a cluster are at about the same distance, meaning differences in apparent brightness translate directly to differences

---

[3]The solar wind refers to the charged particles that continuously flow from the outer atmosphere of the Sun.

in luminosity. Open star clusters are found in the Galactic plane and are associated with the interstellar gas and dust that has collected in the disk component. Therefore, such clusters are part of Baade's Population I. Globular clusters are found in the halo where there is no interstellar gas and dust, and therefore such clusters are part of Baade's Population II. Figure 5 shows HR diagrams for several examples of both types of clusters.

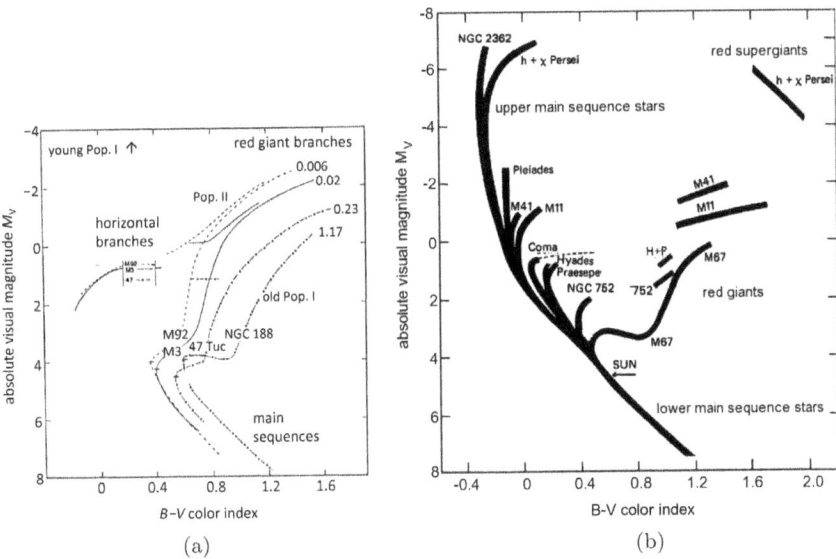

Fig. 5. HR diagrams for star clusters in the Milky Way in the form of absolute visual magnitude versus $B-V$ color index, used as proxies for luminosity and surface temperature, respectively. (a) Diagrams for three globular clusters (M3, M92, and 47 Tucanae), each a population II system more than 10 billion years old, and one old open cluster NGC 188, which is a Population I system about 5 billion years old. The main sequence, red giant, and horizontal branches of these diagrams are indicated. The numbers at upper right are the metallicities. Adapted from R. Buser and I. R. King, eds., *The Milky Way as a Galaxy*, (Mill Valley, California, University Science Books, 1989) p. 23, with permission. (b) The HR diagrams of 10 open star clusters, all Population I systems, made composite by matching their main sequence sections. The peculiar spread of the diagrams is due to the wide range of ages of such clusters, from a few million years in the case of NGC 2362 to about 4 billion years in the case of M67. Adapted from A. R. Sandage, *Astrophysical Journal*, **126**, 326 (1957). © AAS. Reproduced with permission.

The appearance of the HR diagram of a star cluster depends mainly on two factors: the cluster's age, and the metallicity of the cluster's stars. A cluster's age determines how much of the upper part of the main sequence is seen. Since upper main sequence stars are more massive than lower main sequence stars, the upper main sequence stars have much shorter lifetimes. The result is that, as an open cluster ages, stars systematically evolve off the main sequence by turning towards the giant region of the HR diagram. The age of the cluster is the main sequence lifetime of the stars at the turn-off point. High mass stars evolve to red supergiants, while low mass stars evolve to much smaller and much less luminous red giants. Because high mass stars are much rarer than low mass stars, there are usually few red supergiants, if any at all, in the younger clusters. No cluster ever develops a strong "red supergiant branch." However, an old open cluster like M67 or NGC 188 has a strong red giant branch because the stars on the branch are the much more abundant low mass stars.

Metallicity, a measure of the abundance of heavy chemical elements, is usually specified relative to the Sun. A star with the same abundance of heavy elements as the Sun is defined to have a metallicity of 1.0. If the metallicity is equal to or greater than 1/3, we say a star is "metal-rich." If, on the other hand, the metallicity is much less than 1/3, then a star is said to be "metal-poor." Figure 5(b), shows that the main sequences of open star clusters overlap. The absolute visual magnitude and color of the Sun lie on that same composite sequence, indicating that open clusters have the same metallicity as the Sun and thus are metal-rich systems in the Milky Way.

Figure 5(a) shows the drastically different HR diagrams for four very old stellar systems, including three Galactic globular clusters at least 10 billion years old and one Galactic open cluster about 5 billion years old. These old systems characteristically show a strong red giant branch, and in each case, red giants are the brightest stars in the clusters. The numbers at upper right show the wide spread in metallicities of these systems, ranging from 1.17 for old open cluster NGC 188 to 0.006 for the much older globular cluster M92;

thus NGC 188 is metal-rich while M92 is metal-poor. The curves show that, at a given absolute visual magnitude, a metal-poor star is bluer than a metal-rich star. Also, at a given color index (or surface temperature), a metal poor red giant is considerably more luminous than a metal-rich red giant, and a metal-poor main sequence star is considerably fainter than a metal-rich main sequence star.

The reddening of stars through metallicity (as opposed to dust extinction) is due to the fact that metals like iron have many absorption lines in the blue region of the electromagnetic spectrum. These lines act like a "blanket," absorbing energetic blue photons, which are then re-emitted as thermal infrared radiation. Since this weakens the brightness of a star in blue light relative to red light, the net effect is a reddening of the star.

Figure 5(a) also shows that only the metal-poor clusters have a significant horizontal branch. These core-helium-fusing stars are seen in significant numbers in globular clusters because such clusters are old enough to have many of their low mass stars reach that phase of stellar evolution. The term "horizontal branch" applies only to some globular clusters. In M92 and M3, the branch actually curves downward in the HR diagram at its blue end. Some horizontal branch stars can be the hottest stars in a globular cluster. Most open clusters do not hold together long enough for any of their stars to reach this point. In fact, most open clusters disintegrate (that is, lose all of their stars) in less than 200 million years.[4] The stars still evolve of course (like the Sun), but as part of the general field of stars in the Milky Way.

Now we can better understand what Baade and others found out about stellar populations. The most luminous stars in a spiral galaxy will be the hottest, most massive, and youngest upper main sequence stars and any evolved high mass stars, such as red supergiants (or any supergiants/hypergiants regardless of color). Most of the brightest stars will tend to be main sequence stars bluish in color, and since these are usually found lining the spiral arms, they are the population

---

[4]R. Wielen, *Publications of the Astronomical Society of the Pacific*, **86**, 341 (1974).

I stars that we see from great distances, an example being the resolved stars in the OB association NGC 206 in M31 (Figure 3). The older population I stars in a spiral are not easy to resolve in general because old population I, metal-rich red giants are much fainter than Population II red giants. In the bulge of Andromeda, however, there are metal-poor Population II red giants. It is these that Baade found in his pictures taken during the Los Angeles blackout.

Other issues are relevant to the Population I/II dichotomy. Much of the organized motion in a spiral galaxy is contained in the disk component. Stars orbit the center on nearly circular but unclosed paths, as is shown by the Sun's orbit in the plane of the Milky Way (Figure 3 of Chapter 14). But in addition to orbital motion in the plane of the disk, stars also oscillate above and below the disk plane, like horses on a merry-go-round. A typical galactic disk is not infinitesimally thin, but has a clear thickness. Stars that are very young do not wander very far from the plane of the disk as they orbit the center of a spiral galaxy. As time goes on, and after many orbits, the "vertical" motion, that is, the motion in and out of the disk plane, increases, such that older disk stars wander farther from the plane than do the youngest disk stars. The increased motion occurs not because stars encounter other stars as they orbit the galactic center. Star-star encounters have an extremely low probability of occurring because the individual stars in a galaxy are very far apart compared to their typical sizes. For example, in our part of the Milky Way, the nearest star to the Sun, called Rigel Kent or Alpha Centauri, is 4.3 light years away, a distance that is nearly 30 million times the diameter of the Sun. The probability that the Sun will gravitationally encounter another star, and have its orbit significantly changed as a result, is so small that it may take longer than the age of the Universe for an encounter to actually happen.

The source of the disk thickness may instead lie in gravitational encounters with giant clouds of interstellar gas and dust, especially giant molecular clouds (GMCs) which would be large, massive and dense objects scattered throughout the Galactic plane. Encounters with GMCs are far more likely than encounters with other stars would be in a disk-shaped galaxy. As stars interact with GMCs, their orbits get more of a random component, and they can wander further and further from the Galactic plane.

In the bulge and halo components, orbits can have much more of a random component, such that objects in those components do not partake of the motion in the disk. We know that stars in the halo of the Milky Way follow different orbits than do disk stars because globular clusters, the main constituents of the halo, follow a more spherical distribution than do disk stars. Globular clusters in the outer parts of the halo, in fact, do not rotate with the disk component. Instead, we say that their mass in the Galaxy is balanced against gravity by random motions, not by systematic rotation. The halo is said to be a "pressure-supported" subsystem of the Milky Way, much like the Sun balances its mass against gravity due to its intense heat.

The discovery of stellar populations was a monumental development in astronomy, because clues to how the parts of the Milky Way formed and evolved are contained in the movements, colors, and locations of its stars. The current view is that a spiral begins as a large, protogalactic cloud made primarily of hydrogen and helium gas collecting into a seed halo of cold dark matter.[5] As the cloud collapses due its own gravity, a fraction of the mass fragments into smaller collapsing objects that become bulge and halo stars, including globular clusters. A disk will form only if the protogalactic cloud has any initial rotation, which it can get through gravitational encounters with other protogalactic clouds. If there is rotation, then as the cloud collapses, it will rotate faster and faster, and will eventually form a disk plane.

Thick Disks and False Bulges — In addition to the main disk, many spirals also have a feature known as a "thick disk," which is fainter than the main disk and which is several times thicker than the main disk. The thick and thin disks share the same plane, and in edge-on views it is clear that the dust and interstellar gas in a spiral is largely confined to the thin disk; the thick disk itself has little gas and dust in comparison, and is made almost entirely of old stars. In

---

[5]Dark matter is described in Chapter 14.

most discussions of galactic disks, the thin disk is what is referred to as the "disk."

Thick disks were first described in edge-on S0 galaxies by former Arizona State University astronomer David Burstein (1947–2009) in his 1978 PhD thesis. Later, University of Edinburgh astronomers Gerry Gilmore and Neill Reid reported the discovery of a thick disk in our Galaxy.[6]

Many spiral galaxies also have a feature in the center that looks like a bulge, but which is not necessarily made of Population II stars. These bulges can be as flat as the disk and include dust, star formation, spiral structure, even a small inner bar. In the edge-on view, some have a boxy shape. Such bulges are called "pseudobulges" (meaning a false bulge) to distinguish them from the classical bulges seen in galaxies like M31, M81, and especially the Sombrero Galaxy M104. In this book, nearly all of the galaxies illustrated in Chapter 15 have pseudobulges. Pseudobulges were recognized in a remarkable 2004 review article[7] by then University of Texas astronomer John Kormendy and University of Arizona astronomer Robert C. Kennicutt, Jr. These authors concluded that pseudobulges are very common in disk-shaped galaxies, and likely result from the slow migration of interstellar gas towards the center of a galaxy, owing to the gravitational influence of noncircular structures like bars. That is, pseudobulges are thought to be products of secular evolution in galaxy disks. Classical bulges, on the other hand, are believed to have formed early in a galaxy's evolution by a process similar to the formation of elliptical galaxies, through collisions and mergers (Chapter 22). Classical bulges are the type made entirely of Population II stars.

---

[6]G. Gilmore and N. Reid, *Monthly Notices of the Royal Astronomical Society*, **202**, 1025 (1983).

[7]J. Kormendy and R. C. Kennicutt, Jr, *Annual Reviews of Astronomy and Astrophysics*, **42**, 603 (2004); see also E. Athanassoula, *Monthly Notices of the Royal Astronomical Society*, **358**, 1477 (2005).

Chapter 11

Star Factories

On Earth, a "factory" is a place where people or machines manufacture goods or some kind of product. Nothing is produced out of thin air, so a factory needs to have a supply of raw materials to make the items it produces. In order to stay in business, the supply of raw materials has to keep up with demand.

A spiral galaxy is a factory that manufactures stars. The raw material that the galaxy uses is the interstellar gas and dust clouds that are scattered across the disk. The "machine" that manufactures stars is the relentless pull of gravity. Gravity is derived from mass and is an exclusively attractive force, meaning it draws material together. This is different from the electric force, which is derived from charge, and which can be repellant or attractive depending on the sign of any charged particles that come near each other. For example, the electric fields of two objects having charge of the same sign, like protons, will act to cause the objects to repel each other. An electron and a proton have oppositely signed charges, and are attracted to each other as a result.

Star formation is complex and not every aspect of the process is well understood. Nevertheless, the general idea is well established. Stars form by gravitational collapse of a large cloud of interstellar gas and dust. Gravitational collapse takes a large, cold, thinly spread cloud and draws the material together into a much smaller, denser, and hotter object. The heating occurs because collapse converts gravitational potential energy into thermal kinetic energy. Collapse ceases when fusion in the center of the object produces enough heat to balance the mass against gravity. Then we say that a star is born.

The basic requirement for collapse to take place is that gravity has to be able to overcome the random thermal motions within a cloud. Temperature is a measure of the significance of such motions. The higher the temperature, the greater the average speed of the particles in a cloud. These speeds provide an internal pressure that generally opposes gravitational collapse, because a hot gas wants to expand. For most of the interstellar clouds in a given galaxy, the speeds will be high enough to avoid collapse. However, in certain regions, the temperature can be so low and the density high enough that gravity gets the upper hand and overcomes the tendency to expand. This is when collapse begins. The regions where this can happen tend to be the giant molecular clouds, or GMCs, mentioned in the previous chapter. Such clouds can be as massive as a few million suns and can form hundreds to thousands of new stars. Magnetic fields, turbulent motions, and feedback from massive star formation can have an impact on the process. Collapse of a given cloud may require a trigger, such as a shock wave from a supernova explosion or the collision of two clouds.

It was shown in Chapter 8 that star-forming regions can trace the spiral arms of disk-shaped galaxies. This highlights how star formation in spiral galaxies has a significant impact on their appearance. Galaxies that are not forming any new stars can look very different from the ones that are. We can see this in a variety of ways. One way is using color images.

The two stellar populations that Baade identified in spirals were difficult to study and illustrate in his time, but at the present time, not only are these populations well studied, they can be seen directly in color images. This was shown beautifully using UBV photography in James D. Wray's book.[1] The images in this atlas used a dye transfer process designed to give colors that would be scientifically accurate. Some of the best color images based on digital imaging are available from the Sloan Digital Sky Survey. The filters used for

---

[1] J. D. Wray, *The Color Atlas of Galaxies*, (Cambridge University Press, 1988).

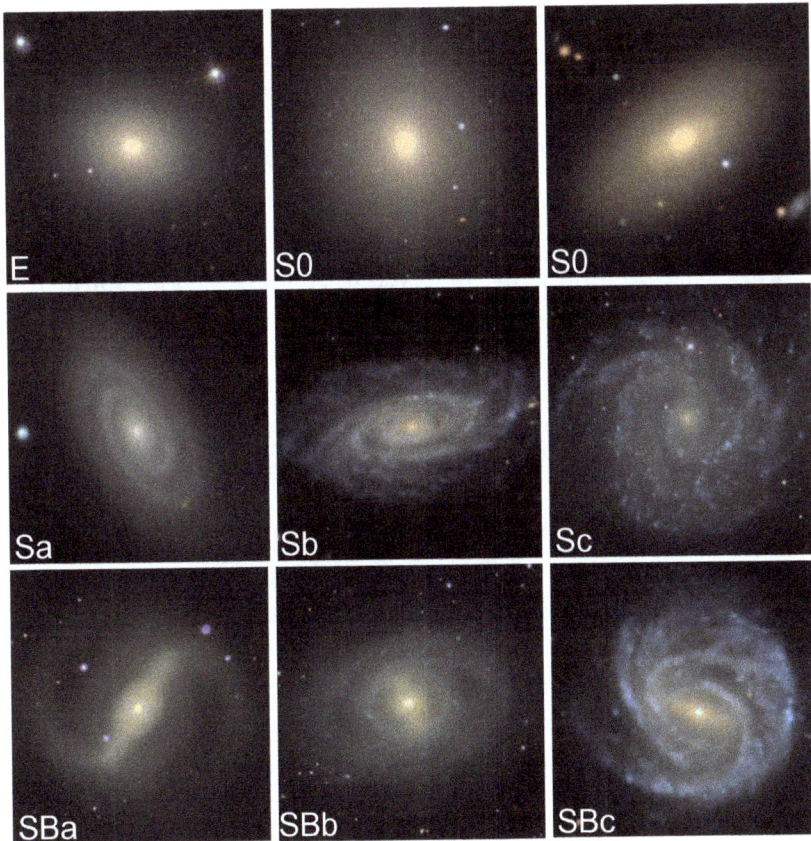

Fig. 1. Color images of nine galaxies of different Hubble types from the Sloan Digital Sky Survey. (top row): NGC 3608, NGC 4203, NGC 6278; (middle row) NGC 4305, NGC 5351, NGC 5668; (bottom row): NGC 4314, NGC 3351, NGC 3367. Images credit: Sloan Digital Sky Survey Collaboration, http://www.sdss.org.

the SDSS are different from the $UBVRI$ filters that were the staple of photometry in the 1980s and 1990s. SDSS blue-green, red, and infrared images can be combined to give images that are a very good approximation to the colors the eye would see.[2]

Figure 1 shows SDSS color images of nine galaxies of different Hubble types. In these images, regions where the light is dominated

---

[2]R. Lupton *et al.*, *Publications of the Astronomical Society of the Pacific*, **116**, 133 (2004).

by young population I stars appear bluish in color, while regions dominated by population II stars appear yellowish-orange in color. These show that elliptical and S0 galaxies are made entirely of old stars. Elliptical galaxies especially are pure population II systems, like bulges without a disk. S0, Sa, and SBa galaxies do have disk components, but also appear to be mainly systems of old stars. In contrast, Sb, Sc, SBb, and SBc galaxies tend to be composite systems, showing bluish spiral arms and yellowish bulges. The spiral arms are clearly dominated by population I stars while the bulges are dominated by population II stars. It is noteworthy also that the bars in the three galaxies shown in Figure 1 are all yellowish in color, indicating that these features are also made of old stars. It was noted in Chapter 6 that bars have a distinctive three-dimensional character that allows them to be recognized in the edge-on or highly-inclined view.

Other sources have also produced excellent color images of galaxies. For example, Figure 2 shows a color image of NGC 1376, a beautiful face-on Sc spiral lying 184 million light years away in the constellation Eridanus. Particularly interesting in this picture, which is based on *Hubble Space Telescope* imaging, is how the inner parts of the disk have a yellowish tint in the inter-arm regions while the inner arms are whitish and the outer arms are bluish. The spiral arms are made of young, blue Population I stars, but the inner arms are whitish probably because of the influence of the yellow disk light. The yellowish inter-arm colors in the inner regions are likely due to a much older stellar population, part of the background exponential disk light (see Figure 13(a) in Chapter 6). The small bulge could be a Population II subsystem in the galaxy, although it could also be a pseudobulge made mostly of disk stars.

There are other ways of displaying star formation in galaxies. It was noted in Chapter 7 that the difference between two filter magnitudes quantifies the color of a celestial object, and we have seen that color can tell us important things about stars, such as how hot they are, how old they might be, and whether they are metal-rich or metal-poor. In conjunction with luminosities, colors can tell us how stars evolve.

Fig. 2.   Color image of NGC 1376, a face-on spiral galaxy 184 million light years away in the constellation Eridanus. Image credit: NASA, ESA, and the Hubble Heritage Team (STScI/AURA). Acknowledgment: R. Thompson (University of Arizona).

Because filters sample relatively small parts of the spectra of stars, a color index like B–V, B–I, or V–I can reveal color in two-dimensional black and white *color index maps*. Such maps, which are shown in Figures 3–7 and in Figures 3, 7, 9, 18, 22, 24, and 32 of Chapter 15, are a more effective tool for studying star formation in spiral galaxies than are color images. All of the color index maps are coded such that blue features are dark and red features are light. The M51 images (Figure 3) show how star formation in the system is confined mostly to the spiral arms. The distribution of star formation in M51 is very complex, with some star-forming regions occurring along interarm "spurs." The map also shows strong dust lanes located on the inner (concave) sides of the two main spiral arms. The companion, NGC 5195, mostly disappears as a distinct object in the color index map, being replaced by a large irregular

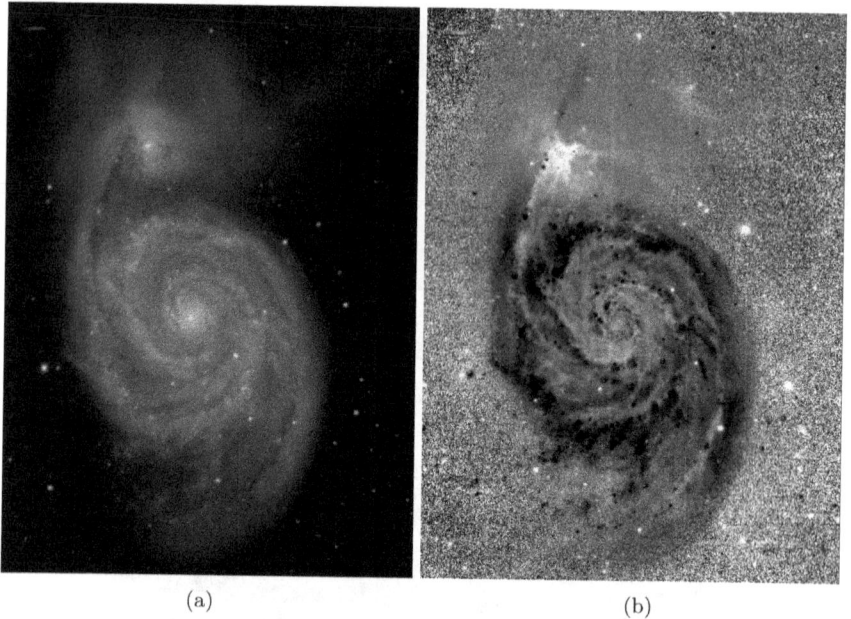

(a)   (b)

Fig. 3.   (a) Blue light (*B*) CCD image and (b) blue minus infrared (*B–I*) color index map for the Whirlpool galaxy M51. Images source: The de Vaucouleurs Atlas of Galaxies published by Cambridge University Press, 2007; reprinted with permission.

patch of dust that is much redder than the concave dust lanes. Star formation in M51 is examined more closely in Chapter 12.

NGC 5364 is the beautiful spiral with a bright ring around the center that was described in Chapter 1. The color index map in Figure 4 shows how star-forming regions follow the spiral arms closely. The map also shows that the inner ring is a zone of enhanced blue colors, and therefore is also a star-forming feature. There are dust lanes both inside and outside the ring, but these are not strongly related to the bright spiral arms.

Figure 5 shows the color index map of NGC 4449, a bright irregular galaxy lying only 12 million light years away in Canes Venatici. NGC 4449 is an example of a galaxy whose light is dominated by population I stars throughout much of its extent, yet the galaxy lacks a coherent spiral pattern like those in M51 and NGC 5364. NGC 4449 is near enough that many individual blue

<table>
<tr><td>(a)</td><td>(b)</td></tr>
</table>

Fig. 4.   (a) Blue light (B) CCD image and (b) blue minus infrared (B–I) color index map for the "beads on a string" spiral NGC 5364. Images source: The de Vaucouleurs Atlas of Galaxies published by Cambridge University Press, 2007; reprinted with permission.

stars are resolved, as are some red supergiants (seen as sprinklings of whitish dots in the blue star-forming areas). The young stars are mostly located in large, stellar associations, as also described for M51 in Chapter 12. Most interesting is the challenge NGC 4449 poses for studies of the impact of spiral structure on galaxies: NGC 4449 has formed massive young star complexes in the absence of a strong spiral pattern. Detailed studies of star formation in low-luminosity irregular galaxies have highlighted the importance of local conditions (like the density of the interstellar gas) and feedback from massive star formation on the process.[3]

In contrast to NGC 4449, NGC 6782 in Pavo (Figure 6) shows extremely well-organized star formation mainly associated with rings. There is a circular nuclear ring of star formation in the central region surrounded by a cuspy oval inner ring of star formation that

---

[3]D. A. Hunter, *Publications of the Astronomical Society of the Pacific*, **109**, 937 (1997); S. Stewart *et al.*, Astrophysical Journal, **529**, 201 (2000).

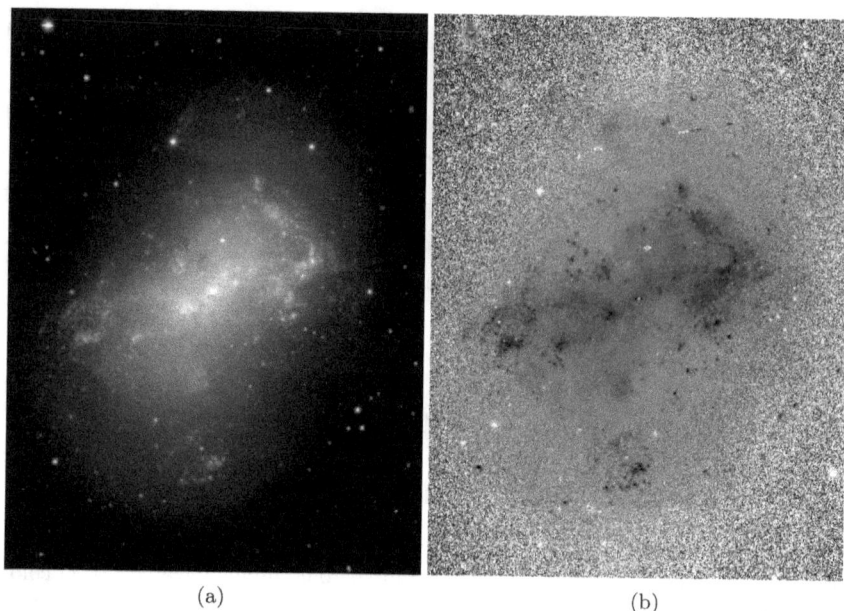

Fig. 5. (a) Blue light (*B*) CCD image and (b) blue minus visual (*B–V*) color index map for the irregular galaxy NGC 4449. Images source: The de Vaucouleurs Atlas of Galaxies published by Cambridge University Press, 2007; reprinted with permission.

strongly resembles the CBS television logo. This ring is bluer at the pointy cusps along its major axis than along its minor axis. Between the two rings are two slightly curved dust lanes associated with a weak-looking bar. Farther out, at about twice the radius of the bar, faint star formation traces a weak outer ring.

Figure 7 shows the "Blackeye Galaxy" NGC 4826 (M64). In this case, there is strong inner dust but little star formation in spite of that dust. The peculiar dust pattern in this famous object, and the recent discovery of nested disks of counter-rotating interstellar gas (meaning an inner disk of gas rotates around the center in the opposite sense to an outer gaseous disk), is thought to signify that the galaxy recently experienced a minor merger with another smaller galaxy.[4]

---

[4]R. Braun *et al.*, *Astrophysical Journal*, **420**, 558 (1994).

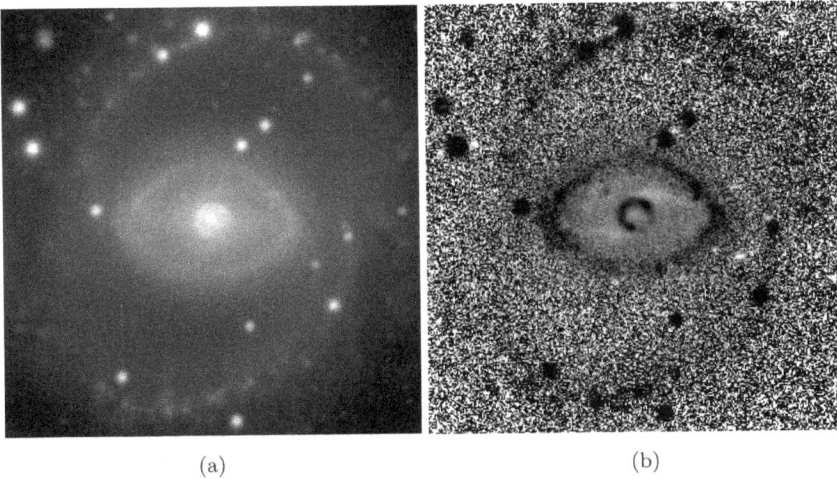

(a)   (b)

Fig. 6. (a) Blue light (B) CCD image and (b) blue minus infrared (B–I) color index map for the "CBS" spiral galaxy NGC 6782. Images source: The de Vaucouleurs Atlas of Galaxies published by Cambridge University Press, 2007; reprinted with permission.

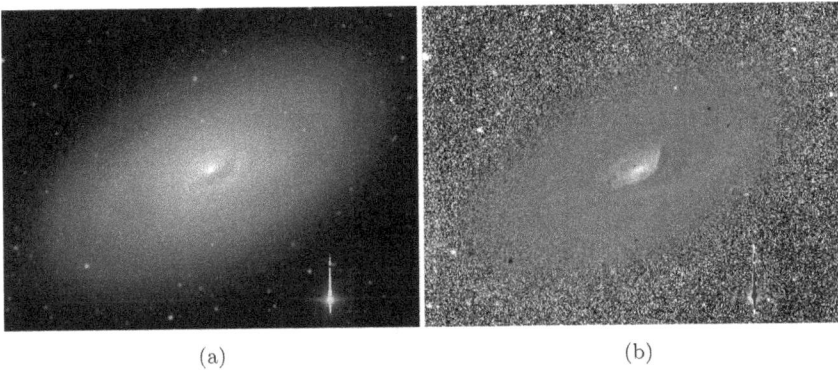

(a)   (b)

Fig. 7. (a) Visual light (V) CCD image and (b) visual minus infrared (V-I) color index map for the "Blackeye" spiral galaxy NGC 4826 (M64). Images source: The de Vaucouleurs Atlas of Galaxies published by Cambridge University Press, 2007; reprinted with permission.

A third way of examining star formation in a galaxy is obtaining an image using a filter that transmits mainly the Hα emission line at 656 nm. As described in Chapter 8, Hα is the most prominent optical emission line seen in HII regions. These often spectacular

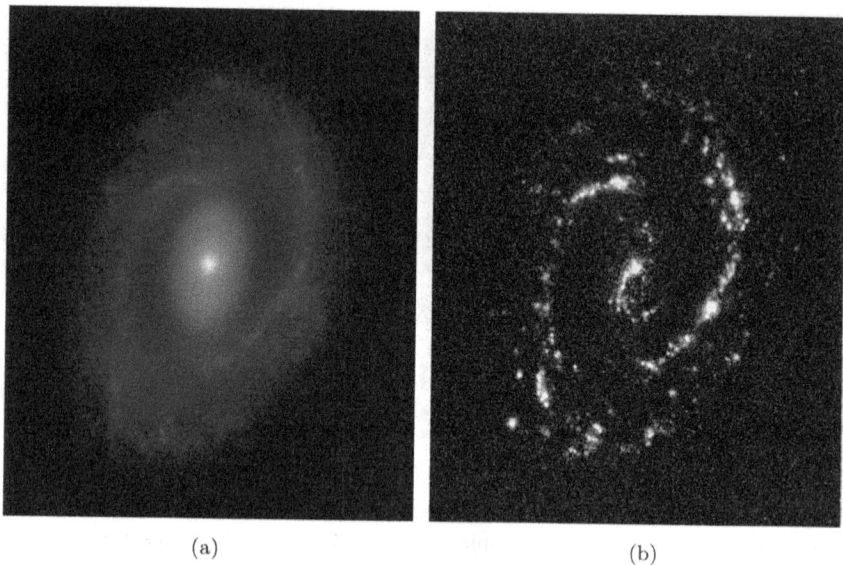

(a)   (b)

Fig. 8.   (a) Red light image and (b) Hα image of the spiral galaxy NGC 210, showing how HII regions trace the spiral arms. From D. A. Crocker *et al.*, *Astrophysical Journal Supplement Series*, **105**, 353 (1996). © AAS. Reproduced with permission.

nebulae always signal the presence of very young, massive upper main sequence stars, because only these kinds of stars are hot enough to photoionize a nebula. A given HII region may include only one or dozens of massive, hot stars.

Figure 8 shows a red image of the bright spiral NGC 210 next to an Hα image of the same galaxy. By "red image," it is meant an image that sampled the red part of the electromagnetic spectrum and which at the same time excluded Hα. NGC 210 lies 75 million light years away in the constellation Cetus, the Whale, and beautifully shows how strongly the spiral arms of a galaxy can be traced by HII regions; there are virtually no HII regions between the main arms and the broad inner structure, which shows a separate spiral pattern of its own.

Chapter 12

M51: The 21st Century View

The previous chapter showed that color images, color index maps, and Hα imaging are very effective tools for revealing the distribution of star formation in spiral galaxies. For most galaxies, these kinds of images do not show much detail in the individual star-forming regions, but when a spiral galaxy is examined up close, what do we see in its structure? Fortunately, thanks to the *Hubble Space Telescope*, we can answer this question well.

In 175 years, the spiral arms of the Whirlpool Galaxy M51 have gone from faint, unresolved, fleeting impressions seen at the eyepiece of the world's largest telescope to a view where the fine details of the arms, and some of the stars that make them up, are on full display. The beauty of M51 is that, at its distance of 24.6 million light years, it is close enough to be well resolved in deep HST images. In 2005, the spiral and its companion were the target of an extensive multi-color imaging survey with the HST, which yielded the image shown in Figure 1.[1] The image is a color representation based on blue, visual, and infrared filters for stars, similar to the blue-green, red, and infrared filters used for the galaxies shown in Figure 1 of Chapter 11, but unlike those images, an Hα filter was also used to reveal the locations of HII regions. The final, fully-processed color image is so spectacular that large-scale prints are required to fully appreciate the information it contains. The discussion here is based on a 0.9 m × 1.2 m poster-sized print.

---

[1] See https://archive.stsci.edu/prepds/m51/.

Fig. 1. Hubble Space Telescope color image of M51 and its companion NGC 5195. Image credit: NASA, Hubble Heritage Team, STScI/AURA, ESA, S. Beckwith (STScI). Additional Processing: Robert Gendler.

For showing the anatomy of a cosmic pinwheel, the HST image of M51 is second to none. The most prominent feature is the large number of HII regions scattered among the spiral arms. These objects, which appear pinkish in the image, range from the size of the Orion Nebula (about 14 light years) to more than 1,000 light years across. Figures 2–4 show closeups of a few areas in the arms.

Fig. 2. Hubble Space Telescope color image of part of M51 showing unresolved massive open clusters, well-resolved OB stellar associations, several giant HII regions, and a faint HII region that gives a good idea as to what the Orion Nebula would look like at the distance of M51. Image credit: NASA, Hubble Heritage Team, STScI/AURA, ESA, S. Beckwith (STScI). Additional Processing: Robert Gendler.

Note: The labeling is solely due to the author and was not a part of the original image. Also, in this and the succeeding two illustrations, the interpretations are mostly suggestive, not necessarily definitive.

The giant HII region labeled in Figure 2 is at least as big as the famous Tarantula Nebula in the Large Magellanic Cloud, a nebula so large itself that it is visible to the naked eye from a distance of 160,000 light years (Figure 7 of Chapter 13). The humble Orion Nebula is a mere 1344 light years away and would match up with the smallest pink nebulae, one of which is highlighted in Figure 2.

As described in Chapter 10, star formation in galaxies is thought to occur when a cold cloud of interstellar gas and dust has enough gravity to overcome its internal pressure. For a cloud of given density and temperature, there will be a mass above which gravitational collapse will ensue. Such clouds are usually dense enough to host pockets of very cold molecular gas, and it is in these pockets that stars will form in associations or clusters. The tendency for star formation to begin as a large, collapsing cloud having a characteristic mass that

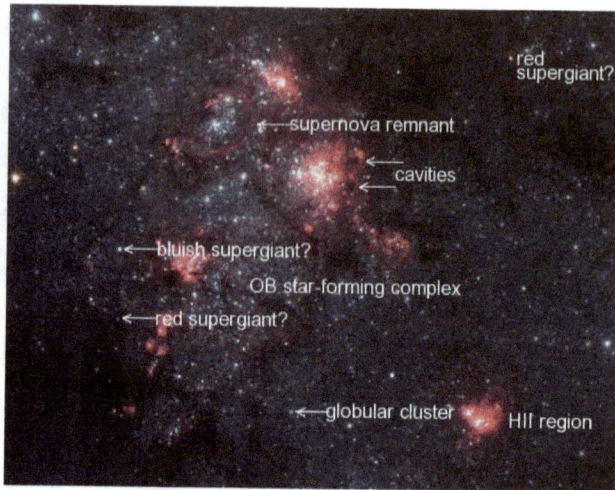

Fig. 3.   Hubble Space Telescope color image of part of M51 showing a star-forming complex with multiple HII regions, a likely supernova remnant, a possible globular cluster, and possible blue and red supergiants. Image credit: NASA, Hubble Heritage Team, STScI/AURA, ESA, S. Beckwith (STScI). Additional Processing: Robert Gendler.

Note: The labeling is solely due to the author and was not a part of the original image.

leads to stars forming in smaller-scale groups like OB associations and even denser open clusters, is usually taken as evidence that star formation in galaxies is *hierarchical* in nature.[2]

Many of the pink nebulae in M51 have obvious resolved bluish-white stars. The smaller pink nebulae may have only one massive star hot enough to photoionize the nebula, while the larger nebulae may involve hundreds of such stars. Although thousands of bluish-white stars are scattered throughout the HST image, the largest concentrations of such stars are in the spiral arms. As shown in Figures 2–4, the resolved stars in the arms are concentrated in large, bluish star complexes that include one or more subgroupings called OB stellar associations because their brightest stars, if they could

---

[2]B. G. Elmegreen, in *Star Formation in the Local Universe*, Charbonnel, Montmerle, eds. (EAS Publication Series Vol. 51, EES2010, EDP Sciences, Les Ulis), p. 19.

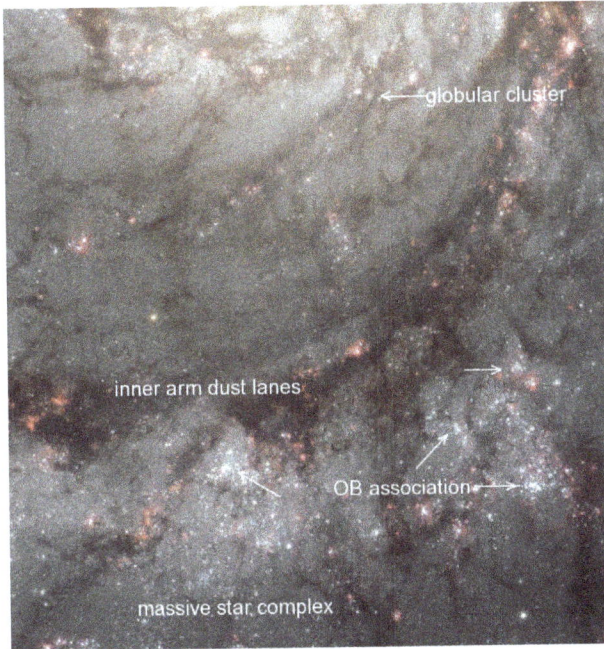

Fig. 4. This field just below the center of M51 includes a possible globular cluster (the largest known in the galaxy), as well as two large star complexes each having several OB associations (arrows). Image credit: NASA, Hubble Heritage Team, STScI/AURA, ESA, S. Beckwith (STScI). Additional Processing: Robert Gendler.

Note: The labeling is solely due to the author and was not a part of the original image.

be spectroscopically observed, would be of spectral classes O and B. There are such groupings near the Sun in our Galaxy. Many of the visible stars in the M51 OB associations are not necessarily hot enough to photoionize a pink nebula. The pink nebulae tell us where the stars are hot and massive enough to do this.

The large, young star complexes are the most obvious groupings of resolved stars in the M51 HST image. Conventional open star clusters (for example, objects like the Pleiades) are less obvious because the stars in such clusters tend to be closer together. Open clusters are generally only a few to a few tens of light years across, and at the distance of M51 would appear less than a millimeter in size on the scale of the 0.9 m × 1.2 m poster. This means real open clusters in M51 should be slightly resolved in the HST image. Two possible cases are identified in Figure 2. Because the brightest stars in the youngest open clusters also tend to be hot, upper main sequence stars, the clusters would have the same color as these stars, but would appear much brighter than any single star since they would have the combined brightness of hundreds of unresolved stars.

Globular clusters are much more massive than open clusters, and as a consequence are much more strongly bound gravitationally. Like open clusters, globular clusters have sizes of only a few to a few tens of light years, and thus appear almost star-like at the distance of M51. However, unlike open clusters, globular clusters are made exclusively of old (10 billion years or older), heavy-element-deficient stars that make the unresolved light of the cluster strongly tinted yellow. Like open clusters, a globular cluster would be brighter than most individual stars. Thus, if we want to find them in the HST images, we have to look for objects in the image having their expected brightness and color. Figures 3 and 4 identify two possible candidates; the latter figure includes the largest example known in the system.[3]

Although the bulk of resolved stars in M51 are OB main sequence stars collected in young stellar associations or scattered along the spiral arms, there are exceptions. The theory of stellar evolution tells us that, at least for stars between one half and 30 times the mass of the Sun, all roads lead to the red giant or red supergiant regions of the Hertzsprung–Russell diagram once a star exhausts its core hydrogen. Low mass stars evolve to red giants that can be a hundred to a thousand times more luminous than when such stars were main sequence stars, but even so they are difficult to detect at the distance of M51. High mass stars, however, become red supergiants with virtually the same luminosity as they had when they were main sequence stars. Such red stars can be more than 10,000 times as luminous as the Sun, and would be easily detected at the distance of M51. Because the giant phase is only a small fraction of the total

---

[3]R. Scheepmaker *et al.*, *Astronomy and Astrophysics*, **469**, 925 (2007).

lifetime of a star, a given red supergiant would have to be relatively young to be seen at all. This suggests that mixed with the OB stars in the resolved stellar associations of M51 should be an occasional red supergiant. There are many candidates for red supergiants in the arms of M51; two possibilities are indicated in Figure 3.

Figure 2 shows an HII region with a conspicuous cavity surrounding a cluster of resolved bluish stars. The cavity is similar to that seen in the Galactic HII region NGC 2237 (the Rosette Nebula) and to the Triangulum Spiral HII region NGC 604, and has likely been "carved" by the winds of massive stars. Some of the nebulae could also be supernova remnants (SNRs), the ejected gases from a long past stellar explosion. Both SNRs and HII regions show, in addition to Hα, emission lines due to ionized sulfur, but the latter are brighter relative to Hα in SNRs than in HII regions. This allows the two phenomena to be distinguished spectroscopically. Visual distinction is likely to be more ambiguous. One possible example is indicated in Figure 3. The two cavities indicated in one HII region in Figure 3 could also have been created in supernova explosions.

There are undoubtedly many planetary nebulae in M51, but being only 1–2 light years in size generally, these objects are not well-resolved in the HST image. Planetary nebulae would be detected with the Hα filter, but would appear as very faint, pink star-like objects.

The HST image of M51 shows how strongly new stars are concentrated along and near the spiral arms. The association is in fact so close that one might deduce that it is the presence of the spiral itself that triggers the formation of large numbers of new stars. That is, the spiral somehow compresses interstellar gas and dust clouds, allowing the self-gravity of the clouds to overcome their internal pressure and collapse into forming new stars. This does not preclude finding pink nebulae between spiral arms, because several large complexes are found far from the arms. Nevertheless, there is a definite preference for the pink nebulae to be found near the spiral arms.

There is also a striking connection between interstellar dust in M51 and its spiral arms (Figure 5). Interstellar dust is manifested in the HST image as dark lanes that are especially strong in the inner

Fig. 5.   Hubble Space Telescope color image of the inner arms of M51 showing the strong and well-organized dust lanes lying on the inner sides of the two main arms as well as interarm dust lanes and complex small-scale lanes in the central region. Image credit: NASA, Hubble Heritage Team, STScI/AURA, ESA, S. Beckwith (STScI). Additional Processing: Robert Gendler.

regions of the galaxy. A high density of atomic and molecular gas is concentrated in these sharp lanes, which are often characterized as being "shocks." A shock is a disturbance which travels faster than the speed of sound in a medium (that is, a region of given temperature, density, and pressure). In this case, the disturbance is the spiral itself, and the medium is the interstellar medium in the galaxy. Shocks can induce star formation, but while this may account for some of the star-forming regions we see, it is thought that much of the effect of the spiral may be in organizing star formation and not in enhancing the rate at which new stars are being formed in the galaxy.[4] This is discussed further in Chapter 16.

---

[4]B. G. Elmegreen, *EAS Publications Series*, **51**, 19 (2011).

Curiously, in the inner regions of M51, the dust distributes along sharp lanes on the *inner parts* of the arms, while the pink nebulae are distributed roughly along the middle or outer parts of the arms. This suggests that the galaxy's interstellar medium encounters the arms while moving in a counterclockwise sense, implying trailing spiral arms.

Figure 5 further shows that there are both arm and interarm dust lanes in M51, the former being more prominent than the latter. Many lanes curve around the center, while some appear to almost point towards the center or just a little ahead of the center. The dust lanes appear to be organized almost to the smallest detectable scale. The main dust lanes on the inner edges of the bright inner arms and the remarkably long arm that starts in this area and ends in front of the companion are tens of thousands of light years long. Figures 1 and 5 show also that if we lived just outside the main arms, virtually any line of sight to the galactic center would intercept multiple clouds of dust, which would likely obscure the galactic nucleus completely as in our Galaxy.

The presence of hot stars in stellar associations tells us that such groupings are not likely to be more than 10–100 million years old. The fate of these star complexes is to disperse over time. The reason for this is that in general, the stars in stellar associations are thinly spread, making the systems very loosely bound gravitationally. As the stars in an association move along their orbits around the galactic center, they spread out into the galactic disk. Over time, a given group will lose its identity. We know this happens because some older associations are found in advanced states of dispersal known as star streams or moving groups. These are groups of stars over a wide swath of sky that appear to be moving together. Studies of infrared images have provided evidence for star streams in many spirals.[5]

The bulk of the stars in the disk of M51 probably formed in a stellar association. The aged and lower mass members of earlier

---

[5]D. L. Block *et al.*, *Astrophysical Journal*, **694**, 115 (2009).

associations are now scattered in the largely unresolved background star clouds which are most easily seen between the bright spiral arms. It is noteworthy that even at the distance of the Andromeda Galaxy, one-tenth as far away as M51, a star like the Sun would be difficult to detect with the HST.

Chapter 13

The Curtis/Ritchey "Novae"

Chapter 3 recounted the role that novae in spirals played in establishing the existence of galaxies. Although most novae in the Andromeda and Triangulum spirals achieved luminosities consistent with Galactic novae, the novae Heber Curtis and G. W. Ritchey found on photographic plates a century ago, in what were obviously much more distant spirals than Andromeda and Triangulum, seemed to be on a luminosity scale very different from Galactic novae.

To understand these peculiar novae, it is necessary to examine more closely how stars change over time. This is not something we can necessarily see happening in real time. The timescales of star evolution are generally much longer than a human lifetime, and even some of the highest mass stars shine for more than 50,000 human lifetimes. Since we cannot observe any star from birth to death, we have to infer what happens from studying many stars. The Hertzsprung–Russell (HR) diagram (Chapter 10) is the primary tool for studying star evolution.

13.1. Low Mass Stars

The evolution of a star begins with its position on the main sequence. A star like the Sun is made of 75% hydrogen, 24% helium, and about 1% all other chemical elements. Flush with hydrogen fuel in its core, a new star will settle on the main sequence for a period of time that we saw in Chapter 10 mainly depends on the star's birth mass. The Sun is about 4.6 billion years old, but a one solar mass star can shine as a main sequence star for about 10 billion years. The Sun is therefore only about halfway through its main sequence lifetime. How a star evolves after the main sequence phase also depends mainly on its mass. The first step in understanding star evolution is to realize that the evolutions of high and low mass stars are different. In general, a star is considered to be "low mass" when its birth mass is between 0.5 and 8 solar masses, and it is considered to be "high mass" when its birth mass is between 8 and 30 solar masses. Stars of very low mass (<0.5 solar masses) have main sequence lifetimes much longer than the age of the universe, while stars more massive than 30 solar masses have main sequence lifetimes of less than 2 million years and are extremely rare.

The red giant branch is the part of the HR diagram extending from near the main sequence to the region of the brightest red giants. Studies have shown that a red giant is an evolved low mass star whose luminosity is powered by fusion of hydrogen into helium in a *shell* around a burned out helium core. The reason this happens is that, in order for the core of a low mass star to reach the temperature of 100,000,000 K needed to power itself by core helium fusion, the core has to shrink so much that the electrons in the core fiercely resist further compression. The act of shrinking, however, compresses unused hydrogen in a shell around the core that, during the main sequence phase, was too cool and too low in density to fuse into helium. Now, in the post-main sequence phase, this gas can fuse and power the star even though the core is not providing any new heat.

Since the core in its unshrinkable condition is producing no new heat, it can't really balance its mass against gravity using ordinary outward thermal gas pressure (that is, the tendency of a hot gas to want to expand). Instead, a peculiar quantum mechanical effect comes into play: the free electrons in the core, which are being forced by gravity to occupy the same locations, increase their speeds enough to balance the mass against gravity. The electrons are said to provide "degeneracy pressure," and any material so balanced is called degenerate matter. Degenerate matter is not something one finds every day. It is an extremely dense form of matter where actual temperature plays no role in providing balance.

As noted in Chapter 10, a horizontal branch star is a star whose luminosity is powered by core helium fusion. These tend to be large, yellow stars which, in globular clusters, occupy a relatively horizontal region of the HR diagram. Low mass stars are thought to eventually overcome degeneracy and explosively begin helium fusion, the byproducts of which are carbon and oxygen. During the horizontal branch phase, a star can become unstable to radial pulsations. This is not a long-lived phase for any star, as the core exhausts its helium at a much faster rate than it consumed its hydrogen. Just as at the end of the main sequence phase, the end of the core helium fusing phase is characterized by contraction of the core in order to raise the central temperature enough to fuse the "ash" of helium fusion, that is, carbon and oxygen. In general, a low mass star will not be able to initiate carbon fusion to power itself because electron degeneracy pressure resists the needed amount of compression. This leads to both shell hydrogen fusion and shell helium fusion powering the star's luminosity, which takes the star up the red giant branch a second time. It is at this point that instabilities in the helium fusing shell can cause the degenerate core and envelope gases of a low mass star to separate, leading to a colorful object known as a planetary nebula (Figure 1).[1] The nebula expands away from the exposed core and eventually disappears, the gases being recycled among surrounding interstellar materials. Once the nebula is gone, the core begins a long process of cooling off as a white dwarf. Thus, a white dwarf is a dead, low mass star made of degenerate matter. In Figure 4 of Chapter 10, white dwarfs are found in the lower left area.

The curious thing about white dwarfs is that they can have masses comparable to the Sun and yet are no bigger than the Earth. They are much denser than any planet and yet would likely have no

---

[1] The remarkable diversity in planetary nebula shapes can be explained in terms of an interacting winds model, where fast-moving gases from the core region interact with more slowly moving gases ejected earlier in the evolution of a star. This is elegantly described in S. Kwok, *Cosmic Butterflies* (Cambridge, Cambridge University Press, 2001).

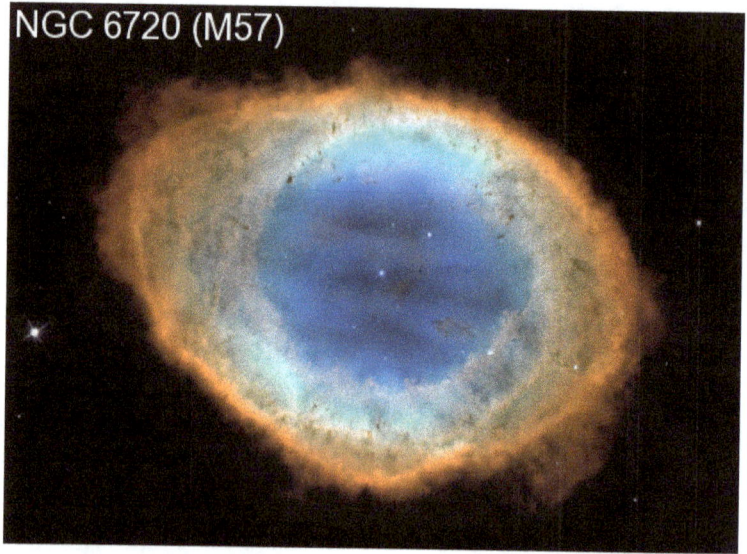

(a)

(b)

(c)

Fig. 1. Three examples of planetary nebulae, the final stage of evolution of low mass, sun-like stars. Essentially dying stars, planetary nebulae represent the non-explosive separation of the core and envelope gases that leaves behind a white dwarf. (a) Image credit: NASA, ESA, and the Hubble Heritage (STScI/AURA)-ESA/Hubble Collaboration. (b) and (c) Image credit: Copyright: Adam Block/Mount Lemmon Skycenter/University of Arizona, reproduced with permission.

surface features. Ordinary objects on Earth would weigh millions of pounds on a white dwarf. White dwarfs are so dim that not a single example is visible to the naked eye.

13.2. High Mass Stars and the Discovery of Supernovae

If we could imagine taking snapshots of the sky every hundred thousand years for 14 billion years, the estimated age of the universe, and then playing the snapshots as a movie where the frames advance one every second, we would gain an interesting perspective on star evolution. Such a movie would compress the entire history of the universe into a period of 39 h, or about a day and a half. In the film, a star like the Sun would shine as a main sequence star for about 28 h, and for most stars in the Milky Way the main sequence lifetime would exceed the length of the film. In contrast, the entire life cycle of a 21 solar mass star like Rigel, whose main sequence lifetime is only about 5 million years, would be over in less than two minutes. This shows that with massive stars, we are dealing with extraordinary star evolution. It is the evolution of massive stars that explains the "novae" that Heber Curtis and others discovered in spiral nebulae on photographic plates nearly a century ago.

As noted in Chapter 4, the puzzle of novae in spirals began with the appearance of a bright example in the Andromeda spiral in 1885. Given the variable star designation S Andromedae, the nova of 1885 was extremely bright and far outshined all of the other stars in that galaxy. This was a visually discovered case; later examples were found photographically. One especially bright example, named Z Centauri, was seen in the peculiar star-forming galaxy NGC 5253 on a Harvard spectrum-plate in 1895.[2] Two more examples were discovered in 1917, when George Ritchey noticed a bright nova in NGC 6946, and Heber Curtis noticed one in NGC 4527 (on a plate taken in 1915). Both NGC 4527 and 6946 are spiral nebulae. Once

---

[2]E. C. Pickering and W. P. Fleming, *Astrophysical Journal*, **3**, 162 (1896).

cued to these objects, Curtis and Ritchey went back to look at other accumulated plates to find other examples.

Novae occur in our Galaxy, one of the most recent and brightest examples being Nova Cygni 1975. This star followed a distinctive light curve where it rose in brightness so rapidly that it was discovered only after it reached maximum brightness, which in this case was second magnitude. It also faded fairly rapidly at first, dropping out of naked eye visibility in only a week, after which it faded much more slowly. If Nova Cygni had occurred in the Andromeda Galaxy during Hubble's time, it would have been easy to detect on photographic plates but would still have been more than 200 times fainter than S. Andromedae.

As previously described in Chapter 4, objects like Nova Cygni are believed to involve non-destructive outbursts on the surface of a white dwarf that is receiving mass from a normal companion star in a very close binary system. A build-up of hydrogen gas on the surface of the white dwarf leads to explosive surface fusion that ejects some mass into space. Because the outbursts do not destroy the white dwarf, it is possible for the same system to have recurring outbursts.

The novae that Ritchey and Curtis discovered occurred in nebulae that looked farther away than the Andromeda Nebula. By 1920, the year of the "Great Debate" on the scale of the Universe, Curtis was convinced that the novae he and others found were on a different scale from those found in our Galaxy and the Andromeda Nebula. Although the nature of novae was uncertain during most of the 1920s, by the 1930s, the existence of galaxies and the large scale of the universe were well-enough established that the issue of novae could be revisited. This was done dramatically in 1934, when Walter Baade and Fritz Zwicky (1898–1974) broke novae up into two categories: *common novae* and *supernovae*. They noted that common novae were seen only in the nearest systems, while supernovae were seen "all over the acceptable range of nebular distances." They also noted that supernovae were so bright that they "emit nearly as much light as the whole nebula in which they originate."

In a second 1934 paper, Baade and Zwicky hypothesized that the peculiar phenomenon of cosmic rays, referring to subatomic particles that regularly strike the Earth's upper atmosphere at near-light speed, is somehow related to the violence of a supernova "flare-up." Given all they knew about cosmic rays and supernovae at the time, Baade and Zwicky stated: "With all reserve we advance the view that a supernova represents the transition of an ordinary star into a *neutron star*, consisting mainly of neutrons." They argue that neutrons can pack more closely than can protons, electrons, or other atomic nuclei, and that this would make a neutron star "the most stable configuration of matter as such."

This brilliant hypothesis is largely correct for some (about half) of all supernovae, and it provides the link needed to massive star evolution. Like low mass stars, high mass stars also become large red stars after they exhaust their core hydrogen fuel. In fact, they get so large that they are called red supergiants rather then mere red giants. A typical red giant is about 100 times the size of the Sun, while a typical red supergiant is more than 500 times the size of the Sun. High mass stars have such high central temperatures that they go through all phases of star evolution without the limitations imposed by core degeneracy. While a low mass star can't power itself by carbon or oxygen fusion because the core becomes degenerate before it reaches the required temperature, a high mass star can power itself not only by these, but by a host of even heavier elemental fusions which, after the main sequence phase, occur in an "onion skin" configuration as multiple fusing shells develop (Figure 2).[3]

The fusion paths lead eventually to a shell where silicon (Si) is fusing into Nickel-56, a form of the element nickel (Ni) that has 28 protons and 28 neutrons, and which is radioactive. This means the atomic nucleus in Nickel-56 is unstable and can spontaneously change to another neighboring element in the periodic table of the elements on a timescale known as the half-life. The half life of any radioactive element is the time its takes for half of the element present in any material to decay into something else. In the case of Nickel-56, one of the protons in the nucleus spontaneously changes into a neutron, leading to the emission of a high energy particle of

---

[3]E. M. Burbidge *et al.*, *Reviews of Modern Physics*, **29**, 547 (1957).

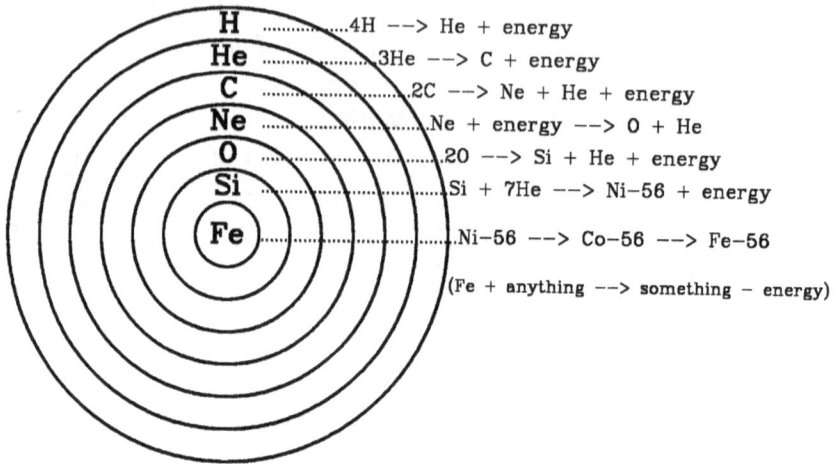

Fig. 2.   Schematic of the tiny core region of a red supergiant. Although no bigger than the Earth, the luminosity of a red supergiant star is powered entirely by multiple shells of fusing chemical elements in this region. The end result of all of these fusions is a degenerate iron core that will eventually collapse violently into a neutron star, followed by the violent ejection of all envelope gases into space. The approximate reactions occurring in the shells are shown (in very abbreviated form) at right. The reaction "Fe-56 + anything → something − energy" means that fusing iron-56 with any other nucleus does not release energy but requires energy. Therefore iron fusion cannot power a star's luminosity.

light called a gamma ray. The half-life of this decay process is only 6 days, so Nickel-56 does not hang around for long in the core. When it decays, it becomes the isotope Cobalt-56, which has 27 protons and 29 neutrons, and which is also radioactive. Cobalt (Co) is next to nickel in the periodic table of the elements (Figure 3). Cobalt-56 decays into Iron-56, again by the spontaneous change of a proton into a neutron, but with a half-life of 77 days. Iron-56 is a stable isotope of iron (Fe) having 26 protons and 30 neutrons. It is the build-up of Iron-56 in the core that leads to the violent core collapse and subsequent explosion of a high mass star. Core collapse is tied directly to the formation of a neutron star. The Iron-56 nucleus is the most tightly-bound atomic nucleus found abundantly in stars. The consequence of this is that iron fusion cannot release energy like hydrogen, helium, carbon, or oxygen fusion can, which makes it impossible for iron fusion to power a star's luminosity. Instead, Iron-56 piles up in the

25	26	27	28	29
Mn	**Fe**	**Co**	**Ni**	**Cu**
manganese	iron	cobalt	nickel	copper

43	44	45	46	47
Tc	**Ru**	**Rh**	**Pd**	**Ag**
technetium	ruthenium	rhodium	palladium	silver

(a)

Iron-56   Cobalt-59   Nickel-58

(b)

Fig. 3. (a) Iron, cobalt, and nickel in the periodic table of the elements. (b) The most abundant stable isotopes of iron, cobalt, and nickel.

center until the core becomes too massive to be balanced against gravity by electron degeneracy pressure. The violent collapse of the core breaks up the iron nuclei into protons and neutrons, and forces the protons to absorb the electrons to form more neutrons. The result would be a ball of neutrons no larger than a city.

Thus, the "novae" that Curtis, Ritchey, and others discovered a century ago were actually supernovae. If Ritchey could have followed the supernova he found in NGC 6946, it likely would have had light curves resembling those of SN 2017eaw (Figures 4 and 5), which occurred in the same galaxy a hundred years later.[4] NGC 6946 is an Sc spiral with a very large population of massive stars (Figure 8 of Chapter 3). Baade and Zwicky originally thought that in any given

---

[4]Modern supernovae are named by the year of discovery followed by a letter or series of letters.

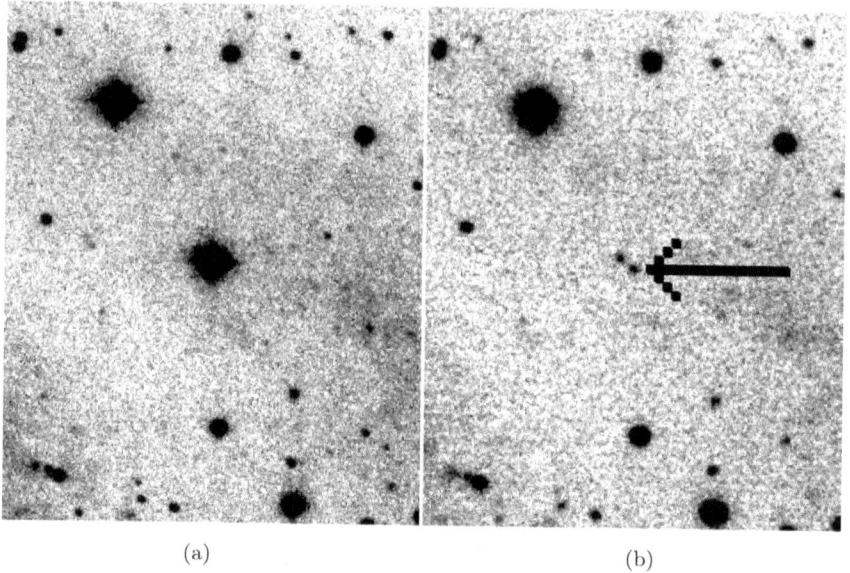

(a)   (b)

Fig. 4.   Close-up images of Supernova (SN) 2017eaw that appeared in the spiral galaxy NGC 6946. (a) shows the supernova on August 17, 2017 UT, 97 days past the explosion date. The arrow in (b) shows the supernova on October 4, 2018 UT, 510 days past the explosion date. This panel reveals a foreground star close to the supernova. From R. Buta and W. C. Keel, *Monthly Notices of the Royal Astronomical Society*, **487**, 832 (2019); Copyright © 2019, Oxford University Press.

galaxy, a supernova would occur only once every several centuries. It is only recently that supernova search programs have been efficient and thorough enough to reveal that some galaxies are "supernova factories," hosting supernovae every few years or decades. NGC 6946 is one such factory. Its high rate of supernova production (10 in 100 years) has led to it being nicknamed the "Fireworks Galaxy."

SN 2017eaw was discovered on May 14, 2017 (universal time),[5] but the explosion itself did not occur on this day. According to the most recent studies, NGC 6946 is 25 million light years away, so the explosion actually occurred 25 million years ago. Only news of the event reached Earth in 2017, and the supernova was noticed near maximum light on May 14.

---

[5]P. Wiggins, *Central Bureau Electronic Telegrams*, **4391**, 2 (2017).

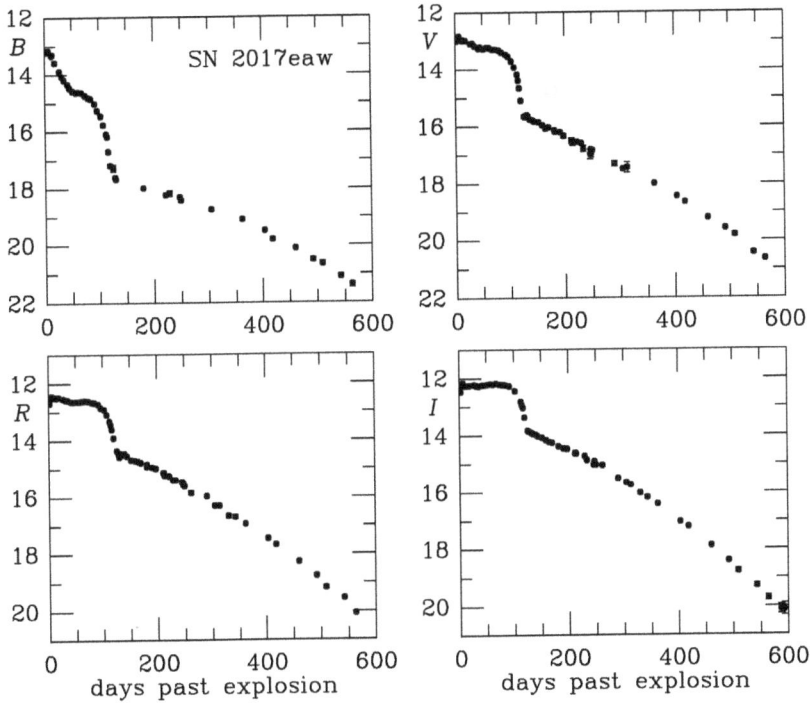

Fig. 5. Light curves of SN 2017eaw in blue, visual, red, and infrared filters. The brightness changes are displayed as the apparent standard magnitude B, V, R, I versus the number of days since the explosion, which has been pinned down to May 13 ± 1, 2017 universal time (UT). The supernova reached its maximum brightness (minimum magnitude) in each filter within 6 days past the explosion. In all filters, the supernova faded by a factor of more than a thousand by the time of the last data point, nearly 600 days after the explosion. From R. Buta and W. C. Keel, *Monthly Notices of the Royal Astronomical Society*, **487**, 832 (2019); Copyright © 2019, Oxford University Press.

It is interesting to follow, in real time, the brightness changes of a supernova, and I had the pleasure of doing so for SN 2017eaw over a period of nearly 600 days (a year and a half), initially with the 16-inch DFM Engineering reflector atop Gallalee Hall on the campus of the University of Alabama, and later (with the collaboration of Prof. W. Keel) with other telescopes at remotely-operated observatories at Kitt Peak National Observatory and the Canary Islands. A CCD camera was used at each observatory to obtain blue, visual, red, and infrared ($BVRI$) images of the supernova, and the observations were all combined to give the well-defined light curves shown in Figure 5.

This was the second supernova I had observed in NGC 6946, the previous one being SN 1980k when I was a graduate student at McDonald Observatory 37 years earlier. It is ironic that, from our vantage point, we were able to know about both supernovae only 37 years apart in time, while in the galaxy itself the two objects could not have known about each other's explosion so soon in relative time because they occurred on nearly opposite sides of the galaxy, some 55,000 light years apart (Figure 6).

Supernovae are classified according to their spectra near maximum light. There are two major recognized types. Type I supernovae are characterized by the absence of lines due to hydrogen in their spectra, while Type II supernovae are characterized by the presence of hydrogen lines in their spectra. Type I supernovae are subdivided further into Ia, Ib, and Ic subtypes according to other differences in their spectra. Of these, only Type Ia supernovae do not involve massive stars. Instead, like common novae, Type Ia supernovae are

Fig. 6.   Blue-light image of NGC 6946 showing where the supernovae 1980k and 2017eaw occurred relative to each other. The two explosion sites are 55,000 light years apart in the plane of the galaxy. Image source: W. Keel and R. Buta.

thought to be related to mass transfer onto a white dwarf in a close binary system. The difference between the two phenomena lies in degree: in a common nova, the mass transfer leads to surface outbursts but does not disturb the balance of the white dwarf against gravity. Electron degeneracy pressure holds on and keeps the white dwarf stable. But the ability of electron degeneracy pressure to balance a white dwarf's mass against gravity has a limit: if the white dwarf mass exceeds 1.4 times the mass of the Sun, degeneracy pressure can't keep the white dwarf stable. In a Type Ia supernova, mass transfer brings a white dwarf over this limit, which ultimately leads to the white dwarf's total destruction. Because Type Ia supernovae involve the destruction of a white dwarf (and not a massive star), Type Ia supernovae can occur in any kind of galaxy: ellipticals, S0s, spirals, and irregulars.

In contrast, Type II supernovae are believed to exclusively involve the explosions of massive stars, and as such they are expected to be found only in star-forming galaxies, like spiral and irregular galaxies. Type II supernovae are known as "core collapse" supernovae, because they result from the violent collapse of the core region of a massive star. It is thought that the minimum stellar mass for this to happen is 8 solar masses. Both SN 1980k and 2017eaw were examples of Type II supernovae.

The life cycle of a massive star usually begins in a cluster like that shown in Figure 7, the cluster associated with the massive HII region known as the "Tarantula Nebula" in the Large Magellanic Cloud. Many of the bluish-colored stars are extremely massive upper main sequence stars, some of which may be slightly evolved. There are so many such stars in the cluster that they photoionize a huge HII region, the inner part of which is known as the Great Looped Nebula NGC 2070. Neither SN 2017eaw nor SN 1980k took place near an area as large or complex as NGC 2070, but they did take place in NGC 6946's spiral arms which are lined by legions of massive, bluish-colored stars. The only supernova known to have occurred in the Tarantula Nebula area is SN 1987A. This was a peculiar Type II supernova that reached naked eye visibility, the only one seen this way in more than 300 years.

Fig. 7.   The massive star cluster in the Tarantula Nebula/Great Looped Nebula NGC 2070. One prominent loop is in the upper left part of the image. Image credit: NASA, ESA, and F. Paresce (INAF-IASF), R. O'Connell (U. Virginia) and the HST WFC3 Science Oversight Committee. Reproduced with permission.

Although both SN 1980k and 2017eaw were Type II supernovae, they were still quite different. SN 2017eaw was a "Type II-P" supernova, where the "P" refers to a prominent plateau in the light curves. In contrast, SN 1980k was a "Type II-L" supernova, a type characterized by a linear decline in apparent brightness past maximum light. Basically, a Type II-L supernova lacks a plateau in the light curves. The difference tells us that the progenitors[6] of the two types likely differed greatly in pre-explosion size.

The *Hubble Space Telescope* allowed the properties of the progenitor of SN 2017eaw to be determined.[7] It is believed that the star was born with a mass of 15 solar masses. When it was a main sequence star, its luminosity was about 15,000 times that of the Sun, and its surface temperature was 26,000 K. Therefore, it was bluish-white in

---

[6]The progenitor of a supernova is the object that actually exploded.

[7]C. D. Kilpatrick and R. J. Foley, *Monthly Notices of the Royal Astronomical Society*, **481**, 2536 (2018).

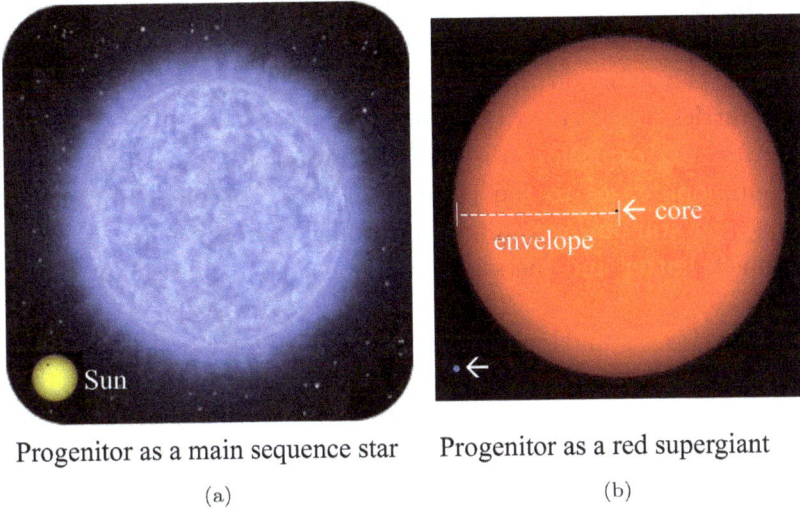

Progenitor as a main sequence star   Progenitor as a red supergiant

(a)   (b)

Fig. 8. (a) The size and color of the progenitor of SN 2017eaw as compared with the Sun. (b) The size and color of the progenitor (white arrow in (b)) of SN 2017eaw as compared to its size as a red supergiant just before it exploded. The definitions of the core and envelope of the red supergiant are indicated. Adapted from wikipedia.

color as shown in Figure 8(a). Its radius was about 6.5 times that of the Sun, and its main sequence lifetime was 11 million years. At the end of its main sequence phase, the star expanded to a red supergiant about 500 times the size of the Sun, or about 80 times its main sequence size (Figure 8(b)). In this state, all the fusion powering the star is taking place in the core, a distinct object which is no bigger than the Earth. The rest of the star makes up the envelope, which is half the size of Jupiter's orbit around the Sun. A red supergiant is like a *Stegosaurus*, a huge dinosaur whose brain cavity was no larger than a hotdog!

The fate of the star after the main sequence phase ends depends on how fast it builds up its iron core. If it accumulates too much iron in its core, the core will collapse violently under its own weight and turn into a tiny ball of neutrons called a neutron star. When this happens, the core becomes incompressible. This leads to "core bounce," where the envelope gases rebound off the tiny, incompressible core. This sends a shock wave through the envelope. The explosion is so violent that it raises the star's brightness to greater than a billion times the Sun in visual light, and the shock wave carries the huge envelope out into space.

After the explosion, the outward-propogating shockwave heats the envelope gases to the point where they become fully ionized, giving the star an expanding and opaque photosphere. The gases are much hotter and the star is no longer red, at least not initially (Figure 9).

The rise to maximum light of SN 2017eaw must have been very rapid, because the object was already close to its maximum brightness the night it was discovered. It is thought that the rise to maximum is due to the rapidly increasing surface area of the envelope gases as they expand away from the center. The light curves further show that the supernova was bluish-white in color when it was discovered. This is because near maximum light, the expanding gases are very hot and emit mostly ultraviolet light. This enhances the brightness in the blue filter as opposed to the redder filters, as shown in Figure 10, where the $BVRI$ light curves have been overlapped by

Fig. 9. Artistic conception of the expanding photosphere of a Type II supernova as it approaches the plateau phase. Image credit: ESO/M. Kornmesser, with permission.

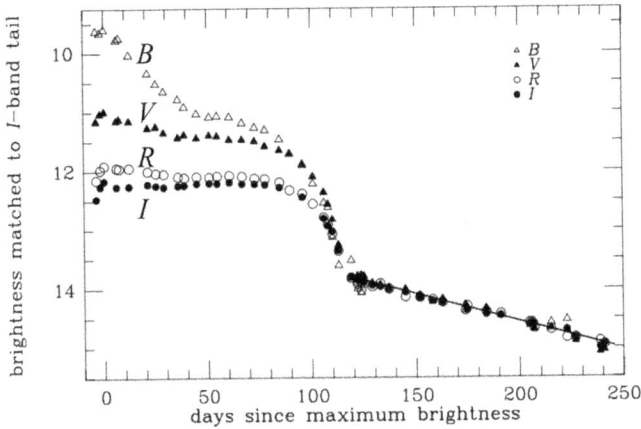

Fig. 10. The *BVRI* light curves of SN 2017eaw are matched in the "tail" region to show that the brightness near maximum light is enhanced in the blue and visual filters as compared to the red and infrared filters, relative to the "plateau." The fact that the difference decreases (that is, blends in with the plateau) with time tells us the gases are cooling at these times.

matching them on the post-plateau "tail." As the pre-plateau gases cool, the color gets redder.

The "plateau" is a result of the fact that the star that exploded was a red supergiant with a deep hydrogen-rich envelope.[8] A red supergiant is a huge star hundreds of times bigger than the Sun. They have only a tiny core and are practically all envelope (Figure 8). Up to the time of maximum light, the expanding photosphere is opaque because it is still very hot and mostly fully ionized, but eventually the gases begin to cool. When the photosphere cools to a temperature of about 6000 K, ionized hydrogen nuclei (free protons) recombine with free electrons into ordinary hydrogen atoms. The gas then becomes more transparent, and we start being able to see deeper into the expanding material (Figure 11). As the envelope continues to expand and cool, the recombination boundary propogates deeper and deeper into the star. Because the boundary involves a well-defined

---

[8]W. D. Arnett, *Supernovae and Nucleosynthesis* (Princeton, Princeton University Press, 1996); see also D. Branch and J. C. Wheeler, *Supernova Explosions* (Berlin, Springer-Verlag, 2017).

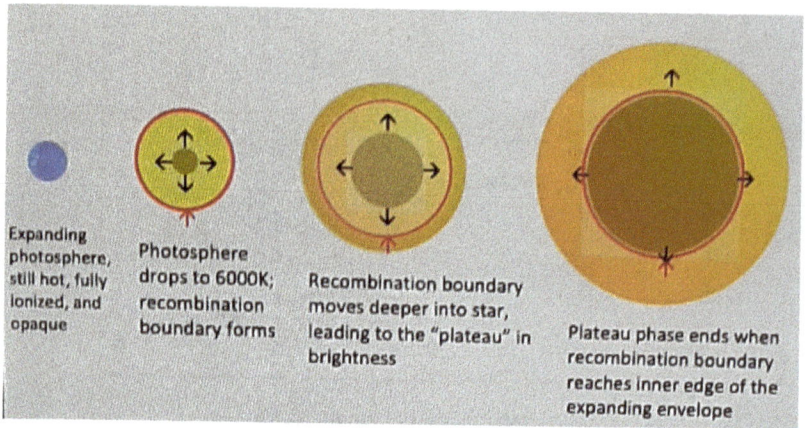

Fig. 11. Illustration of how the "plateau" in the light curves of a Type II-P supernova comes about. It arises from the increasing transparency of the shell of gases as it expands further into space. Adapted from https://astronomy.swin.edu.au/cosmos/.

temperature, the star maintains a constant brightness. This is seen as the "plateau."

The plateau phase ends when the recombination zone passes through the entire envelope, which is now largely in the form of a shell. At this point, the supernova becomes a relatively transparent expanding nebula too small to resolve, and the brightness drops drastically. In SN 2017eaw, this sharp drop in brightness occurred about $3\frac{1}{2}$ months after the explosion, and it would have continued to the point of invisibility were it not for the fact that another process took over to power the light curves. The post-plateau drop in brightness suddenly halted, 120 days after the explosion, and for the next 6 months, SN 2017eaw faded linearly in the light curves at a rate of about 1 magnitude every 100 days in all of the filters. This "tail," as it is called, is seen in all supernovae, and has been interpreted in terms of a radioactive decay process involving none other than Ni-56, the same isotope of nickel that led to the buildup of the iron core that was the death knell of the star to begin with. While nucleosynthesis is what powers the luminosities of stars during their normal cycles, the explosion itself can also induce nucleosynthesis in the expanding gases. Ni-56 is the most important of these explosively-synthesized materials. The amount of Ni-56 produced is proportional to the brightness of the supernova on the tail, summed over all wavelengths of the light. In the case of SN 2017eaw, it has been estimated that 120 Jupiter masses (about a tenth of a solar mass) of Ni-56 was produced in the explosion.[9]

We have already seen that Ni-56 is involved in a chain reaction where radioactive Ni-56 decays into radioactive Co-56 which then decays into stable Fe-56. Both steps in the chain release gamma rays into the surrounding gases. However, the gamma rays are largely confined (that is, do not escape directly into space) due the opaqueness of the material, and the energy trickles out mainly as visible light. Because the half-life of Ni-56 is only 6 days, it quickly leaves Co-56 in its wake. The decay of Ni-56 into Co-56 mostly occurs during the plateau phase. It is the decay of Co-56 that takes over the light curve as the brightness drops at the end of the plateau phase. We can tell that the decay of Co-56 into Fe-56 powers the tail because the rate of decline of the apparent brightness is consistent with this interpretation. The half-life of Co-56 is 77 days, which translates into a decline of 1 magnitude per 100 days on the tail, exactly what is found for the tail of SN 2017eaw and for supernovae in general. This is actually true only for the early part of the tail. In R and I especially, the slope of the tail in SN 2017eaw increases over time, most likely because the confinement assumption of the gamma rays begins to fail.

Finally, as SN 2017eaw faded into oblivion, it likely left behind a neutron star which from some directions would appear as a pulsar. A pulsar is thought to be a rapidly rotating, highly magnetized neutron star. Several thousand pulsars are known in our Galaxy, but the one left behind by SN 2017eaw would not be easy to detect from a distance of 25 million light years. If we could have seen SN 2017eaw 1,000 years after the explosion, it might have resembled the Crab Nebula in Taurus, one of the nearest supernova remnants (Figure 12). The gases of the Crab Nebula are still expanding, and the appearance

---

[9]R. Buta and W. C. Keel, *Monthly Notices of the Royal Astronomical Society*, **487**, 832 (2019).

Fig. 12. *Hubble Space Telescope* image of the Crab Nebula M1 in Taurus, a supernova remnant known to harbor a pulsar. If SN 2017eaw could have been seen a thousand years after the explosion, it likely would have looked something like this. Image credit: NASA/ESA and Allison Loll/Jeff Hester (Arizona State University). Acknowledgment: Davide De Martin (ESA/Hubble). Reproduced with permission.

of the object as complex filaments betrays the violence with which it was created, If we could see what is left of SN 2017eaw today, 25 million years post-explosion, the expanding gases likely have mixed with the interstellar gases that were present in the supernova's vicinity, enriching them with the heavy chemical elements created not only in the star's core, but also in the explosion itself. This underscores the important role supernovae play in enriching the interstellar gases of a galaxy like NGC 6946 with heavy chemical elements.

Chapter 14

Rotating Pinwheels

An interesting aspect of spirals is their rotation. Spirals have to be rotating because their shapes are so highly flattened. To achieve such shapes, spirals must have formed from gravitationally collapsing proto-galactic clouds that had some initial rotation. By the law of conservation of angular momentum, a slowly rotating, large proto-galactic cloud will spin faster as it collapses into a smaller, highly-flattened disk shape. How fast the final object rotates will depend on the total mass of the system, with more massive cases having higher rotation speeds.

One of the first astronomers to measure the rotation of a disk galaxy was F. G. Pease who, in 1916, used the Mount Wilson Observatory 1.5-m telescope to detect the rotation of the Sombrero Galaxy M104. It was a long time afterward that extensive rotation studies of spirals could be performed. This depended on the availability of red-sensitive plates that could enhance the detection of Hα in emission, and on the development of high quality spectrographs that could provide spectra with a long slit, that is, a slit long enough to sample the outer parts as well as the inner parts of a galaxy.

The main goal of long-slit spectroscopy is to determine the rotation curve of a system. A rotation curve is a graph of the rotation speed versus galactocentric distance across a galaxy. Rotation curves are derived by measuring Doppler wavelength shifts of spectral lines, usually emission lines due to interstellar gas. Rotation curves are fundamental to understanding spiral galaxies, because such curves can tell us something about how matter in a galaxy is distributed, and that could tell us something about the nature of spiral structure. Many early rotation studies of galaxies were made by Geoffrey Burbidge (1925–2010) and Margaret Burbidge (1919–2020) using the 82-inch telescope of McDonald Observatory, but the most extensive studies with a long slit were made by Vera C. Rubin (1928–2016), an astronomer at the Department of Terrestrial Magnetism in Washington, D. C., and her coworkers using spectrographs available at Kitt Peak National Observatory and Cerro Tololo Inter-American Observatory.[1]

A slit spectrograph has three main parts: a narrow slit which allows selected light into the instrument, a dispersing element which breaks this light into its component colors, and a detector to record the spectrum. When I used the CTIO long-slit spectrograph in 1981, the dispersing element was a grating, and the detector was a photographic plate. Figure 1 shows the rotation curve of the highly-inclined spiral NGC 7531, based mainly on Doppler wavelength shifts of the Hα emission line.[2] The rotation curve of NGC 7531 is on one hand ordinary and typical for a spiral, but on the other hand is also intriguing. The intriguing aspect is that the point in the rotation curve where the rotation speed changes from rapidly rising to a near constant value (the "turnover point") corresponds to the location of the galaxy's bright inner ring. This cannot be a coincidence, and was unknown before the rotation curve was obtained. It has not yet been determined whether turnover rings are common or relatively rare.

The ordinary aspect of the rotation curve of NGC 7531 is how the rotation speed remains about the same at points beyond the location of the ring. The rotation curve is said to be "flat." In the 1970s and 1980s, Vera Rubin and her coworkers obtained rotation curves of many different spiral galaxies, and found that most of the curves were relatively flat at large distances from the center.[3]

---

[1]V. C. Rubin, W. K. Ford and N. Thonnard, *Astrophysical Journal*, **225**, L107 (1978).

[2]R. Buta, *Astrophysical Journal Supplement Series*, **64**, 1 (1987).

[3]V. C. Rubin, *Publications of the Astronomical Society of the Pacific*, **112**, 747 (2000).

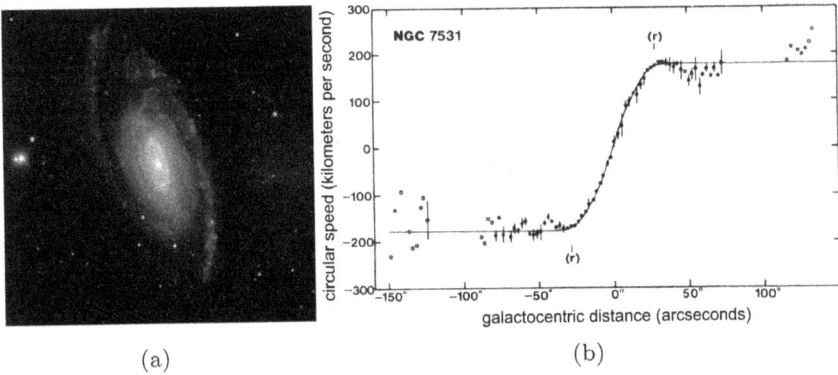

(a)   (b)

Fig. 1. (a) Blue light image of NGC 7531; image source: The de Vaucouleurs Atlas of Galaxies published by Cambridge University Press, 2007; reprinted with permission; (b) rotation (or circular orbit) speed in kilometers per second versus distance from the center of NGC 7531. In this display, the receding (positive speed) and approaching (negative speed) halves of the rotation curve are shown separately, and the galactocentric distance is in arcseconds. The location of the bright inner ring is indicated by "(r)." From R. Buta, *Astrophysical Journal Supplement Series*, **64**, 1 (1987). © AAS. Reproduced with permission.

The significance of flat rotation curves comes through when we ask: What determines the rotation speed at a given galactocentric radius in a spiral galaxy? This speed is determined by the amount of mass enclosed within the radius. In NGC 7531, rotation speeds from the center to the ring rise because the galaxy is bright inside the ring and each increase in radius encloses more mass than the previous radius did. In the outer parts, we do not expect the rotation curve to rise this way, because there is much less light out there. To check this, we can assume a certain amount of light corresponds to a certain amount of mass. When this is done, it is found that the predicted rotation speeds for NGC 7531 should drop off from a maximum at the position of the ring. In general, for any disk-shaped galaxy, we expect that the rotation speeds should drop off in the same manner as the orbital speeds of planets decrease with increasing distance from the Sun. The orbital speeds in NGC 7531 nevertheless do not decrease, but instead remain as high as the speeds reached in the ring.

It was rotation studies like these that led to the idea that spiral galaxies harbor a large amount of material known as *dark matter*.

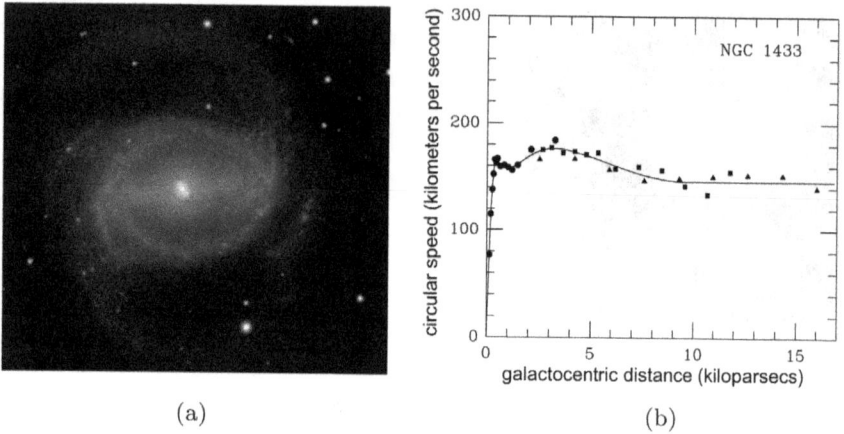

(a)   (b)

Fig. 2.   (a) Blue light image of NGC 1433; image source: The de Vaucouleurs Atlas of Galaxies published by Cambridge University Press, 2007; reprinted with permission; (b) rotation speed versus distance from the center of NGC 1433. Points beyond 10 kpc (32,000 light years) are based on radio observations at 21 cm wavelength, and show that the rotation curve remains flat well beyond the light of the visible disk. From R. Buta *et al.*, *Astronomical Journal*, **121**, 225 (2001). © AAS. Reproduced with permission.

Rotation curves remain flat at large radii when they should decline. This is especially true for rotation curves obtained at radio wavelengths, because these usually extend much farther into the outer disk than do Hα rotation curves like that for NGC 7531. An example is shown by the rotation curve of NGC 1433 in Figure 2.[4] The rotation curve in this case is displayed with the receding and approaching halves folded (that is, averaged across the major axis), and the galactocentric distance R is in kiloparsecs, not arcseconds. This rotation curve was not based on long-slit spectroscopy, but on the more sophisticated technique known as Fabry–Perot interferometry. The method involves an optical element known as an etalon, which consists of two closely-spaced, highly reflective flat surfaces facing each other. When light of a given wavelength passes into the etalon, a pattern of fringes is created on a detector at locations depending on the etalon spacing and the wavelength of that light. When focussed on H-alpha emission from a tilted, rotating spiral galaxy, this method

---

[4]R. Buta *et al.*, *Astrophysical Journal*, **121**, 225 (2001).

can give the line of sight speed across the whole face of the galaxy, not just along different slit positions as in a slit spectrograph. The resulting two-dimensional map of line of sight speeds is called the *velocity field*. Such a map is shown for NGC 1433 in Figure 5 of Chapter 15; this was used for the inner part of the rotation curve shown in Figure 2. The outer part of the rotation curve is based on 21-cm radio observations. The combined rotation curve is nearly flat out to 50,000 light years galactocentric distance. The broad inner bump is due to the considerable mass of the bar and inner ring region.

Because dark matter mostly affects the rotation speeds of interstellar clouds at large galactocentric radii, it is thought to be concentrated in the halo of a spiral galaxy. To account for its effects, the dark matter has to make up at least 90% of the mass of a disk-shaped galaxy. Thus, the features that define galaxy morphology — spiral arms, bars, rings — are in a way like the frosting or decoration on a cake. The bulk of the matter in a galaxy is invisible.

The nature of dark matter has been a prime research objective of astronomers now for several decades. We know that dark matter has mass through its gravitational influence, just as ordinary matter does. The difference is that ordinary matter emits, absorbs, and reflects light while dark matter does not do any of these things, so far as is currently known.

14.1. Non-circular Motions in Spiral Galaxies

Even if the visible matter seen in a spiral is only a small fraction of the total amount of mass present, this does not mean that the "frosting on the cake" has no significant impact on the movement of stars and gas clouds in a galaxy. In the bright parts of a galaxy where there is considerable light, the visible mass can dominate over the influence of the dark matter. It is mostly at large distances from the center where dark matter has its main impact.

The common assumption astronomers make in deriving rotation curves is that the motion detected is perfectly circular. However, this is only true if a galaxy has no non-circular structures, such as spiral arms or a bar. The presence of such structures can cause deviations from circular motion. The study of such motions has led to important

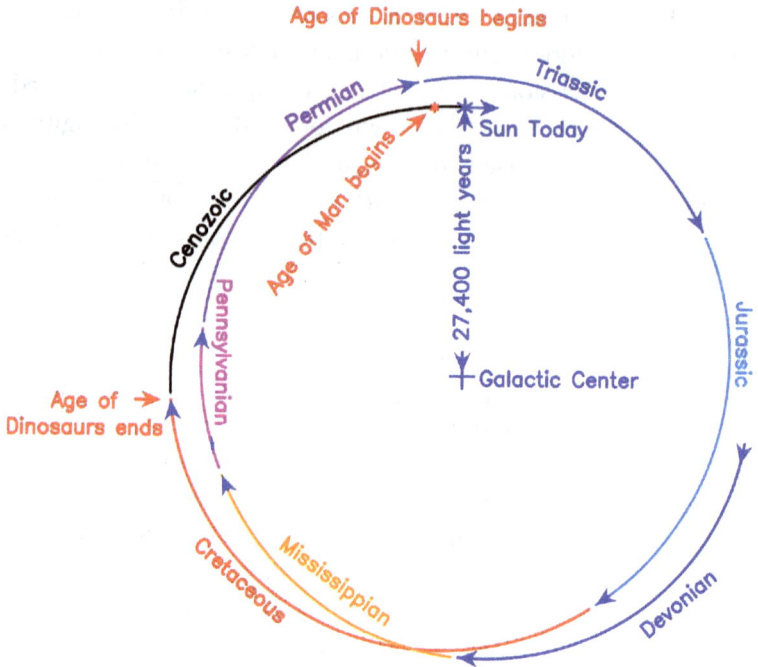

Fig. 3.   The orbit of the Sun around the Galactic center, assuming the Milky Way has no non-circular structures. The orbit is labeled with important periods in Earth history to highlight the extreme length of time, 240 million years, it takes for the Sun to execute a single orbit. Adapted from R. J. Buta and D. C. Kopaska-Merkel, *Footprints in Stone: Fossil Traces of Coal-Age Tetrapods* (Tuscaloosa, © University of Alabama Press, 2016); reproduced with permission.

insights into the impact of non-circular structures on the evolution of disk-shaped galaxies.

Like planets orbiting the Sun, the stars and gas clouds in a spiral galaxy are following orbits around the center. In the absence of any massive non-circular structures, the orbit of a star around the galactic center is nearly circular but never repeats itself. If the Milky Way were such a galaxy, the orbit of the Sun would look like that shown in Figure 3.[5] It has been known for nearly a century that the Sun orbits very far from the center, one of the more recent estimates placing it

[5]R. J. Buta and D. C. Kopaska-Merkel, *Footprints in Stone: Fossil Traces of Coal-Age Tetrapods* (Tuscaloosa, University of Alabama Press, 2016).

27,400 light years out. This is so far from the center that it takes the Sun nearly 240 million years to orbit once. How this astronomical timescale connects to different periods in Earth history is shown by the labeled sections in Figure 3. These show that if one were to go back in time a single orbit (or 1 "galactic year"), the age of dinosaurs would be just beginning.

Rotation curves typically show that spiral galaxies do not rotate like solid objects. A rotating solid object, like a record, would display a rotation curve such that the rotation speed increases directly in proportion to the distance from the center, which would mean that all points on the object would have the same rotation period. In a galaxy, the orbital period of a star or gas cloud depends on the distance from the center in the sense that the farther a star is from the center, the longer its orbital period. In the Milky Way, a star 1/10th as far from the center as the Sun would take less than 30 million years to orbit once, while a star twice as far from the center as the Sun would take nearly half a billion years to do the same.

The presence of a bar can have a significant impact on the way material moves in a galaxy. This can be seen in NGC 6300,[6] one of the brightest barred spirals in the sky (Figure 4 of Chapter 1). The galaxy is inclined about 52° to the line of sight, the tilt being such that the bar is not being viewed down its long axis or its short axis, but along an intermediate axis. A schematic of NGC 6300 outlining the bar, the spiral structure, and an ellipse characterizing the projected shape of the disk is shown in Figure 4.

In a typical disk-shaped galaxy, the long dimension is called the major axis while the short dimension is called the minor axis. If the galaxy has no non-circular structures, then placing the long slit of a spectrograph along the major axis directly gives the rotation curve (multiplied by the sine of the inclination). The long red arrows in Figure 4 show the projected speeds on the approaching and receding halves of the major axis. It is only on the major axis that the full rotation speed at any given location projects to the line of sight. In contrast, placing the slit along the minor axis should give a spectrum

---

[6]R. Buta, *Astrophysical Journal Supplement Series*, **64**, 383 (1987).

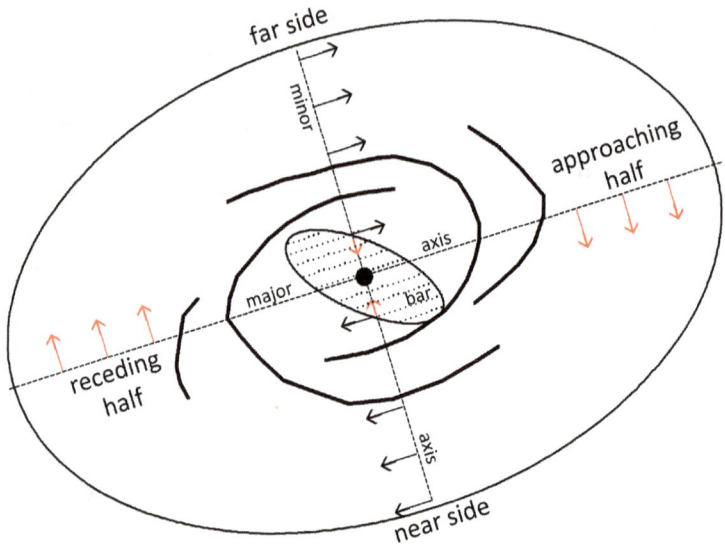

Fig. 4.   A schematic of the spiral galaxy NGC 6300 showing how rotation and non-circular motions are detected along the galaxy's major and minor axes. Red arrows indicate points where motions project along the line of sight. Black arrows indicate points where motions project onto the plane of the sky.

where all points have a line of sight speed equal to the systemic speed.[7] This is because along the minor axis no part of the rotation projects along the line of sight (long black arrows in Figure 4). The major axis spectrum shows the maximum effect of rotation, while the minor axis spectrum reveals no effect of rotation.

Figure 5 shows the actual rotation profiles along the major and minor axes of NGC 6300. As expected, the major axis shows the rotation curve of the galaxy, revealing rotation speeds that rise slightly with increasing distance from the center. The rise occurs across the bright inner pseudoring of spiral structure, indicating that the ring actually has significant mass. It turns out that NGC 6300 is a case where the visible light traces the mass well even to the farthest optical points. However, when the optical rotation curve is extended to larger galactocentric distances by 21-cm observations of

---

[7]The systemic speed is the speed at which the whole galaxy is participating in the universal expansion. For NGC 6300, this speed is 1107 km/s.

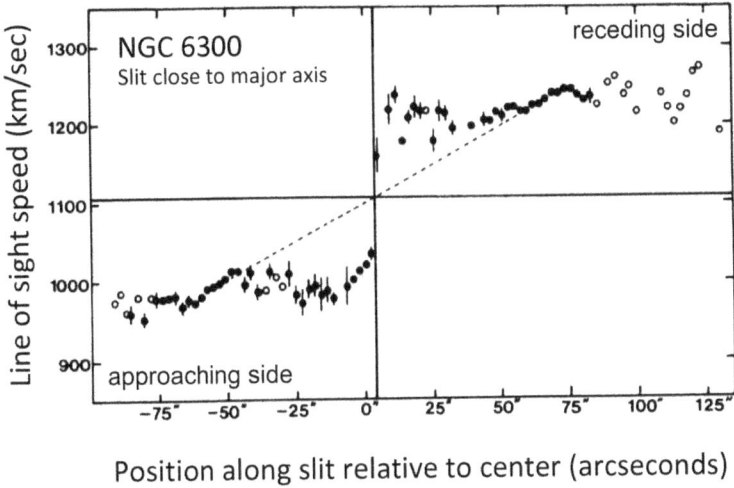

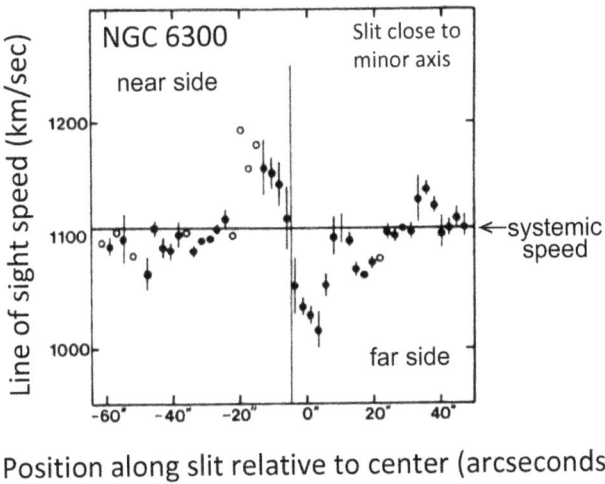

Fig. 5. Rotation profiles close to the major and minor axes of NGC 6300. By "rotation profiles," it is meant that the graphs show line of sight speed versus galactocentric distance, uncorrected for tilt. The dashed line in the top panel highlights a narrow zone of solid body rotation, where the rotation speed is proportional to the distance from the center. The lower profile shows the clear "zig-zag" attributable to non-circular motions in the bar region. These are detectable only when the slit is placed nearly or along the galaxy minor axis. The line of sight profiles are not symmetric around "0" arcseconds galactocentric distance owing to heavy internal dust in the bar of NGC 6300. From R. Buta, *Astrophysical Journal Supplement Series*, **64**, 383 (1987). © AAS. Reproduced with permission.

cold atomic hydrogen, the rotation speeds remain flat far from the center, still indicating the presence of dark matter.[8]

Non-circular motions are not detectable along the major axis of NGC 6300, but along the minor axis they are glaringly obvious. The minor axis rotation profile in Figure 5, bottom panel, shows a peculiar "zig-zag." If the motion were purely circular, there would be no zig-zag and the rotation profile along the minor axis would be the systemic speed, 1,107 km/s, at all points. The zig-zag instead tells us that there are significant non-circular motions in NGC 6300, on the order of 100 km/s in amplitude. When the location of the zig-zag speeds is examined, it is found that they are occurring mostly in the bar, which has diffuse ionized gas. The orbits in the bar region are apparently highly-elongated ovals. One such (idealized) oval is shown in Figure 4. At any point along the orbit, the speed can be broken up into two components: a tangential speed in the direction of rotation, and a radial speed towards or away from the galactic center. It is only along the galaxy minor axis that the full radial speed projects into the line of sight, This is shown by the short red arrows in the bar in Figure 4. Internal dust in NGC 6300 tells us which side of the minor axis is the near side, and with that knowledge we can deduce that the zig-zag seen in the observed minor axis profile is due to an inward (towards the galactic center) motion. Apparently, when a slit is placed along the minor axis of NGC 6300, it intersects the oval bar orbit at points where the gas is flowing inward, owing to the fact that the bar is being viewed at an intermediate angle. Had the bar instead been oriented end-on or broadside-on, no zig–zag would have been seen in the minor axis profile.

This all highlights how non-circular motions in the inner parts of the Milky Way support Gérard de Vaucouleurs's conclusion, made nearly 60 years ago, that our Galaxy is barred.

14.2. Dark Matter Today

The current field of dark matter research is rich in its scope but still has not been conclusive in determining the nature of the material.

---

[8]R. Buta, *Astrophysical Journal Supplement Series*, **121**, 225 (2001).

It is thought that some of the dark matter is baryonic[9] in nature, that is, is made of ordinary matter whose light is hard to detect owing to extreme faintness. A black hole is an example of real dark matter, since it is by definition an object that has considerable mass and yet can emit no light. Nevertheless, a black hole originates from ordinary baryonic matter and there are too few black holes in any given galaxy for them to account for 90% of the total mass of a galaxy. Small, but common massive objects that are not very self-luminous, such as planets and substellar objects known as brown dwarfs, are detectable through microlensing events, which refers to a temporary brightening of a background star when one of these objects (called massive compact halo objects, or MACHOs) by happenstance eclipses it. An extensive search for these events confirmed the existence of MACHOS, but not in the numbers required to account for all the dark matter needed to explain flat rotation curves.[10]

Another popular explanation of dark matter is that it is largely in the form of exotic particles that have mass but which only weakly interact with ordinary matter through the weak nuclear force, one of the four fundamental forces of nature. These particles, called weakly interacting massive particles or WIMPs,[11] are massive enough to be held by gravity in a galaxy's halo, well beyond the extent of most of a galaxy's stars.

---

[9]The term "baryon" refers to subatomic particles like protons and neutrons. Stars, planets, and related objects are all baryonic in nature.

[10]C. Alcock, *et al.*, *Astrophysical Journal*, **542**, 281 (2000).

[11]R. Agnese, *et al.*, *Physical Review D*, **92**, 7, 72003 (2015).

Chapter 15

Rings of Fire

The rings of Saturn are one of nature's true marvels. More than than 280,000 km in diameter, the rings are thinner relative to their diameter than an 8.5 × 11-inch sheet of paper and are comprised of dozens of narrow ringlets. Although the rings have never been photographed close enough to be resolved, they are thought to be made of billions of small fragments of ice, each orbiting the planet like a tiny moon. Saturn's rings may be thought of as "rings of ice."

As described in Chapter 1, spiral galaxies also often have rings as part of their structure. Rings in spirals are not as well-known as Saturn's rings in part because many of the brightest and best-known spirals, such as M51, M81, M100, NGC 1300, NGC 1365, and M83 (Figure 1), do not have such features. These objects, known as (s)-shaped "grand design" spirals,[1] have two major arms which break from a bright central region, or the ends of a bar, and then wind outward in a fairly symmetric pattern lined with star-forming regions. Ringed spirals are different. In addition to a spiral pattern, these galaxies show remarkable "rings of fire," that is, ring-shaped patterns dominated by recent star formation. In some cases, the main spiral arms break directly from a ring pattern, as in NGC 2523 (Figure 7(b) of Chapter 6), or the arms *may define* the pattern, that is, may close into a ring-like pattern (Figure 2). Multiple star-forming ring patterns in the same galaxy are also seen, including in some cases as many as four distinct features. Although ringed galaxies are not

---

[1]See Figure 7 of Chapter 6 and Chapter 16.

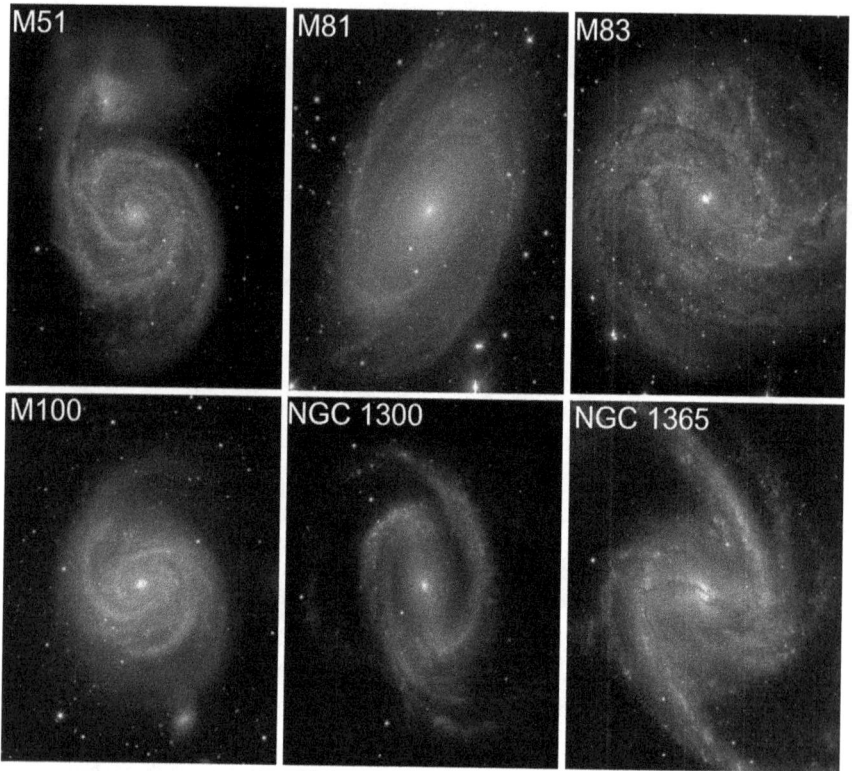

Fig. 1.   Blue-light images of six classic nearby "grand-design" spiral galaxies. Images source: the de Vaucouleurs Atlas of Galaxies published by Cambridge University Press, 2007; reprinted with permission.

as common as non-ringed galaxies, they are not rare by any means. The mystique of these patterns begs the question: what do the rings represent?

It is interesting to focus on galactic rings because of distinctive aspects of their morphology, not the least of which is how much rings resemble orbits, especially orbits predicted by models of galaxies. Add to this the fact that rings may be visible tracers of specific aspects of galaxy dynamics, and the features become very important for understanding the structure and evolution of galaxies in general.

One of the nearest and most spectacular ringed galaxies is NGC 1433 in the southern constellation Horologium, the Clock. NGC 1433 is 39 million light years away from the Milky Way, and happens

Fig. 2. Blue light image of NGC 1433, a ringed spiral galaxy lying 39 million light years away in the constellation Horologium, the Clock. Image source: The de Vaucouleurs Atlas of Galaxies published by Cambridge University Press, 2007; reprinted with permission.

to be very nearly face-on. Figure 2 presents a blue light (B-band) image of NGC 1433 which shows the inner ring and prominent bar of the galaxy. Figure 3 shows a blue minus infrared (B–I) color index map of the inner regions of the galaxy. As described in Chapter 11, a color index map is an effective way of displaying the distribution of star formation and interstellar dust in a spiral galaxy. The map is coded such that blue features are dark and red features are light. The most prominent features in color index maps are usually star-forming regions (which appear dark) and dust features (which appear light).

The schematic in Figure 4[2] highlights the most important aspects of NGC 1433, which has been characterized in the de Vaucouleurs

---

[2]From R. Buta, *Astrophysical Journal Supplement Series*, **61**, 631 (1986).

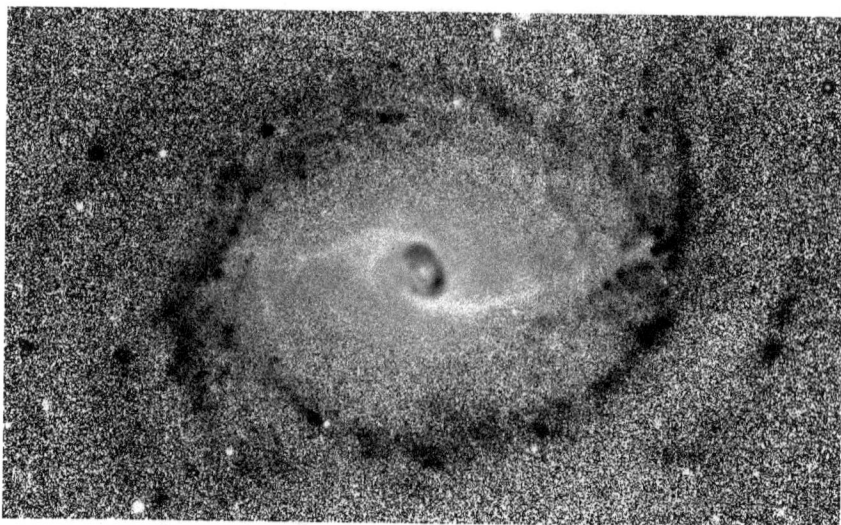

Fig. 3.   Blue minus infrared ($B-I$) color index map of NGC 1433, showing star formation and spiral structure in the large inner ring, and dust lanes in the bar region. The dust lanes are on the leading edges of the bar, that is, the sides of the bar that are ahead of the middle part of the bar, in the direction of rotation (assuming trailing arms). The map is coded such that bluish features are dark and reddish features are light. Image source: The de Vaucouleurs Atlas of Galaxies published by Cambridge University Press, 2007; reprinted with permission.

Atlas of Galaxies as having "virtually every morphological feature a galaxy can have and still be considered normal." The galaxy shows a very strong bar (b_1, b_2) and two rather dim outer spiral arms (a_1, a_2). The inner ring is the strong oval feature around the bar. The color index map highlights the star formation in the ring, which is elongated like an ellipse and appears to be made of tightly-wrapped spiral arms (r_1–r_4) in a pattern that looks largely independent of the two outer arms. The nearly closed spiral ring also has another characteristic: its longer (or major) axis coincides almost exactly with the long axis of the bar. That is, the ring is elongated parallel to the bar, an apparent alignment that can neither be an artifact of inclination nor a coincidence. Gérard de Vaucouleurs depicted this alignment in his famous sketch of spiral galaxy types in Figure 8 of Chapter 6.

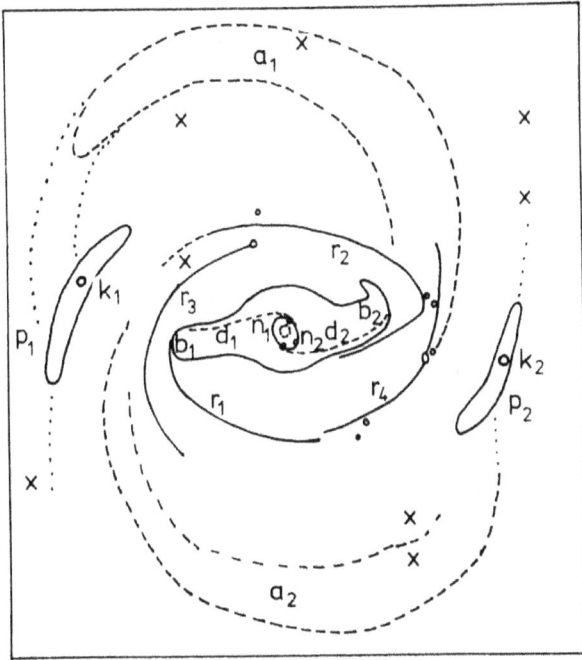

Fig. 4. A schematic of NGC 1433 identifying different features of interest. The features labeled are the: bar: b_1, b_2; outer arms: a_1, a_2; inner ring: r_1 to r_4; nuclear ring: n_1, n_2; plumes: p_1, p_2; knots (star-forming regions) in plumes: k_1, k_2, and bar dust lanes d_1, d_2. The spiral arms are assumed to open outward opposite (or "trailing") the direction of rotation, which would make the galaxy turning in a clockwise sense. The dust lanes would then be on the leading edges of the bar. Other small circles are other knots. X symbols refer to foreground stars. From R. Buta, *Astrophysical Journal Supplement Series*, 61, **631** (1986). © AAS. Reproduced with permission.

Definitive proof of the alignment is provided by the velocity field of NGC 1433 shown in Figure 5.[3] This is based on a special spectroscopic instrument, a Fabry–Perot interferometer, which was described in Chapter 14; it measures Doppler wavelength shifts of spectral emission lines across the face of a whole galaxy (unlike long-slit spectroscopy which is also described in Chapter 14). These Doppler shifts give the line of sight speed at any point in the galaxy

---

[3]From R. Buta *et al.*, *Astronomical Journal*, **121**, 225 (2001).

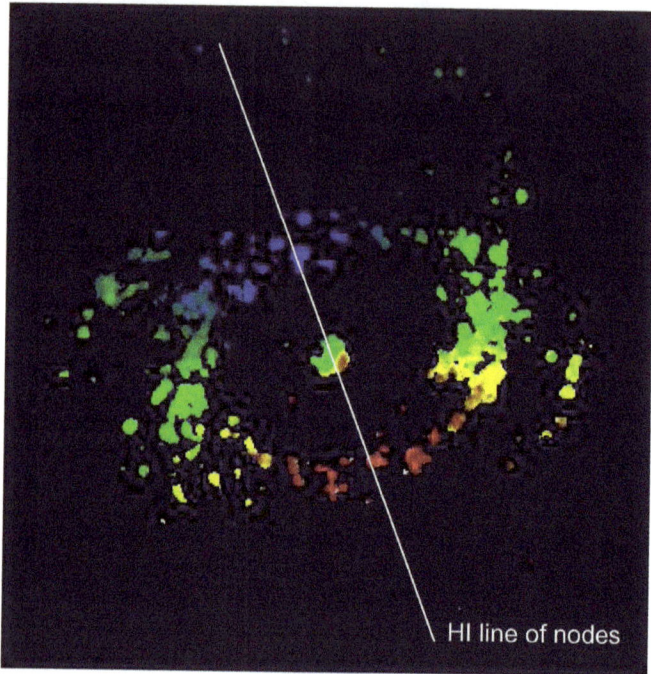

Fig. 5.   Line of sight speeds of ionized gas clouds across the face-on spiral NGC 1433. The clouds are detected mainly in the central region, the inner ring, and the plumes. The HI line of nodes refers to the line of intersection between the plane of the sky and the plane of the galaxy's disk as determined in a study of the interstellar atomic hydrogen in the galaxy. The color coding is such that blue shows the clouds approaching us while red shows the clouds receding from us, with green intermediate. From R. Buta *et al., Astronomical Journal,* **121**, 225 (2001). © AAS. Reproduced with permission.

where Hα is detected in emission. The galaxy is tilted around the line of nodes[4] shown in the illustration, which is nearly perpendicular to the major axis of the inner ring. This means the inner ring is more intrinsically oval than it looks. If the ring were simply a tilted, circular feature, the line of nodes would be along the ring's major axis. The fact that it is not verifies the bar-inner ring alignment. From the color coding, the north side of the galaxy is approaching

---

[4]The line of nodes is the line of intersection between the plane of the galaxy and the plane of the sky.

us while the south side is receding. Note that much of the galaxy has no Hα emission and so Fabry–Perot interferometry can provide no information on rotation speeds in those regions.

Most interesting in Figure 4 is the presence of two short spiral arcs (p_1, p_2) located just off opposing sides of the inner ring. The symmetry of these star-forming arcs (known as plumes), and their isolated nature (not clearly connected to the ring), is remarkable to the point that each arc even has its own major star-forming region (k_1, k_2).

Inside the bright ring of NGC 1433 are other interesting features. The bar, for example, shows two thin dust lanes (d_1, d_2) that begin near its ends and extend all the way to the center, where they suddenly curve more sharply inwards into a small central ring of star formation (n_1, n_2). This central ring is partly defined by a few distinct star-forming regions. The dust lanes appear to lie mainly on the leading sides of the bar (that is, on the sides in the direction of rotation, assuming the spiral arms are trailing) and terminate at the position of this small ring. Some recent star formation is also seen on the leading side of one of the bar dust lanes. The dust in the lanes has reddened the starlight in the bar, which is why the lanes appear light in the color index map.

It was noted in Chapter 6 that inner rings were made a part of galaxy classification by Allan Sandage, who recognized them with the symbol (r). The ring around the bar of NGC 1433 is called an inner ring because it lies within the main outer arms and is approximately the same size as the bar. The ring is a very large feature, almost 40,000 light years in diameter. The small ring in the center of NGC 1433, which is most conspicuous in the B–I color index map in Figure 3, is known as a "nuclear ring." This type of feature was not recognized in galaxy classifications until very recently. Since NGC 1433 is nearly face-on to us, the apparent elongated shapes of the inner and nuclear rings are not artifacts of tilt, but are close to the actual shapes in the galaxy plane.

Some galaxies show an even larger scale of ring. NGC 1291, located 32 million light years away in Eridanus, is a beautiful face-on spiral showing a spectacular *outer ring* (Figures 6 and 7). The ring

Fig. 6.   Blue-light (*B*-band) image of NGC 1291 in Eridanus showing the prominent outer ring and spiral structure. The central region has a small bar misaligned with a larger, fainter bar by about 30°. The larger bar is oriented nearly vertically in the image. Image source: The de Vaucouleurs Atlas of Galaxies published by Cambridge University Press, 2007; reprinted with permission.

was first described by Charles D. Perrine in 1922,[5] and was examined in more detail much later by Gérard de Vaucouleurs.[6] The ring is 65,000 light years in diameter and is sharply defined by active star formation on its outer edge. Instead of breaking from near the ends of a bar, or directly from the bright central bulge, the spiral arms break from the outer edges of the outer ring. There is no nuclear ring or inner ring. There is a bar, but it is much weaker-looking than that in NGC 1433. Most remarkable is that the central region of NGC

---

[5]C. D. Perrine, *Monthly Notices of the Royal Astronomical Society*, **82**, 486 (1922).
[6]G. de Vaucouleurs, *Astrophysical Journal Supplement Series*, **29**, 193 (1975).

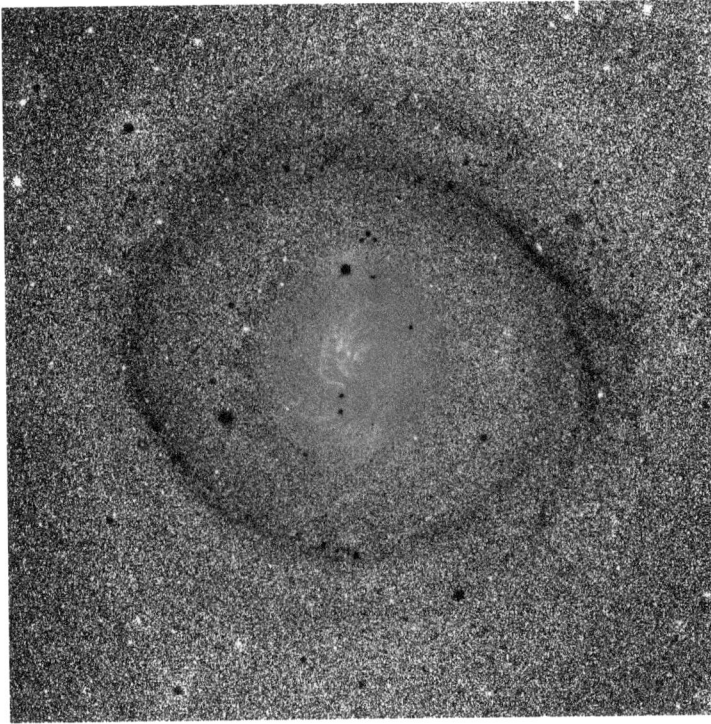

Fig. 7.   Blue minus infrared $(B-I)$ color index map of NGC 1291 showing star formation in the large outer ring and spiral arms, and peculiar dust lanes in the inner region. As in Figure 3, the greyscale coding in the color index map is such that bluish features are dark and reddish features are light. Image source: The de Vaucouleurs Atlas of Galaxies published by Cambridge University Press, 2007; reprinted with permission.

1291 has a secondary or "nuclear bar." That is, NGC 1291 has two bar features. The smaller bar is misaligned with the larger bar by about 30°. The color index map reveals peculiar dust lanes in this same area, likely signifying a past minor merger with another galaxy. NGC 1291 was part of de Vaucouleurs's survey of southern galaxies with the Reynolds 30-inch reflector of Mount Stromlo Observatory. Surprisingly, NGC 1291 is described erroneously as being an "easily resolvable" globular cluster in the NGC.

NGC 4736, also known as M94, is another remarkable case of galactic rings. Lying 16 million light years away in the constellation Canes Venatici, M94 is a spiral galaxy, but is distinctly different from

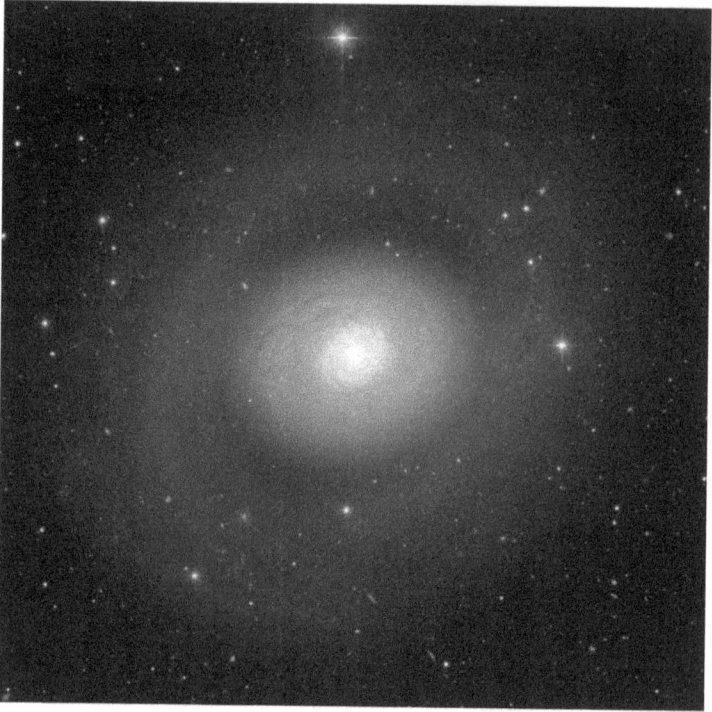

Fig. 8.   Blue-light (*B*-band) image of M94 (NGC 4736), located 16 million light years away in Canes Venatici. The inner ring is the small, knotty feature near the center, while the outer ring is the large, outer diffuse feature. Between the two rings is an oval zone defined in part by multiple spiral dust lanes, and a dark "gap" which separates the outer ring from the inner structures. A few "islands" of star formation are found in the oval zone. Image source: The de Vaucouleurs Atlas of Galaxies published by Cambridge University Press, 2007; reprinted with permission.

most spirals in that the range in surface brightness from the inner to the outer parts of the galaxy is very large. The galaxy has two ring features: an inner pseudoring that is bright and full of structure, and an outer ring which is very faint, diffuse, and largely featureless except for subtle spiral structure. In between these rings is a dusty bar-like structure known as an oval. These structures are highlighted in Figures 8 and 9.

The "fire" in the inner pseudoring of NGC 4736 is spectacularly displayed in the picture shown in Figure 10, which is based on a set of filter images obtained with the *Hubble Space Telescope*,

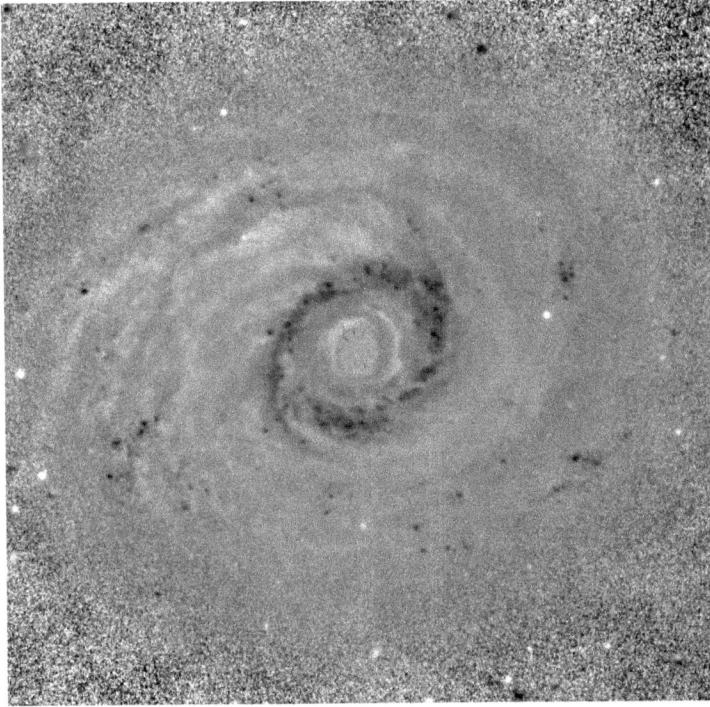

Fig. 9. Blue minus infrared $(B–I)$ color index map of the inner ring and oval zone of M94, coded such that blue, star-forming features are dark, and red, dusty features are light. This highlights how the inner ring is a sharp pseudoring of enhanced star formation in a tightly-wrapped spiral pattern. The map also shows the extensive dust pattern in the massive oval zone. Note: the colors in the central area with the short serrations are not reliable owing to slight overexposure of the original I-band image. The outer ring is outside the field of the image. Image source: The de Vaucouleurs Atlas of Galaxies published by Cambridge University Press, 2007; reprinted with permission.

including a filter that selected the Hα emission line. The color index map in Figure 9 gives the impression that the ring is a pattern of "knots" or discrete star-forming regions, but the HST image shows a feature lined by probably several thousand individual and relatively uniformly spread bluish-colored stars that occupy a broad radial zone. The appearance of knots in the color index map is partly an artifact of the complex dust lane pattern in the ring region. There are many individual HII regions (pink nebulae), but most of the blue stars look like they are part of a large, ring-shaped star-forming

Fig. 10.   *Hubble Space Telescope* color image of the inner pseudoring of NGC 4736 showing literally thousands of young, blue stars and a complex inner dust pattern. The image also has Hα included, and thus also shows the distribution of HII regions (pink objects). Image credits: ESA/Hubble and NASA; processing and copyright: Leo Shatz, used with permission.

complex. The ring is nearly 8,000 light years in diameter and is not so much a collection of star-forming entities than it is a special type of hierarchical star formation structure, like a stellar OB association in the form of a ring. Although not evident in the color picture, the ring is filled with diffuse ionized gas[7] that is not associated with specific HII regions but is due to the collective effect of all the hot, blue stars

---

[7]P. C. van der Kruit, *Astronomy and Astrophysics*, **52**, 85 (1976).

in the ring. This diffuse gas in not present inside the ring, nor is it strongly present outside the ring. Figure 10 also shows the complex dust pattern inside the ring. The yellowish-orange color inside the ring indicates that unresolved old stars dominate this region. The ring itself is dominated by young population I stars.

15.1. Galactic Rings as Orbital Resonances

An orbital resonance is a special location in a system of orbiting bodies where objects are moving in step with each other, meaning there are periodic repeats of gravitational configurations. These can lead to slow, cumulative changes in orbits that could affect the evolution of the system. Orbital resonance is one of the most important physical phenomena that can impact the long-term behavior of a dynamical system like a galaxy, solar system, moon system, or planetary ring system.

In the 1960s, theoretical studies of spirals suggested that some, if not most, galactic rings could be visible manifestations of orbital resonances in galactic disks. The galaxy NGC 5364 was held up as an example where a spiral pattern breaks directly from a "resonance ring" (Figure 11).[8] When a large number of galaxies are inspected, it is found that about a fifth of all disk-shaped systems have a ring, and an additional one-third have a pseudoring made of spiral arms.[9]

The possibility that rings in galaxies are somehow related to orbital resonances is interesting. Orbital resonance is connected to orbital dynamics, and orbital dynamics is tied to the overall mass distribution in a galaxy. Mass distribution refers to how the matter in a galaxy is distributed with distance from the center. In normal cases, visible matter is concentrated at the center of a galaxy, and spreads more thinly with increasing distance from the center. Orbital dynamics is, however, influenced by more than the distribution of visible matter — it is sensitive to the distribution of all matter, including the dark matter described in the previous chapter.

---

[8]C. C. Lin, in W. Becker, G. I. Kontopoulos, eds., *The Spiral Structure of our Galaxy* (Reidel, Dordrecht, 1970), p. 377.
[9]R. Buta and F. Combes, *Fundamentals of Cosmic Physics*, **17**, 95 (1996).

Fig. 11.   Blue-light image of NGC 5364, a spiral 54.5 million light years away in Virgo. In the 1960s, the galaxy was proposed as an example having a "resonance ring," referring to the narrow, elliptical ring close to the center. Image source: the de Vaucouleurs Atlas of Galaxies published by Cambridge University Press, 2007; reprinted with permission.

The mass distribution within a galaxy sets the orbital speed every object in the galaxy needs to maintain its orbit. As was described in Chapter 14, orbital speeds are relatively constant with increasing galactocentric distance in most disk-shaped galaxies. The constant orbital speed means that objects farther from the center take more time to orbit than do objects closer to the center.

Orbital resonance can be understood most easily in the context of Saturn's rings. These rings are a physical laboratory for studying resonance phenomena in disk-shaped celestial objects: a disk made of billions of much smaller objects (likely snowballs) revolving around Saturn with orbital periods dependent on distance from the center of Saturn. The planet also has more than 60 known moons, some of which orbit just outside the outermost edge of the rings. Orbital resonances in Saturn's rings are set up when the orbital period of a moon outside the rings, and the orbital period of ring objects, are related by a rational number, such as 2/1, 3/2, 6/5, etc.

For example, the Cassini Division is a broad gap in Saturn's rings that is easily visible with small telescopes.[10] Ring particles that once orbited near the inner edge of the gap went around Saturn at exactly half the orbital period of Saturn's innermost moon Mimas. This means that ring particles that used to be in the Cassini Division were in 2:1 orbital resonance with Mimas, since those particles orbited Saturn twice for every single orbit of Mimas. Mimas acts as a perturber in this case because although it is a relatively small moon, it is much more massive than any ring snowball or particle. After every two orbits, ring particles in the Cassini Division would feel the gravitational tug of Mimas from the same distance and direction they had two orbits ago. These repeating tugs perturb the orbits of the ring particles, making those orbits more elliptical and enhancing collisions at the resonance.[11] Because there would have been many particles in 2:1 resonance with Mimas, the incessant repeating of exact gravitational configurations would have caused ring particles to collectively move away from the resonance region, leaving a gap which goes all the way around the planet. The Cassini Division is only the most conspicuous gap in Saturn's rings; other lesser gaps and fine structure (ringlets) are attributable to gravitational effects with other moons or higher order resonances with Mimas.

The main point is that in Saturn's rings, orbital resonances can clear gaps. Galactic rings, however, are not dark gaps but are bright, often star-forming features. How does orbital resonance produce a bright ring in a galaxy when a similar process produces a dark gap in Saturn's rings?

First, a galaxy and a planetary ring system are two different phenomena on hugely different scales. While Saturn's rings are made

---

[10]The Cassini Division was discovered by Italian French astronomer G. D. Cassini (1625–1712) at the Paris Observatory in 1675.

[11]F. Franklin *et al.*, in *Planetary Rings*, R. Greenberg, A. Brahic, eds. (Tucson, The University of Arizona Press, 1984), p. 581.

of innumerable small objects, each orbiting the planet like a tiny moon, a galaxy is made of stars and interstellar gas and dust clouds. Second, most of the mass within the Saturnian system is in Saturn itself, meaning the particles follow Keplerian orbits[12] with their orbital speeds declining inversely with the square root of their distance from the center of Saturn. In contrast, much of the mass of a galaxy is well-outside the visible disk.

Spiral galaxies have a great deal of ordered motion. Stars and interstellar clouds follow roughly circular orbits in the galactic disk. When a system like a galaxy has much of its total kinetic energy in the form of ordered motion, it can become unstable to forming a bar.[13] A bar is a massive linear feature generally made of old stars in normal galactic disks, and can begin to form by aligning stars on highly-elongated orbits in the inner regions. As more stars are trapped, the gravitational influence of the feature gets stronger, allowing more stars to be trapped. The end result is a rotating, highly-elongated pattern.[14] Although made of stars orbiting the center at different rates, the bar is a uniformly rotating pattern having its own rotation period. This makes a bar analogous to Saturn's innermost moon Mimas. Since stars at different distances from the galactic center have different orbital periods, the presence of a rotating bar in the disk sets up orbital resonances with disk stars and gas clouds. Any star or gas cloud that moves in step with the bar will experience repeating gravitational configurations. The bar's gravity perturbs the orbits of resonant objects, causing orbits to cross in resonance regions. This sets up the possibility of collisions in those regions.

It is at this point that the difference between a galaxy and the rings of Saturn leads to a different outcome for a resonance

---

[12]Keplerian orbits are orbits, like those in the solar system, that obey Kepler's laws of planetary motion.

[13]See also Chapter 20.

[14]D. Lynden-Bell, *Monthly Notices of the Royal Astronomical Society*, **187**, 101 (1979); J. Sellwood and A. Wilkinson, *Reports on Progress in Physics*, **56**(2), (1993).

region. While orbital resonance can clear a gap in Saturn's rings due to an enhanced collision rate and a resonant-induced migration of solid particles out of the resonance region, in a typical disk galaxy collisions of interstellar gas and dust clouds, which are not solid, can occur in resonance regions leading to gravitational collapse and star formation. Thus, we expect a galactic resonance ring to be a bright feature "lit up" with star-forming regions, as is observed in many cases.

15.2. How do Galactic Rings Form?

The fact that galactic rings tend to be narrow zones of active star formation tells us that interstellar gas is an element essential to ring formation. A bar or bar-like feature also seems essential to the process because most ringed galaxies are barred.[15] A bar sets up the resonances needed to collect the interstellar gas into rings. The study of spiral galaxies with rings therefore provides an opportunity to learn more about galactic dynamics and evolution than might be provided by galaxies which lack such features.

To see how resonances are set up, we need to first examine the orbits that are characteristic of a disk-shaped galaxy with no noncircular structures (that is, no bar or spiral). Figure 12 shows the two parts of such an orbit: a circular component representing the main part of the rotation, and a small ellipse called an epicycle, which represents the first-order deviation[16] from pure circular motion. The sense of motion along the epicycle is retrograde, or backwards, with respect to the circular motion. The epicycle is also periodic, and at most locations in a galactic disk, the period of the radial excursion represented by the epicycle is not related by a rational number with the period of the circular rotation. The net effect of the two motions

---

[15]A non-negligible fraction of disk galaxies have a ring but no clear bar; an example is NGC 5364 (Figure 11).

[16]Motion in the plane of a galactic disk depends on the gravitational field of the disk. Without knowing the exact characteristics of this field, the first-order deviation is the lowest-order correction to pure circular motion in the plane.

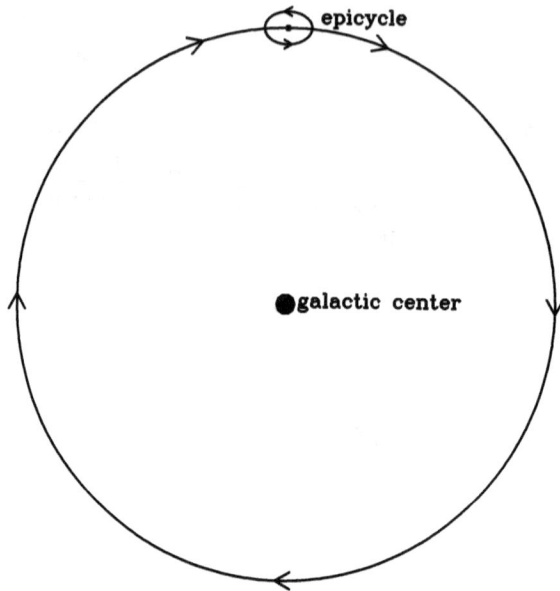

Fig. 12.   Orbits in a galaxy with no noncircular structures are the combination of two motions, a circular part and an elliptical, backwards-circulating "epicycle" representing a small radial excursion around the main circular orbit. Figure 3 of Chapter 14 shows how the combination of these two motions leads to an orbit that never repeats itself.

is an orbit which is nearly circular in shape but which never repeats itself.[17]

Resonances are set up when a bar is added to the disk gravitational field. Motions relative to the bar then become important. Resonances will occur at the locations in the disk where the orbital period of a star or gas cloud relative to the bar is related by a rational number with the period of the radial excursion. By "relative to the bar," it is meant the orbit as seen from a reference frame rotating with the bar. Many resonances are possible, because the bar pattern rotates like a solid object while the disk of stars and gas clouds does not. However, only a few of the possible resonances are relevant to ring formation. These are:

---

[17]For example, see Figure 3 of Chapter 14.

(1) The Inner and Outer Lindblad Resonances (ILR and OLR, respectively), where stars and gas clouds execute two radial excursions for every orbit relative to the bar. These resonances are named after Swedish astronomer Bertil Lindblad (1895–1965), who first examined their possible role in galactic dynamics.
(2) The Inner and Outer 4:1 resonances (I4R and O4R, respectively), where stars and gas clouds execute four radial excursions for every orbit relative to the bar.

In addition to these, another important resonance is called corotation, (CR), where the angular speed of rotation of a star or gas cloud equals the angular speed of rotation of the bar (also known as the bar "pattern speed"). The characterization of a resonance as "inner" or "outer" comes from its position relative to the CR: the ILR and I4R lie inside CR, while the OLR and O4R lie outside CR.

Since the 1970s, astronomers have been able to use computer simulations to study barred galaxy dynamics and in particular to examine how the presence of a bar in a galactic disk can lead to ring formation. A computer simulation is a program used to simulate the formation or evolution of a phenomenon, such as a galaxy, using physics to move particles representing stars and gas clouds through time. Computer simulations can shed considerable light on a process or phenomenon, and is a major technique used by astronomers.

Computer simulations of barred galaxies have shown that rings form by interstellar gas accumulations at resonances, under the continuous influence of gravity torques due to the bar. This was first shown by former Mt. Stromlo Observatory student Mark Philip Schwarz in a remarkable 1979 PhD dissertation.[18] Gravity torques are actions on the motions of stars and gas clouds in a galaxy that occur because a bar is a massive, elongated feature, not a spherical object in the center. The torques serve to collect interstellar gas into resonant orbits aligned either parallel or perpendicular to the bar. The shape of the resulting ring pattern will depend on the

---

[18]M. P. Schwarz, *Astrophysical Journal*, **247**, 77 (1981); see also R. Buta and F. Combes, *Fundamentals of Cosmic Physics*, **17**, 95 (1996).

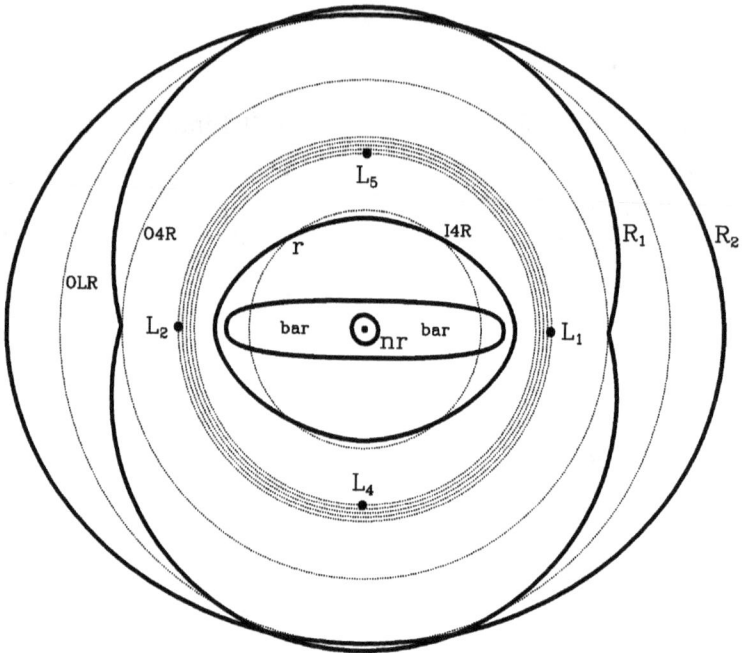

Fig. 13.   Resonant orbits in a model barred galaxy that have been linked to specific galactic rings. The R_1 and R_2 orbits (labels due to author) are actually families of periodic orbits that occur near the OLR. Gravity torques redistribute interstellar gas into these orbits whose alignment parallel or perpendicular to tbe bar causes the torques to eventually vanish, leaving a clear outer ring or pseudoring having the shape of one or both of the orbits (M. P. Schwarz, *Astrophysical Journal*, **247**, 77 (1981)). Observed rings having these shapes define outer resonant subclasses (Figures 14 and 15). The "r" orbit is an inner ring orbit near the I4R. The "nr" inside the bar refers to a nuclear ring; its existence depends on periodic orbits not shown in the illustration. The set of dotted circles show the "corotation region." The locations of Lagrangian points L_1, L_2, L_4, and L_5 define this region.

resonance involved. The locations of these resonances depend only on the pattern speed of the bar and the shape of the rotation curve.

Figure 13 shows examples of some of these orbits, with the locations of the OLR, the O4R, and I4R indicated, based on the assumption of a flat rotation curve (Chapter 14). The ILR was once thought to play a significant role in ring formation, but its location is strongly affected by the shape of the rotation curve in the inner regions, and as a consequence it cannot be specified in a general way like the other resonances.

In Schwarz's models, a key factor in ring formation is that orbits in a barred galaxy change orientation by 90° across a major resonance. Orbits just inside the ILR are elongated perpendicular to the bar. Between the ILR and the CR, orbits are elongated parallel to the bar, while between the CR and the OLR, orbits are once again oriented perpendicular to the bar. Finally, outside the OLR, orbits become parallel to the bar again. These orientation changes can be seen in Figure 13.

Schwarz's simulations showed that before gas clouds can settle into these orbits, they will trace out a spiral pattern because tracing out a spiral is a gradual way for gas clouds to react to the sudden change in orbit orientations by 90°. Gravity torques will cause interstellar clouds to slowly drift away from the CR outwards to the OLR. The material will stop drifting when it has settled into an orbit aligned either parallel or perpendicular to the bar. Near the OLR, Schwarz identified two families of periodic orbits having these characteristics. These are the orbits labeled R_1 and R_2 in Figure 13. The perpendicular-aligned R_1 orbit was favored in the simulations when there was little interstellar gas beyond the OLR; the parallel-aligned R_2 orbit was favored when more gas was outside the OLR.

The R_1 and R_2 orbits are generally taken to be linked with outer rings and can account for some of the distinctive spiral morphologies associated with such rings.[19] This is shown schematically in Figure 14 and with a few examples in Figure 15. An R_1 ring has two main characteristics: an intrinsic elongation perpendicular to the bar, and a peculiar "dimpled" shape, referring to the way the orbit dips in towards the ends of the bar. An especially good example of a dimpled R_1 feature is seen in UGC 12646, a nearly face-on spiral located 367 million light years away in Pegasus (Figure 15, top). Dimpling is often more subtle than this, and may even be absent, but even without dimpling, an R_1 ring can be identified from its alignment perpendicular to the bar.

---

[19]R. Buta and D. A. Crocker, *Astronomical Journal*, **102**, 1715 (1991).

NGC 2665 R$_1'$ ESO 509-98 R$_1$R$_2'$

ESO 325-28 R$_2'$ ESO 507-16 R$_1$R$_2'$

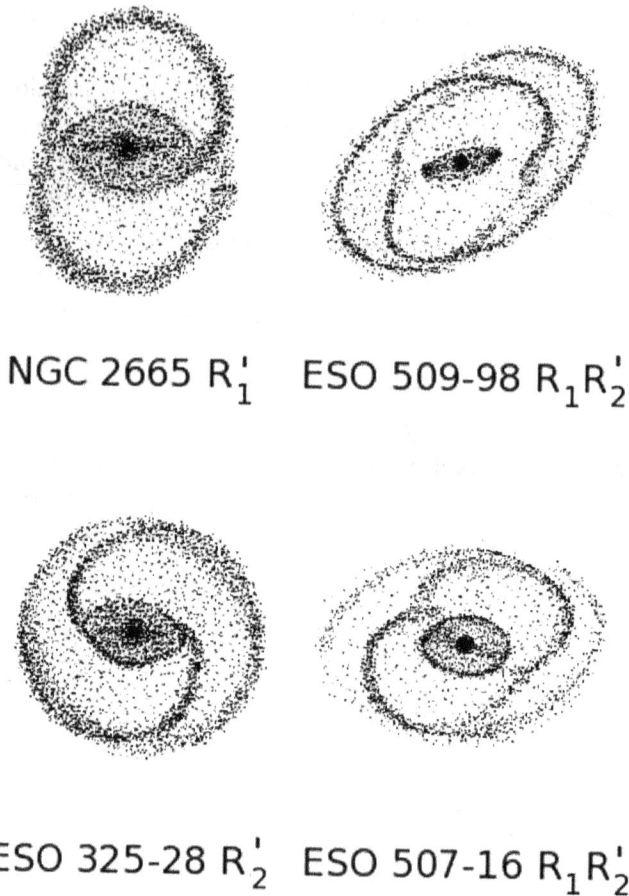

Fig. 14.   Schematics of four examples of galaxies showing outer spiral structure that resembles outer resonant structures. From R. Buta and D. A. Crocker, *Astronomical Journal*, **102**, 1715 (1991). © AAS. Reproduced with permission.

An R$_1$ ring may be found in closed form (classified as type R$_1$), or it may be in the form of a pseudoring (classified as type R$_1'$) where each of two arms breaks from one end of the bar and winds 180 degrees to the other end. UGC 12646 is actually a good example of an R$_1'$ outer pseudoring, as is NGC 2665, another face-on example lying 69 million light years away in Hydra. In contrast, an R$_2$ ring is slightly elongated parallel to the bar and may appear as a detached, closed ring or as a pseudoring (classified as type R$_2'$) where each of

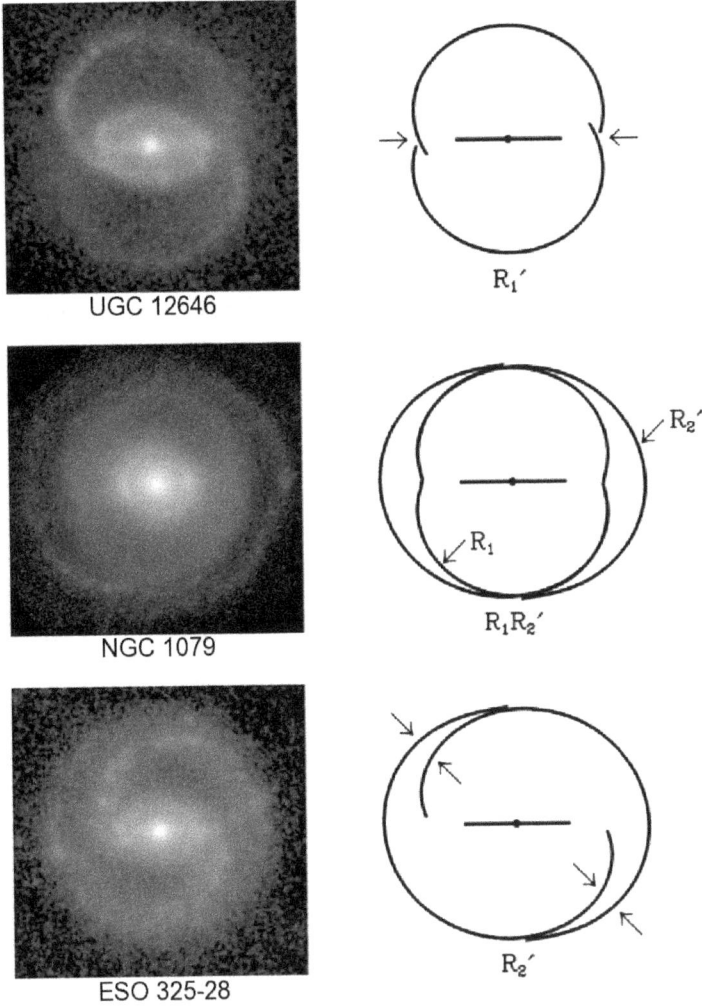

Fig. 15. Three galaxies showing ring patterns resembling the R_1 and R_2 orbits in Figure 13. The schematics at right highlight the distinctions. The arrows point to characteristics that define the features: dimpling in R_1' pseudorings and arm-doubling in two opposing quadrants for R_2' pseudorings. The pattern called R_1R_2' is a combination of the other two. From R. Buta, *Monthly Notices of the Royal Astronomical Society*, **470**, 3819 (2017). Copyright © 2017, Oxford University Press.

Note: ESO stands for European Southern Observatory.

two arms winds about 270° with respect to the bar. A good example of an R'_2 outer pseudoring is seen in ESO 325-28 (Figure 15, bottom), lying 319 million light years away in Centaurus. An R'_2 pseudoring can be simulated by combining the R_1 and R_2 orbits in Figure 13 and leaving out two quadrants of the R_1 part.

Most interesting is the combined morphology, classified as $R_1R'_2$, in which the complete influence of both the R_1 and R_2 orbits is seen. NGC 1079 (Figure 15, middle), located 61 million light years away in Fornax, is an excellent example where the R_1 component is strong and relatively circular in shape; only subtle dimpling is seen in this case, perhaps because it has a relatively weak bar. Other examples are ESO 509-98, lying 352 million light years away, and ESO 507-16, lying 317 million light years away, both in Centaurus. These are shown in schematic form in Figure 14.

The R_1 and R_2 orbit ring/pseudoring features are important because they were predicted by Schwarz's models well before they were actually found in significant numbers. This is remarkable for late 1970s numerical simulations. Although easily identifiable, R_1 and R_2 rings are nevertheless not common in the disk galaxy population and examples of outer rings and pseudorings can be found that do not match either type. Only about 10% of bright galaxies have an outer ring/pseudoring of any kind. All of the R_1 and R_2 types of features are known as outer resonant subclasses, and most of the illustrated examples were identified in a major survey called the *Catalogue of Southern Ringed Galaxies* (CSRG).[20] More examples were identified in the *Galaxy Zoo 2 — Catalogue of Northern Ringed Galaxies* (GZ2-CNRG).[21]

The orbit labeled "r" in Figure 13 refers to inner rings. As for R_1 and R_2 rings, gravity torques could play a role in the formation of these structures by causing interstellar gas to drift inwards from the CR region. The typical inner ring is intrinsically elongated parallel to the bar and may be slightly cuspy (or pointy-ended) on its major axis. Dimples are never seen in the shapes of inner rings. The average

---

[20]R. Buta, *Astrophysical Journal Supplement Series*, **96**, 39 (1995).
[21]R. Buta, *Monthly Notices of the Royal Astronomical Society*, **471**, 4027 (2017).

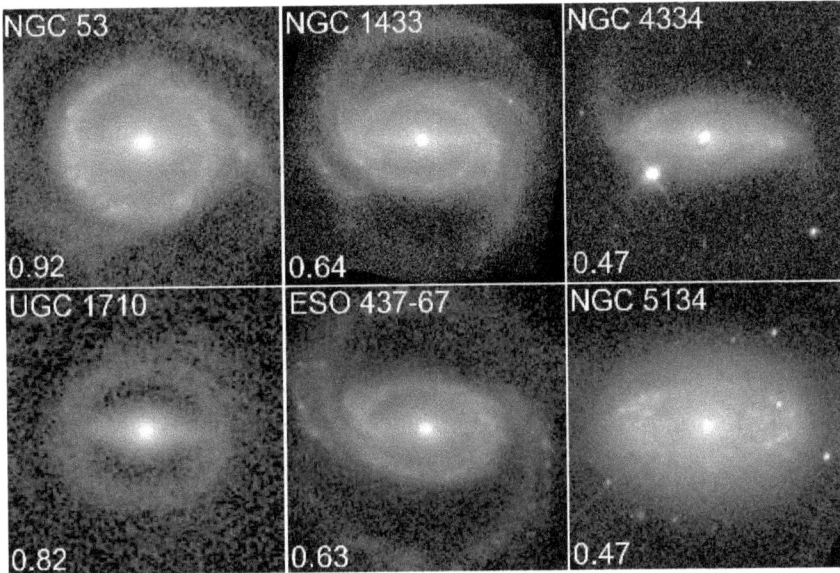

Fig. 16.   Six galaxies showing how the shapes of inner rings in barred galaxies range from nearly circular to highly elongated. The six galaxies have been computer-deprojected (although none is highly inclined) and the images have been rotated to make the bar in each case horizontal. Foreground and background objects have been removed from several of the images. The number at lower left in each image is the face-on inner ring minor-to-major axis ratio. Image sources: The de Vaucouleurs Atlas of Galaxies (NGC 53, NGC 1433, ESO 437-67, and NGC 5134); Sloan Digital Sky Survey Collaboration, http://www.sdss.org (UGC 1710 and NGC 4334). Copyright © 2019, Oxford University Press.

barred galaxy inner ring has a face-on minor to major axis ratio of about 0.8 ± 0.1,[22] but individual cases can have shapes in the range 0.5–1.0. This is shown by the six examples illustrated in Figure 16. These also show that the bar usually fills the inner ring along its major axis, in the form of a flattened Greek letter theta (θ) shape.

The feature labeled "nr" in Figure 13 refers to nuclear rings. A typical nuclear ring is either circular or slightly elongated. If elongated, a nuclear ring is usually oriented at a significant angle ($\approx 60°$ or more) to the bar.

---

[22]R. Buta, *Astrophysical Journal Supplement Series*, **96**, 39 (1995).

Figure 13 interprets inner rings in terms of the I4R, which accounts for the cuspy or oval shapes of the rings and their typical alignment parallel to the bar.[23] Because the I4R lies inside the CR, the formation of a ring near this resonance depends mainly on the shapes of the 4:1 resonant orbits, which can have a four-fold symmetry with peculiar loops in the corners. These orbits can overlap with other orbits, and a ring can form from interstellar gas that collects into one of these orbits that does not loop or overlap any other orbits.[24]

The most popular interpretation of nuclear rings has been in terms of the ILR region.[25] However, the ILR is likely most relevant to ring formation only for weak bar-like ovals or nonbarred spirals. For example, the inner pseudoring of NGC 4736 (Figure 10) lies within a massive oval, and the sharp oval inner ring of NGC 5364 (Figure 11) lies within a bright spiral; both features are good candidates for ILR rings. In barred galaxies, what is most important to nuclear rings is the existence of a family of periodic orbits of sufficient extent in the inner regions that are aligned perpendicular to the bar.[26] These orbits intersect those in the bar itself. A nuclear ring can form from the interaction of gas clouds moving along these orbits, which can lead to a slightly oval, misaligned shape for the ring as depicted in Figure 13. There also has to be a mechanism for interstellar gas flow into the central regions, and this could be provided by a bar's offset dust lanes (like those seen in NGC 1433; Figure 3). These dust lanes are believed to be shocks that carry interstellar gas from near the ends of the bar to the area where the ring forms.[27]

Because the I4R and the ILR are inner resonances, it is possible for their existence to be avoided if the bar pattern speed is high

---

[23]M. P. Schwarz, *Proceedings of the Astronomical Society of Australia*, **5**, 464 (1984).

[24]M. Regan and P. Teuben, *Astrophysical Journal*, **600**, 595 (2004).

[25]F. Combes, in *N-body Problems and Gravitational Dynamics*, F. Combes, E. Athanassoula, eds. (Aussois, Maurienne, France, 1993), p. 137; R. Buta and F. Combes, *Fundamentals of Cosmic Physics*, **17**, 95 (1996).

[26]M. Regan and P. Teuben, *Astrophysical Journal*, **582**, 723 (2003).

[27]E. Athanassoula, *Monthly Notices of the Royal Astronomical Society*, **259**, 345 (1992).

enough. Thus, we might expect that some barred galaxies will have R_1 or R_2 (or both) features, but no inner or nuclear ring. However, the absence of a given ring may not necessarily be due to the absence of a given resonance. Instead, it could be due to a lack of interstellar gas. In the case of nuclear rings, the absence of such a feature could also be due to the lack of the required orbits and not the absence of the resonance.[28]

Figure 13 shows the location of the corotation resonance (CR) as the dividing zone between the inner and outer resonances. When viewed in the context of gravitational dynamics, the corotation resonance has something special to offer to interpretations of spiral galaxy structure: five *equilibrium* points called Lagrangian points, named after the French-Italian mathematician Joseph-Louis Lagrange (1736–1813), who recognized the significance of the points in the 1770s while examining the so-called "restricted problem of three bodies." The same points occur in the gravitational field of a bar because of the extended nature of the mass distribution. According to Geneva Observatory dynamicist Daniel Pfenniger, "the effect of these [Lagrangian] points on the dynamics of a galaxy is crucial."[29]

The four most important points are labeled L_1, L_2, L_4, and L_5 in Figure 13, and are known for their stability or lack thereof.[30] The way to think of these points is that an object can stay indefinitely near an equilibrium point if the point is stable, but cannot do so if it is unstable. In a barred galaxy, the L_4 and L_5 points can be stable or unstable depending on the strength of the bar, while the L_1 and L_2 points are always unstable.

The L_1 and L_2 points are located at corotation near the ends of the bar, while the L_4 and L_5 points lie at corotation along the line perpendicular to the bar. The pairs of points are not necessarily at

---

[28]M. Regan and P. Teuben, *Astrophysical Journal*, **582**, 723 (2003).

[29]D. Pfenniger, *Astronomy and Astrophysics*, **230**, 55 (1990).

[30]The fifth Lagrangian point, L_3, would be at the center of the galaxy. In the restricted problem of three bodies, this point is usually unstable, but in a galaxy can be stable.

exactly the same galactocentric distance,[31] and an annulus defined by them (dotted zone in Figure 13) is called the "corotation region."[32]

Because gravity torques will cause gas clouds to drift outwards outside of CR and inwards inside of CR. we do not expect to see a ring at the CR radius itself. What is predicted to be seen are gaps in the luminosity distribution associated with the L_4 and L_5 Lagrangian points. If the bar is weak, these points are stable, and stars can circulate around the points on periodic orbits that blend in with regular orbits centered on the galactic nucleus. However, if the bar is sufficiently strong, the L_4 and L_5 points become unstable, and the orbits around these points become chaotic, causing stars to leave the vicinity of the points.[33] This leads to a deficiency of mass around the points, which then leads to dark gaps in surface brightness lying on the minor axis line of the bar between inner and outer rings. The fact that we can see such deficiencies in real galaxies provides a way of locating the CR. How the Lagrangian points can be used to further interpret galaxy dynamics is described later in this chapter and in Chapter 16.

As previously noted, models show that across the ILR, CR, and OLR, orbits change their orientation relative to the bar by 90°. These changes are essential to the process of ring formation,[34] but they also highlight something else: a bar cannot extend beyond its corotation resonance. This was concluded in a 1980 paper[35] by University of Athens professor George Contopoulos, who noted that the main orbits in the gravity field of a bar support the bar only between the ILR and the CR. In order to "support the bar," the orbits must be elongated along the bar axis. But outside the CR and up to the OLR, the orbits align perpendicular to the bar, and therefore do not

---

[31] In a model bar described by D. Pfenniger, *Astronomy and Astrophysics*, **230**, 55 (1990), L_4 and L_5 are about 9% closer to the center than L_1 and L_2.

[32] J. Binney and S. Tremaine, *Galactic Dynamics* (Princeton, Princeton University Press, 2008).

[33] D. Pfenniger, *Astronomy and Astrophysics*, **230**, 55 (1990).

[34] M. P. Schwarz, *Astrophysical Journal*, **247**, 77 (1981).

[35] G. Contopoulos, *Astronomy and Astrophysics*, **81**, 198 (1980).

support the bar. Figure 13 shows a bar whose CR is at about 1.3–1.4 times the bar radius, which galaxy theorists consider to be a "fast" (rapidly rotating) bar.[36]

15.3. Special Examples

In the remainder of this chapter, examples of ringed galaxies are described that are either dynamically or morphologically special, and have something to offer to further our understanding of the structure and evolution of spiral galaxies. These are interesting cases to put the spiral galactic phenomenon into perspective.

15.3.1. *Dead and Dying Rings*

The first highlight is that although the title of this chapter is "Rings of Fire," not all rings are active zones of star formation. Some rings stopped forming new stars billions of years ago. For example, Figure 17 shows IC 1438, a spiral 120 million light years away in Aquarius. IC 1438 has an R_1R_2' outer structure, but with a peculiar twist: the R_2' feature is most prominent in a blue filter image while the R_1 feature is most prominent in an infrared filter image. This dichotomy suggests that the R_1 part formed first and left behind a stellar remnant (meaning a ring made mainly of very old stars). IC 1438 also has star-forming inner and nuclear rings.

NGC 1211 in Cetus is a spectacular related example of a face-on spiral having three well-defined rings: an inner ring (r), an R_1 outer ring, and an R_2' outer pseudoring (Figure 18(a)). Except for the absence of a nuclear ring, NGC 1211 bears a strong resemblance to the schematic in Figure 13. The R_1 ring has only subtle dimpling. Surprisingly, only the R_2' feature harbors active star formation in this case (Figure 18(b)). The inner ring is only subtly enhanced in color while the R_1 feature is as dead as a ring can be. What likely happened to this galaxy is that, for a long time after it formed, its inner ring and its R_1 ring had sufficient interstellar gas to fuel star

---

[36]For example, V. P. Debattista and J. A. Sellwood, *Astrophysical Journal*, **543**, 704 (2000).

(a)

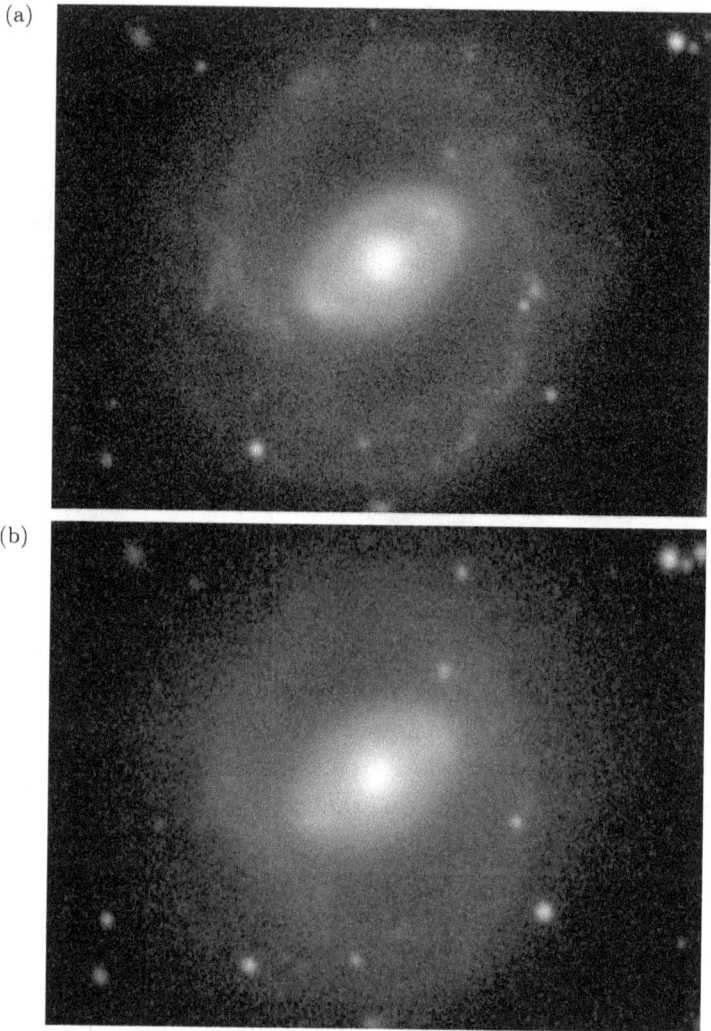

(b)

Fig. 17.   (a) Blue-light (*B*-band) and (b) near-infrared (*I*-band) images of IC 1438, a galaxy showing an R_1R_2' morphology. The R_2' component has recent star formation and as a result is most prominent in the blue-light image. In contrast, the R_1 component is most prominent in the infrared image. The difference suggests that the R_1 component formed first and left behind a stellar remnant. Images source: the de Vaucouleurs Atlas of Galaxies published by Cambridge University Press, 2007; reprinted with permission.

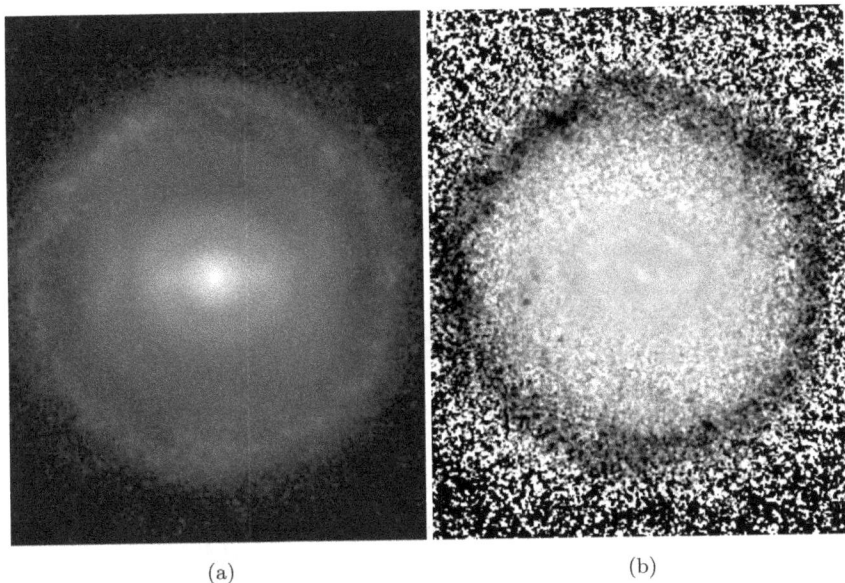

(a)   (b)

Fig. 18.   (a) Bluegreen-light (*g*-band) and (b) *g–i* color index map of NGC 1211, a face-on early-type spiral located 143 million light years away in the constellation Cetus, the Whale. The color index map is coded such that blue, star-forming regions are dark, while redder areas are light. Images source: Sloan Digital Sky Survey Collaboration, http://www.sdss.org.

formation long enough to build up stellar remnants. Eventually, the gas was used up and no new stars could form in either feature, leaving the stars that formed in the rings to age. The gas that fueled these rings could have been left over from the time the galaxy formed. There may not have been enough gas in the outer regions of the galaxy to fuel an R'_2 component during this period. The interstellar gas currently fueling the star formation in the R'_2 outer ring most likely was accreted more recently from *outside* the galaxy.

15.3.2. *Traffic Pileup*

Images taken in Hα bring attention to another interesting aspect of star formation in inner rings: a sensitivity of the distribution of HII regions around inner rings to the face-on shape of the ring. This is shown by the four examples in Figure 19, which compare

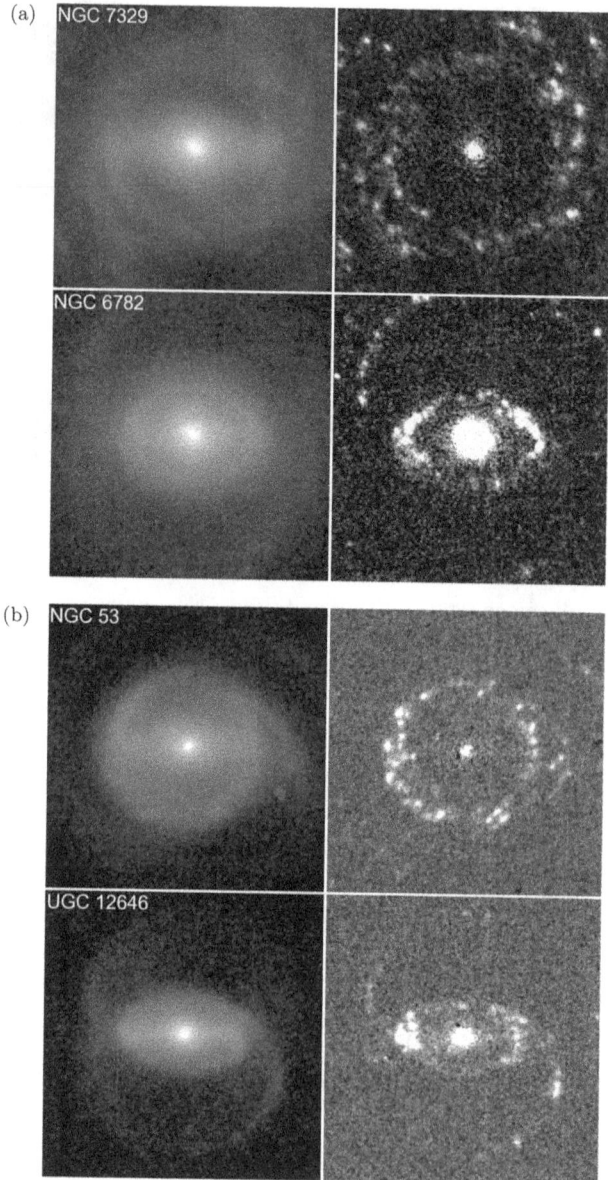

Fig. 19.   (a) Red light and (b) Hα images for four galaxies having intrinsically round or extremely elongated inner rings. The images have been computer-deprojected and show that photoionized clouds spread uniformly around circular inner rings, but "bunch up" around the major axis points of highly-elongated rings. Images source: The de Vaucouleurs Atlas of Galaxies published by Cambridge University Press, 2007; reprinted with permission.

computer-deprojected[37] red light images in the left panels with deprojected Hα images in the right panels for two circular inner rings (NGC 53 and 7329) and two highly-elongated inner rings (NGC 6782 and UGC 12646).[38] The images show that HII regions are strongly concentrated around the major axis points of the highly-elongated inner rings, but are more evenly distributed around the more circular inner rings. The likely reason for this is that in a reference frame rotating with the bar pattern speed, the HII regions in the inner rings of NGC 6782 and UGC 12646 are forming from clouds which are moving more slowly near the ring major axis as compared to the ring minor axis. This leads to a "pile-up" around the ends of the bars, enhancing star formation in those zones.[39] The effect is similar to the orbital behavior of a comet: when it is close to the Sun, it moves faster, but when it is far from the Sun, it moves more slowly.

15.3.3. *Misaligned Bar and a Super Nuclear Ring of Fire*

It is well-established that most inner rings in barred spiral galaxies are elongated parallel to the bar. The galaxy ESO 565-11 (Figures 20 and 21),[40] located 202 million light years away in the constellation Hydra, is an unusual spiral that peculiarly violates this rule. The galaxy has several features of interest: a conspicuous bar, an inner pseudoring made of tightly wrapped spiral structure, an outer pseudoring made of two very faint outer arms, and a spectacular nuclear ring made of about a dozen "hotspots" or star-forming "knots." A small but distinct central bulge is also seen. What makes this galaxy stand out is that it is being viewed in a nearly face-on orientation, yet the bar is misaligned with the major axis of the

---

[37]Computer deprojection means that a galaxy is deprojected with a computer program that uses the shape and orientation of faint outer isophotes to tell the inclination of the galaxy and the line in the plane of the sky around which the galaxy is tilted, that is, the line of nodes.

[38]D. A. Crocker, P. D. Baugus and R. J. Buta, *Astrophysical Journal Supplement Series*, **105**, 353 (1996).

[39]For example, G. Byrd *et al.*, *Astronomical Journal*, **131**, 1377 (2006).

[40]R. Buta, G. B. Purcell and D. A. Crocker, *Astronomical Journal.* **110**, 1588 (1996).

(a)   (b)

(c)

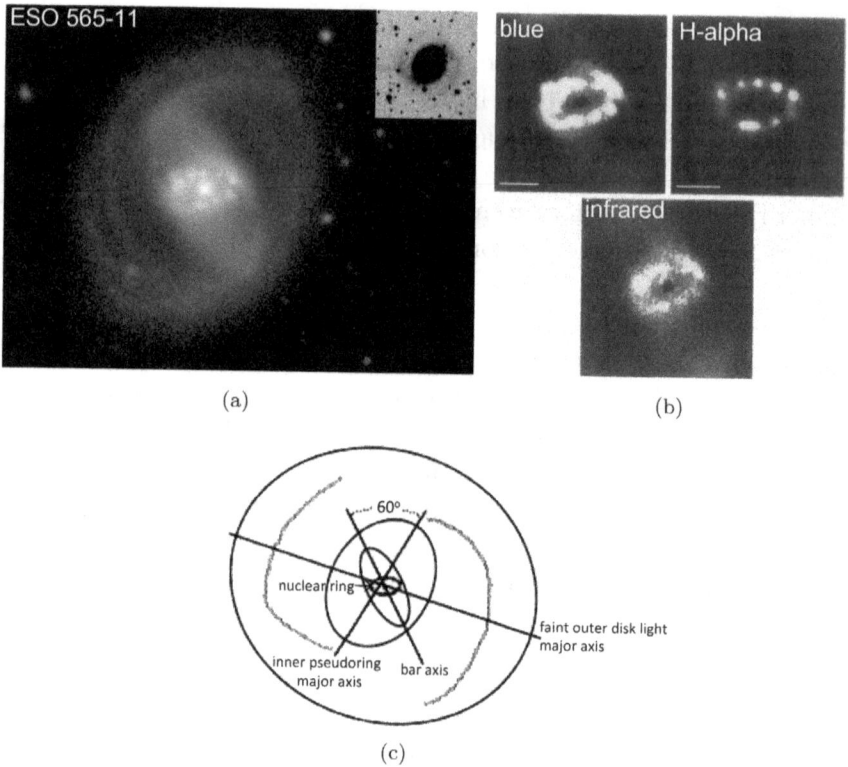

Fig. 20.   ESO 565-11. (a) Blue light image (inset shows the two faint outer spiral arms); (b) images of the center showing how the orientation of the super nuclear ring changes with increasing wavelength; (c) schematic highlighting the galaxy's structure. Adapted from R. Buta *et al.*, *Astronomical Journal*, **110**, 1588 (1996). © AAS. Reproduced with permission.

inner pseudoring by 60° (see schematic diagram Figure 20(c)). This is highly unusual and very rare. To add to the mystique of the galaxy, the nuclear ring turns out to be a huge feature, 16,600 light years in diameter, when the average nuclear ring is only about 5,000 light years in diameter. The nuclear ring of ESO 565-11 is also a highly-elongated oval; most nuclear rings are either circular or slightly oval in intrinsic shape.[41] The galaxy has no major companions, so its

---

[41]S. Comerón *et al.*, *Monthly Notices of the Royal Astronomical Society*, **402**, 2462 (2010).

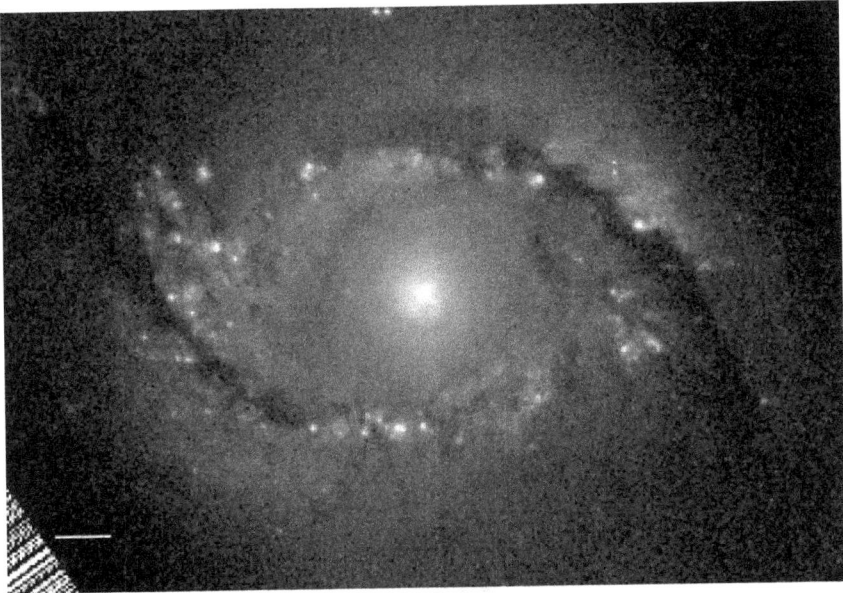

Fig. 21.   The super nuclear ring in ESO 565-11, as seen in the V-band filter at high resolution with the *Hubble Space Telescope*. R. Buta *et al.*, *Astronomical Journal*, **118**, 2071 (1999). © AAS. Reproduced with permission.

peculiar structure is unlikely to have been caused by a gravitational interaction.

The star formation in the nuclear ring is so intense that it can only be described as a "starburst." A starburst is a star-forming event where a large number of new stars are born in a relatively short amount of time. Starbursts are often triggered by a galactic merger, but nuclear rings are also prone to such events. The nuclear rings in NGC 1512 (Figure 22) and NGC 1097 (Figure 23) are exceptional examples, but in both of these cases the feature is nearly circular and not as large as the one in ESO 565-11.

The top right panels in Figure 20 show images of the nuclear ring after removal of the background galaxy. The three frames, Hα, blue, and infrared are an age sequence: Hα shows where the photoionized clouds are, and these are always associated with the youngest stars. The infrared image alternatively shows mainly the older stars, while blue light shows stars in between what these two filters show.

(a)

(b)

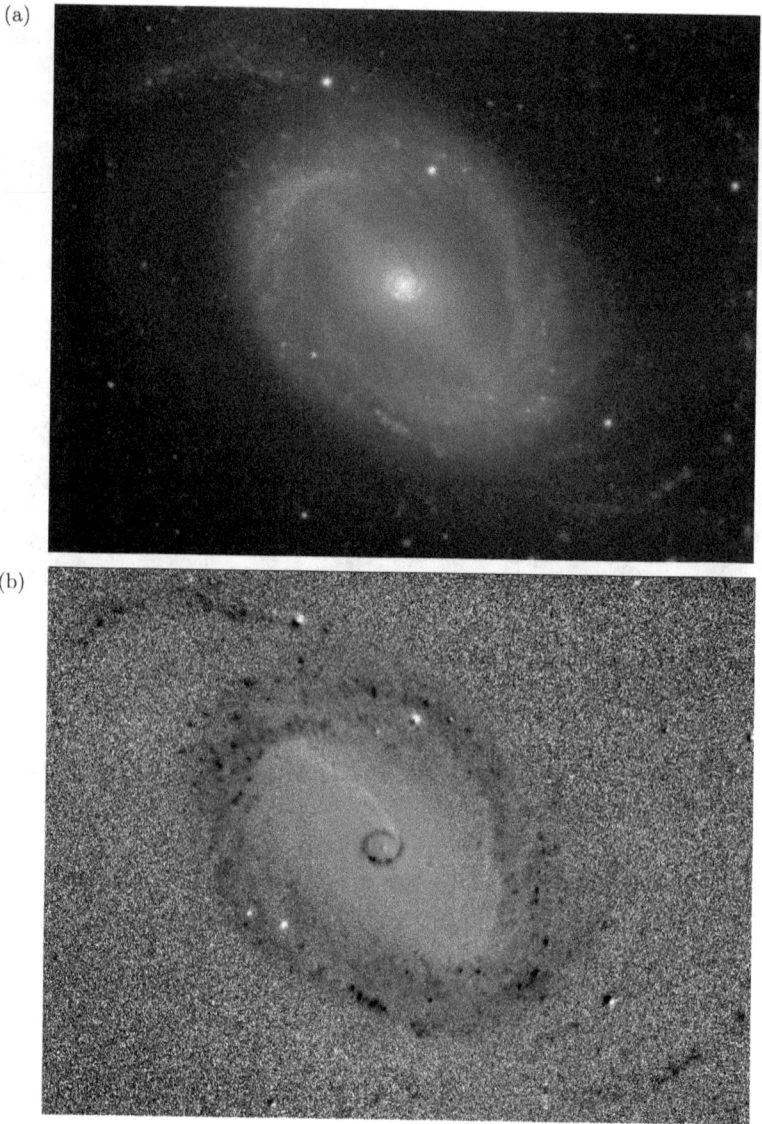

Fig. 22.   NGC 1512 is a beautiful ringed spiral with a strong bar. (a) Blue-light image; (b) Blue minus visual (B–V) color index map. The color index map shows that the inner ring and the nuclear ring are zones of active star formation. The region between the two rings is largely devoid of star formation. The galaxy is strongly interacting with a companion, NGC 1510, which is just outside the field. Color index map coding: bluish features are dark, and reddish features are light. Images source: the de Vaucouleurs Atlas of Galaxies published by Cambridge University Press, 2007; reprinted with permission.

Fig. 23. NGC 1097 is a strongly interacting ringed, barred spiral lying 53 million light years away in the constellation Fornax. In this blue-light image, a large, nearly circular starburst nuclear ring is seen. This blue-light image is from the de Vaucouleurs Atlas of Galaxies published by Cambridge University Press, 2007; reprinted with permission.

The long axis of the nuclear ring seems to turn slightly from Hα to infrared; in the latter image, the ring is oriented more nearly perpendicular to the bar. The cause of this effect is likely gravity torquing due to the bar, perhaps suggesting that the ring has existed for more than one generation of star formation.

ESO 565-11 is so distant that groundbased images do not give a clear view of the structure of the starburst nuclear ring. Figure 21 shows a visual (V-band) image of the feature obtained with the *Hubble Space Telescope*.[42] At HST resolution, we can see individual (but largely unresolved) clusters of massive stars, some of which are so luminous that they are called "super star clusters." Interestingly,

---

[42]R. Buta, D. A. Crocker and G. Byrd, *Astronomical Journal*, **118**, 2071 (1999).

in spite of the extreme elongation of the ring, the star-forming regions are not preferentially found around the major axis points of the ring, as they would be for inner rings (Figure 19). The HST image reveals strong dust lanes cutting into the nuclear ring region, in a manner that is typical of such features.

What is the nature of ESO 565-11? One clue may be provided by the location of the inner pseudoring: it appears to lie on the rim of a massive oval structure that could be interpreted as an evolved bar. In a 1979 paper, John Kormendy (then at U. Cal. Berkeley) hypothesized that interactions internal to a galaxy could destroy a bar by making its star orbits more circular.[43] These interactions can heat up a galactic disk, that is, increase random stellar motions, which can destroy a bar because its longevity will depend on how effectively it can maintain order in the motions of the old stars that make it up. A galaxy will likely remain cured of a bar unless something happens that can cool the disk and reduce the level of random stellar motions. In 2002, French astronomers Frederick Bournaud and Francoise Combes showed that cold interstellar hydrogen gas falling into a galaxy from outside can cool a disk enough for a new bar episode to occur in the same galaxy.[44] Perhaps what we are seeing in ESO 565-11 is a second episode of bar formation: the oval is the remnant of the first episode, while the currently visible bar is the second episode. The second bar may have formed out of alignment with the first bar owing to a different pattern speed. If the observed bar in ESO 565-11 developed only recently, then the super nuclear ring may be what a newly formed nuclear ring looks like.

15.3.4. *The "Quintessential Resonance Ring Galaxy"*

NGC 3081, located 98 million light years away in the constellation Hydra, is a remarkable barred spiral galaxy whose morphology is dominated by star-forming rings more than by spiral arms (Figure 24). The set of ring patterns in NGC 3081 bears an uncanny

---

[43] J. Kormendy, *Astrophysical Journal*, **227**, 714 (1979).
[44] F. Bournaud and F. Combes, *Astronomy and Astrophysics*, **392**, 83 (2002).

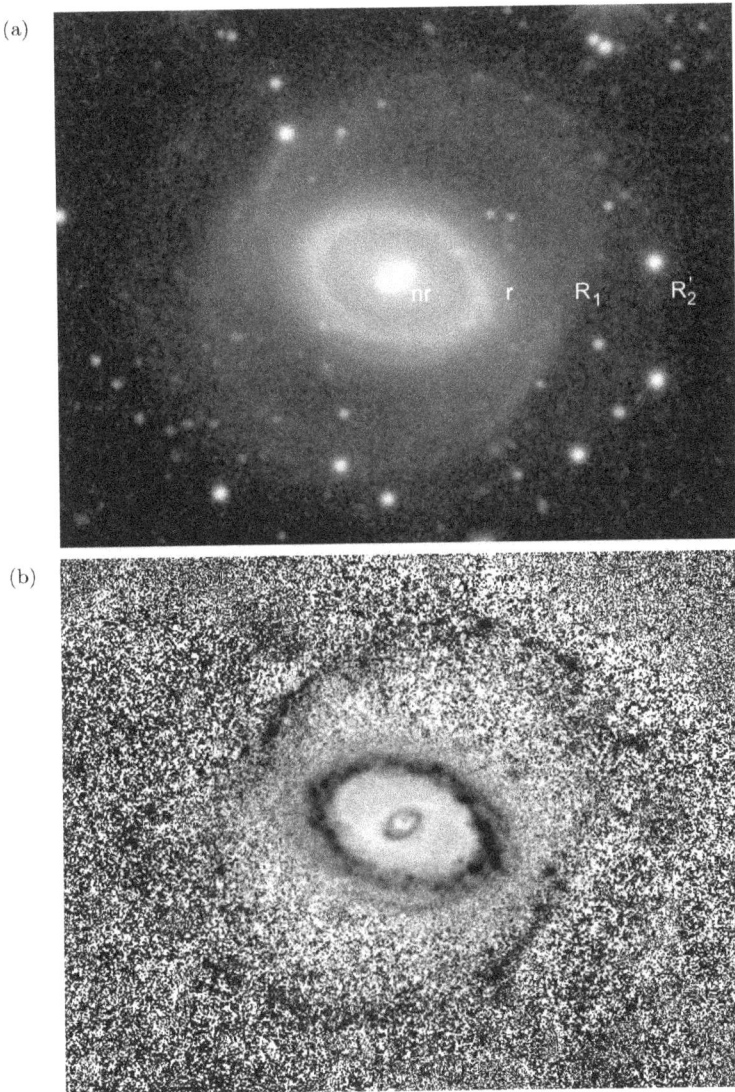

Fig. 24. The resonance ring galaxy NGC 3081 in Hydra, 98 million light years away. This galaxy shows four ring features: a nuclear ring (nr), an inner ring (r), an R_1 outer ring, and an R_2' outer pseudoring (all labeled on the top image). The top image is in blue light (B-band) while the bottom image is a blue minus infrared (B–I) color index map (foreground and background objects have been removed). The color index map coding is such that bluish features are dark and reddish features are light; the map shows that all four rings are zones of enhanced star formation. Images source: adapted from the de Vaucouleurs Atlas of Galaxies published by Cambridge University Press, 2007; reprinted with permission.

resemblance to the resonance patterns in Figure 13. Few galaxies seem to slide into a dynamical concept as easily as this one does.

The galaxy has four distinct ring features: a nuclear ring, an inner ring, an R_1 outer ring, and an R_2' outer pseudoring. These are all shown and identified in Figure 24(a). In contrast to NGC 1211, each ring is a zone of enhanced star formation, as is seen in the blue minus infrared color index map in Figure 24(b). When computer-deprojected, the rings alternate in orientation: the nuclear ring is elongated nearly perpendicular to the bar, the inner ring is elongated parallel to the bar, the R_1 ring is elongated nearly perpendicular to the bar, and the R_2' pseudoring is nearly intrinsically circular. Again, the resemblance to the behavior of orbits in a barred galaxy model is astonishing.

The highly organized nature of the star formation in NGC 3081 is also remarkable, especially in the inner ring which is significantly non-circular in its face-on shape. This is highlighted in Figure 25, which shows a color image of the inner and nuclear rings of NGC 3081 as seen with the *Hubble Space Telescope*. At the high resolution of the HST, the star formation in the inner ring is resolved into myriads

Fig. 25. *Hubble Space Telescope* color image of NGC 3081. ESA/Hubble and NASA; Acknowledgment: R. Buta (University of Alabama).

of mostly compact blue star clusters. In the manner of Figure 19, an enhancement in the star formation around the major axis points is clearly seen, and the image also highlights how the ring appears to be a very tightly wrapped spiral pattern. It is interesting also that, in spite of having enhanced blue colors as well, the nuclear ring star formation is largely unresolved compared to the inner ring.

The bar of NGC 3081 is a broad, weak-looking feature that in the HST image is yellow in color, implying the feature is made entirely of very old stars. Both the HST image and the $B-I$ color index map reveal weak dust lanes in the bar region. Also, a secondary (or nuclear) bar (nb) is found inside the nuclear ring.

Surprisingly, the modified tuning fork (Figure 10 of Chapter 6) classification of NGC 3081, given in Allan Sandage and John Bedke's *Carnegie Atlas of Galaxies*, is SBa(s), meaning that none of the observed rings is formally recognized.[45] While the nuclear ring, nuclear bar, and the two outer rings of the galaxy might be left unrecognized within the adopted classification system, the inner ring is very bright and one of the best-defined examples of such a feature. Why classify the galaxy as (s)-variety? Sandage and Bedke explain that the reason for their choice of (s) variety is because the inner ring of NGC 3081 appears to be a tightly wrapped spiral. It is a "near-ring" in their view, not a "ring." While this is confirmed by the HST observations, the feature is much more of a ring than it is a spiral. Nevertheless, Sandage and Bedke note: "NGC 3081 has one of the most complex morphologies of the (Revised Shapley–Ames Catalogue) galaxies. There are rings within rings at the edges of disks within disks, as intricate a structure as in nesting, concentric Russian dolls."

An important question is: how much mass makes up a galactic ring? For example, the inner ring of NGC 3081 is very bright in blue light. This means the ring is dominated by young, massive, bluish-colored stars as shown in Figure 25. It does not mean, however, that the ring has much mass. If we could map the mass instead of

---

[45] The Comprehensive de Vaucouleurs revised Hubble-Sandage (CVRHS) classification is $(R_1R_2')SAB(r,nr,nb)0/a$.

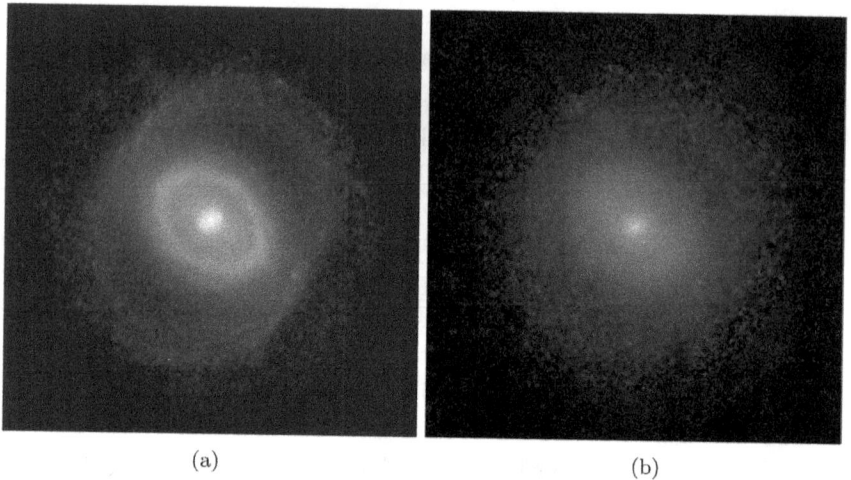

(a)   (b)

Fig. 26.   Light and mass in the rings of NGC 3081: (a) blue-light image; (b) the
distribution of stellar mass in NGC 3081 using star color as an indicator of mass.
The images have been computer-deprojected, and foreground and background
objects have been removed. From R. Buta, presented at: *Galactic Rings: Signposts
of Secular Evolution in Disk Galaxies*, a meeting held at the University of
Alabama May 27–June 1, 2018.

the light, what would the ring look like? We can get such a map
using color as an indicator of mass.[46] When this is done, the ring
virtually disappears, as shown in Figure 26. Of the four rings in
NGC 3081, three (the nuclear ring, the inner ring, and the outer R'_2
ring) disappear in the mass map; only the R_1 feature is still seen,
implying it has a stellar remnant.

The simultaneous star formation in four rings separated by
significant distances poses an interesting problem. Each ring in NGC
3081 has a different associated timescale, because each is at a different
location. Based on known rotation speeds in the galaxy,[47] the orbital
period at the positions of the nuclear, inner, outer R_1, and outer
R'_2 rings is about 30, 150, 360, and 450 million years, respectively.
(Compare these with the known orbital period, 240 million years,
for the Sun in the Milky Way.) The timescale for the formation of

[46]E. F. Bell *et al.*, *Astrophysical Journal Supplement Series*, **149**, 289 (2003).
[47]R. Buta and G. B. Purcell, *Astronomical Journal*, **115**, 484 (1998).

these rings is expected to be proportional to the orbital period at their location. The large difference between the orbital periods from the nuclear ring to the outer R'_2 poses a question: why do we see all four rings engaged in star formation at the same time? Why haven't all the rings developed a stellar remnant? This likely all boils down to determining where the interstellar gas in rings actually comes from: inside or outside the galaxy? Paris Observatory astronomer Francoise Combes has suggested that the co-existence of multiple star-forming rings at very different locations in a galaxy implies that interstellar gas continually falls into the disk from the outside to fuel the simultaneous star formation in the rings.[48]

The rings of NGC 3081 highlight the likely secular evolutionary nature of galactic resonance rings in general. The term "secular" refers to a process that occurs only slowly with time. Rings are not necessarily short-lived phenomena that form quickly and then disappear. They can be the product of long-term gas flows that work to build up a central mass concentration and at the same time spread out the outer disk. Because outer rings and pseudorings tend to develop far from the center, it can take more than a billion years for such features to fully form. This is in the domain of secular evolution. Secular evolution is described further in Chapter 20.

Finally, if we were to ask why NGC 3081 has such regular, symmetric structure, it could be because the galaxy has no major companions, at least nothing within about a million light years. This has allowed it to evolve in relative isolation, and perhaps the way it looks is the way a disk-shaped galaxy looks when resonances are allowed to run their full course.

15.3.5. *Ghost Companion*

The galaxy NGC 7531 is a bright, tilted spiral 70 million light years away in the constellation Grus, the Crane. It was one of the six galaxies that I studied for my PhD dissertation work, and

---

[48]F. Combes, in *N-body Problems and Gravitational Dynamics*, F. Combes and E. Athanassoula, eds. (Paris, Paris Observatory, 1993), p. 137.

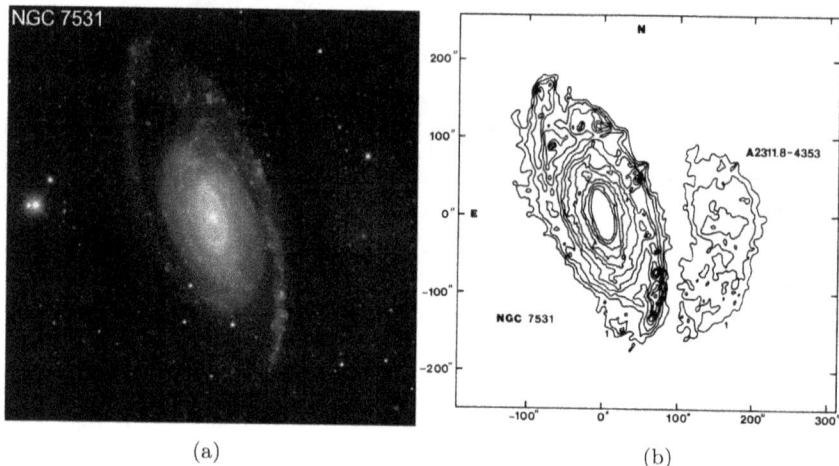

(a)   (b)

Fig. 27.   The southern spiral galaxy NGC 7531 in Grus: (a) blue-light image (source: The de Vaucouleurs Atlas of Galaxies published by Cambridge University Press, 2007; reprinted with permission.); (b) a contour map showing NGC 7531 and its ghostly companion Anonymous 2311.8−4353 (PGC 70787). From R. Buta, *Astrophysical Journal Supplement Series*, **64**, 1 (1987). © AAS. Reproduced with permission.

is interesting for its bright star-forming inner ring.[49] But equally interesting about the galaxy is that a few arcminutes to the west is a large, diffuse object of very low surface brightness known as PGC 70787.[50] The object is so faint that the best way to display it is using a contour plot (Figure 27). At the faintest surface brightness level, the object is about two-thirds the size of NGC 7531, which is a massive spiral of high luminosity. The object could be a foreground dwarf galaxy in which case its proximity to the spiral would be a chance alignment. The main argument against this is the strong asymmetry in the spiral arms of NGC 7531. The arm closest to PGC 70787 has a higher level of star formation than does the opposing arm, which could signify a gravitational interaction. An alternative interpretation to chance alignment could then be that PGC 70787 is

---

[49]R. Buta, *Astrophysical Journal Supplement Series*, **64**, 1 (1987).
[50]"PGC" stands for "Principal Galaxy Catalogue;" reference: G. Paturel *et al.*, *Catalogue of Principal Galaxies* (Lyon, Lyon Observatory, 1989).

a physical companion of NGC 7531 that has been tidally disrupted in the spiral's halo.[51] It has not been possible to distinguish between the two possibilities since no one has yet successfully measured the line of sight speed of PGC 70787.

15.3.6. *Wing Tip Bar*

NGC 7098 (Figure 28) is a bright spiral galaxy lying 101 million light years away in the far southern constellation of Octans, the Octant. The galaxy is noteworthy for its bright, somewhat broad bar having distinct linear enhancements at the ends. These enhancements, which are more easily seen in the unsharp masked image[52] shown in Figure 28(c), were pointed out by David S. Evans in his 1957 southern galaxy atlas,[53] where he colorfully named them "wing tips" because of their resemblance to structures sometimes found at the tips of airplane wings. Today, enhancements such as these are called "ansae," meaning "handles of the bar," and they are most frequently seen in SB0 and SBa galaxies.[54] Ansae can appear as wing tips, short arcs, or round spots. The cause of wing tips in particular, and ansae in general is uncertain, but recent numerical simulation studies have suggested that a merger between two disk-shaped galaxies could lead to a disk-shaped remnant with an ansae-type of bar.[55]

A blue minus infrared color index map reveals that recent star formation is occurring only in the pseudorings of NGC 7098. The bar, including the "wing tips," is made only of old stars.

15.3.7. *X Marks the Center*

NGC 7020 (Figure 29) is a stunningly beautiful southern galaxy lying 139 million light years away in the constellation Pavo, the Peacock.

---

[51]D. Martínez-Delgado *et al.*, *Astrophysical Journal*, **140**, 962 (2010).

[52]A type of image processing where a heavily smoothed version of an image is subtracted from the unsmoothed version, revealing the fine details in an image without a strong background gradient.

[53]D. S. Evans, *Cape Atlas of Southern Galaxies* (1957).

[54]I. Martinez-Valpuesta *et al.*, *Astronomical Journal*, **134**, 1863 (2007).

[55]E. Athanassoula *et al.*, *Astrophysical Journal*, **821**, 90 (2016).

(a)

(b)

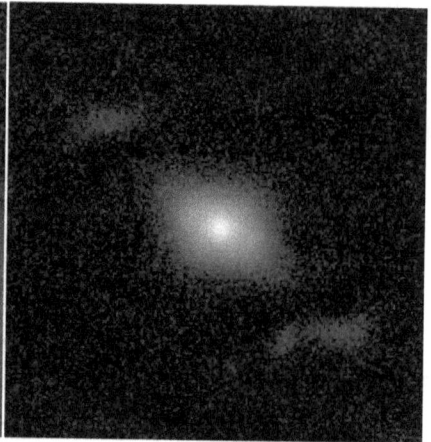

(c)

Fig. 28.   The tilted spiral galaxy NGC 7098. The upper frame is a blue-light image of the galaxy showing the spiral pattern which is mostly in the form of pseudorings. The lower panels are infrared images of the bar region. The lower right panel is an unsharp mask image where a heavily-smoothed version of the lower left image has been subtracted, revealing the "wing tips." Foreground stars have been removed in the lower panels. Top image from R. Buta, *Astrophysical Journal Supplement Series*, **96**, 39 (1995). © AAS. Reproduced with permission.

(a)

(b)

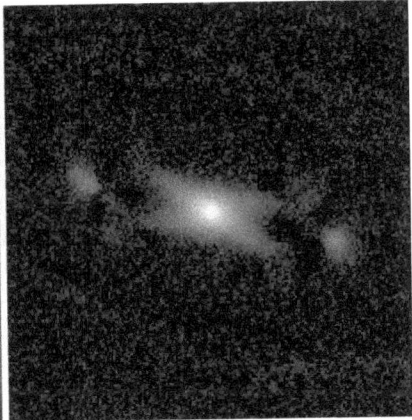

(c)

Fig. 29.   The peculiar X-galaxy NGC 7020. (a) Blue light image; (b) and (c) Infrared images of the inner hexagonal zone, (c) showing an unsharp masked residual image to highlight the X pattern and the "spots." Foreground stars have been removed in (b) and (c). The blue-light image is from the de Vaucouleurs Atlas of Galaxies published by Cambridge University Press, 2007; reprinted with permission.

The galaxy has two noteworthy features: a bright outer ring, and a peculiar central region showing a strong hexagonal pattern. Although not an obvious cosmic pinwheel, NGC 7020 nevertheless shows short spiral segments all around the outer ring, in the manner of what is described as a "flocculent" spiral in Chapter 16. The outer ring and the inner hexagonal zone are separated by a wide gap. In a blue minus infrared color index map, only the outer ring appears as a zone of enhanced blue colors. This means that the outer ring is the only place in NGC 7020 where recent star formation is occurring. Both the inner hexagonal zone and the gap region are made entirely of old stars.

In spite of the apparent ring-like appearance of the inner zone in NGC 7020, there is no ring in this area. The zone consists of isophotes which are circular near the center, then boxy at larger distances from the center, followed by the hexagonal isophotes that stand out so well at the inner edge of the gap. The pointed tips of the hexagonal zone almost look like distinct "spots." The structure of the zone, however, becomes clear when we see it in an unsharp mask image. Figure 29(b) shows the inner zone in infrared light, while the lower right panel shows the same image after subtraction of a heavily smoothed version of the lower left image. When this is done, the resulting unsharp masked image shows a peculiar X pattern crossing the center, in addition to the two distinct round spots that were at the tips of the hexagon.

The appearance of the residual X pattern across the center of NGC 7020 tells us that the inner hexagonal zone is a bar with considerable thickness perpendicular to the plane of the galaxy and viewed at fairly high inclination. Motion perpendicular to the plane of a galaxy is called vertical motion. The X is believed to be caused by three-dimensional orbits where the vertical motion is in step with the pattern speed of the bar in the plane. Both the X and the spots are line-of-sight illusions through the three-dimensional bar structure. This has been shown by Finnish astronomers Eija Laurikainen and Heikki Salo, who used sophisticated numerical simulations that could model the three-dimensional structure of bars as seen from various vantage points. In the case of NGC 7020, their models suggest that we are viewing the bar nearly broadside-on at an inclination between 60° and 75°. The actual inclination of NGC 7020 is 69°.

15.3.8. *False Barred Spiral*

The southern galaxy ESO 235-58 looks for all the world like a barred spiral galaxy. This, however, is an illusion, as shown in Figure 30 which is based on a blue light CCD image. The apparent bar of ESO 235-58 is unusual in that it appears to be split practically down the middle by a dust lane, something which is almost never seen in normal, obvious bars. The issue isn't just the presence of dust in the bar, as many normal bars have dust lanes. Only, instead of cutting through the center, normal bar dust lanes are usually offset to the leading edges of the bar. The apparent bar of ESO 235-58 is

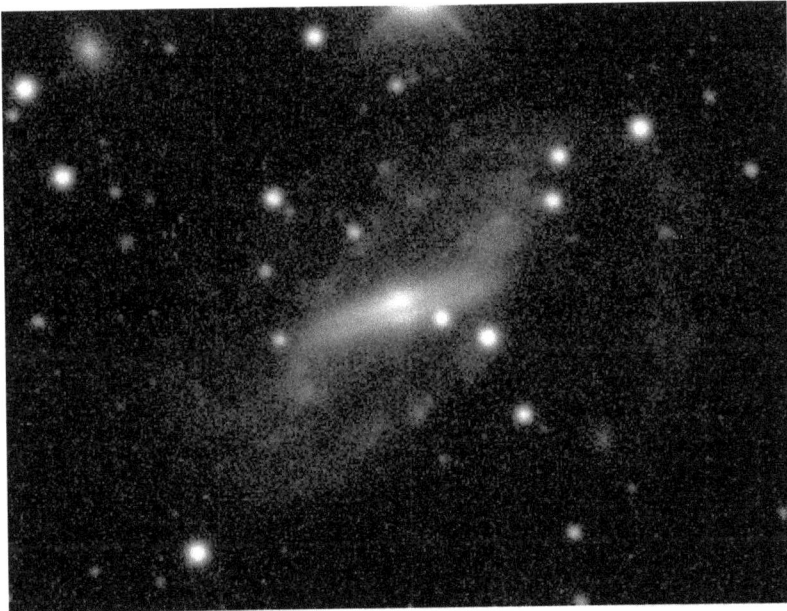

Fig. 30. Blue-light image of ESO 235-58, a possible polar-ring-related system that deceptively looks like a barred spiral. The dark band dividing the central bulge into two halves suggests that the apparent bar is actually an edge-on disk galaxy. Image source: The de Vaucouleurs of Galaxies published by Cambridge University Press, 2007; reprinted with permission.

more likely to be an edge-on disk galaxy like NGC 4565 (Figure 1 of Chapter 10) and NGC 891 (Figure 2 of Chapter 10).

If the apparent bar of ESO 235-58 is an edge-on disk galaxy, then the material making up the apparent spiral arms is not orbiting the center in the same plane as the disk galaxy's stars. This would place the system into the domain of objects known as *polar ring galaxies* (PRGs).[56] In a PRG, a small satellite galaxy has been disrupted into a polar orbit around a disk-shaped galaxy. A polar orbit is simply one which passes through the poles of a disk at an orientation $\approx 90°$ to the plane of the disk. Generally, the disk is more massive than the companion galaxy that was disrupted. Because forming a polar ring requires a very special type of gravitational encounter, PRGs are very rare. In the most obvious cases, the main disk and the polar ring are both seen edge-on.[57] Nevertheless, there must be cases where only one component is seen edge-on and the other is nearly face-on. ESO 235-58 is likely one of these cases. The disrupted satellite is seen as a faint spiral pattern, and the spiral may have been generated because the main disk "feels like" a bar to the polar disk.

ESO 235-58 is part of a small galaxy group including the large edge-on disk galaxy ESO 235-57. Compared to this one, the edge-on disk in ESO 235-58 is relatively small and of low luminosity.

15.3.9. *Four-leaf Clover and Feathers*

The symmetry of structure in some galaxies is remarkable. This is the case in the southern galaxy ESO 566-24, a barred spiral 148 million light years away in the constellation Hydra. The galaxy shows an unusually symmetric four-armed spiral that, in a computer-deprojected view, resembles a four-leaf clover (Figure 31(a)). Each arm appears to break from a different part of the bright inner ring, and there is a trace of an outer ring just outside the four-armed pattern. This can be compared with NGC 1433, which instead of showing a four-armed pattern, shows two outer arms and two

---

[56]F. Schweizer *et al.*, *Astronomical Journal*, **88**, 909 (1983).
[57]B. C. Whitmore *et al.*, *Astronomical Journal*, **100**, 1489 (1990).

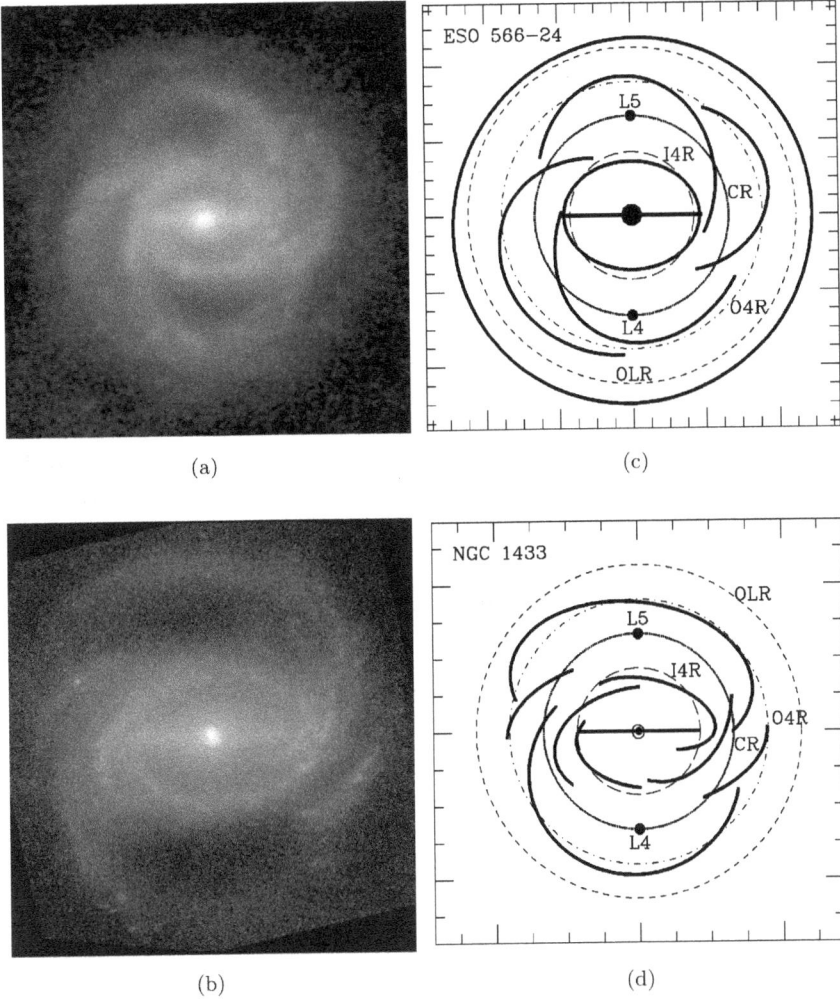

Fig. 31. Two barred spiral galaxies showing a symmetric, four-armed spiral pattern around an inner ring. (a) Computer-deprojected blue-light image of ESO 566-24; (b) computer-deprojected blue-light image of NGC 1433. Both images have been cleaned of foreground and background objects. (c) and (d) Schematic diagrams showing the locations of resonances in these galaxies, assuming a flat rotation curve. The corotation resonance has been assumed to be in the dim region between the inner ring and outer spiral arms. L_4 and L_5 are special gravitational points that would lie in these regions. From R. Buta, *Monthly Notices of the Royal Astronomical Society*, **470**, 3819 (2017). Copyright © 2017, Oxford University Press.

secondary arms ("plumes") which appear like feathers just outside the inner ring (Figure 31(b)).

The best interpretation is that these patterns are due to the influence of dynamical 4:1 resonances with the bar pattern speed. Finnish astronomers Pertti Rautiainen and Heikki Salo used numerical simulations to show that the four-armed pattern in ESO 566-24 is largely confined between the inner and outer 4:1 resonances.[58] Although NGC 1433 is not a strong four-armed case, it seems likely that its two secondary arms are related to the pattern seen in ESO 566-24, except that the bar is stronger in NGC 1433 than in ESO 566-24. As was noted earlier in this chapter, the corotation resonance lies between the inner and outer 4:1 resonances, and in the presence of a strong bar, orbits near the L_4 and L_5 Lagrangian points can be unstable (meaning they become chaotic). This would make it difficult for the secondary arms in NGC 1433 to connect to the inner ring as they might for a weaker bar.

The diagrams in Figures 31(c) and 31(d) show how the structures of NGC 1433 and ESO 566-24 connect to resonances.[59] These diagrams utilize the low surface brightness gaps between the inner rings and outer spiral arms as tracers of the location of the CR. The gaps are taken to be centered on the L_4 and L_5 points (Figure 13). Models have suggested that when the bar is strong enough, these points are unstable and leave a minimum of surface brightness between the inner ring and the outer arms. The corotation circles in Figures 31(c) and 31(d) were placed at those minima, and once the CR is set, the locations of the other main resonances follow if we assume a flat rotation curve. The graph for ESO 566-24 confirms the result of Rautiainen and Salo, that the four-armed pattern lies mainly between the inner and outer 4:1 resonances. The analysis for NGC 1433 places the secondary arcs between corotation and the outer 4:1 resonance.[60]

---

[58] P. Rautiainen, H. Salo and R. Buta, *Monthly Notices of the Royal Astronomical Society*, **349**, 933 (2004).

[59] R. Buta, *Monthly Notices of the Royal Astronomical Society*, **470**, 3819 (2017).

[60] See also P. Treuthardt *et al.*, *Astronomical Journal*, **136**, 300 (2008).

Figure 31 also shows that the bars in both ESO 566-24 and NGC 1433 may not extend to the CR. Instead, they could extend to the I4R, a not unreasonable interpretation given that both galaxies have a bar enveloped by an inner ring. In this circumstance, the L_1 and L_2 Lagrangian points would lie a little beyond the ends of the bar (Figure 13).

15.3.10. *The "Quintessential Non-barred Ringed Galaxy"*

NGC 7217 is a remarkable example of a nonbarred ringed galaxy. The *B*-band image in Figure 32 shows a very symmetric system with a clear flocculent pattern of inner spiral structure, a small central ring, and a conspicuous outer pseudoring made of tightly wrapped arms. Most interesting is what the color index map reveals about the system: the outer pseudoring is a spectacular ring of star formation. There is also star formation in the inner flocculent zone, but the small ring seen in the blue light image is not a star-forming feature. Instead this feature is a *dust ring*, a much rarer phenomenon than a typical star-forming ring.

The nature of NGC 7217 becomes clear when we view its faint outer isophotes and examine the prominence of its bulge. The isophotes become perfectly circular at large galactocentric distances and the bulge is almost twice as bright as the disk. This suggests that NGC 7217 is a nearly face-on version of the Sombrero Galaxy M104 that was shown in Figure 1 of Chapter 4. In Chapter 18, a 3.6-micron[61] image shows that M104 has an outer ring, which strengthens this interpretation. The fact that the isophotes of NGC 7217 become circular at large radii suggests that the disk of the galaxy formed in a pre-existing E0 system.

The nature of the rings in NGC 7217 is unclear. It has been suggested that they are resonance features like those seen in barred spirals, but in response to only a weak oval in the mass distribution.[62]

---

[61] A micron corresponds to one millionth of a meter, or 1,000 nm.

[62] R. Buta *et al.*, *Astrophysical Journal*, **450**, 593 (1995).

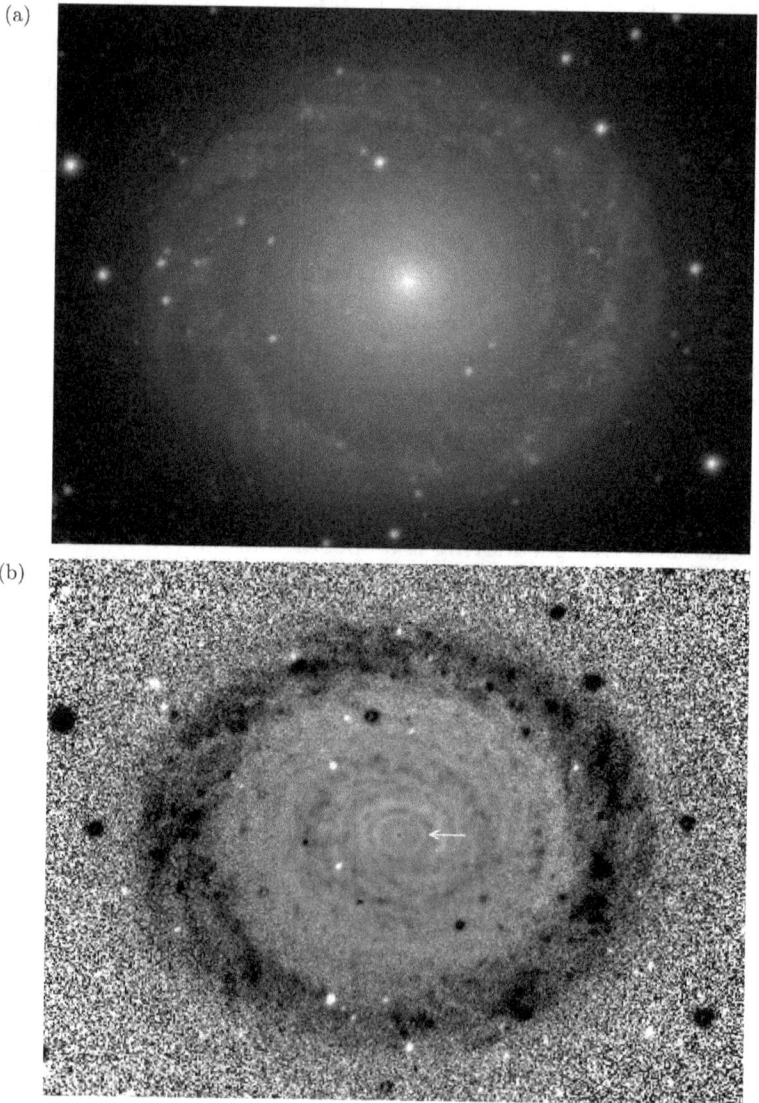

Fig. 32.   (a) Blue light (*B*-band) image; (b) blue-minus infrared (*B–I*) color index map of NGC 7217, a nearby nonbarred ringed galaxy lying 52 million light years away in Pegasus. In the color index map, the small white arrow points to the central dust ring. The arc of dust around this ring indicates that the top half of the galaxy is the near side. The dark spot in the center of the *B–I* map is an artifact of saturation in the *I*-band image, which can happen when the center is very bright. Images source: The de Vaucouleurs Atlas of Galaxies published by Cambridge University Press, 2007; reprinted with permission.

Chapter 16

Waves of Density

The previous chapter showed that the interpretation of ringed spirals in terms of specific orbital resonances with a bar or oval has had some success in explaining the shapes, alignments, and morphologies of galactic rings. However, the majority of spirals do not have rings, or have only a trace of a pseudoring, and the idea of "seeing galactic dynamics" in the form of an orbit is not characteristic of spirals in general. The dominant spiral variety is (s) and the dominant outer ring type is no outer ring or pseudoring. Without the ring features, how can we interpret what we see in a typical spiral galaxy?

The nature of the pure spiral patterns in disk-shaped galaxies, especially those lacking bars, is still elusive. Since the 1960s, several theories have been put forward, but each has limitations and cannot necessarily explain everything we see. One way to examine the problem is to first recognize that there are three basic kinds of spiral patterns, known as *arm classes*[1]:

(1) **Grand design spirals (G)**: The most beautiful spirals, described using a term popularized as early as 1970 by MIT astronomer Chia Chiao Lin (1916–2013).[2] In a grand design spiral, two arms extend as a pattern across the entire disk of a galaxy, that is, the pattern is said to be *global*. The arms are

---

[1] D. M. Elmegreen, *Astrophysical Journal Supplement Series*, **47**, 229 (1981); D. M. Elmegreen *et al.*, *Astrophysical Journal*, **737**, 32 (2011).
[2] C. C. Lin in *The Spiral Structure of our Galaxy*, IAU Symposium No. 38, W. Becker, G. Ioannou, eds. (Dordrecht, Reidel, 1970), p. 377.

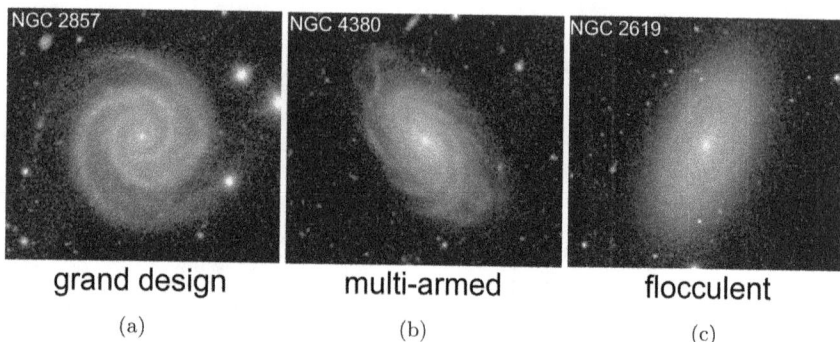

grand design   multi-armed   flocculent

(a)   (b)   (c)

Fig. 1.   The three basic kinds of spiral structure (arm classes) found in disk-shaped galaxies (D. M. Elmegreen *et al.*, *Astrophysical Journal*, 737, **32**, (2011).) Images source: Sloan Digital Sky Survey Collaboration, http://www.sdss.org.

generally long and symmetric as in NGC 2857 (Figure 1(a); see also Figure 2 of Chapter 17), which is an exceptional face-on example. In addition, all of the galaxies shown in Figure 1 of Chapter 15 are examples of grand design spirals.

(2) **Multi-armed spirals (M)**: In some spirals, the pattern is two-armed only in the inner regions. A multi-armed spiral also has a global spiral pattern, but the number of arms is greater than 2 (Figure 1(b)).

(3) **Flocculent spirals (F)**: A flocculent spiral is characterized not by a global spiral pattern, but by a large number of small pieces of ill-defined arms (Figure 1(c)). Some galaxies that show a flocculent spiral pattern in blue light show a grand design spiral pattern in infrared light, owing to the effects of excessive internal extinction on the blue-light image. An excellent example of this is the highly-inclined spiral NGC 253 (Figure 2).

These categories are due to Debra and Bruce Elmegreen, who made extensive studies of spirals, both nearby and very distant.[3] In addition to spiral types, there are other issues:

(1) Spirals are very common, but only among high mass, high luminosity disk galaxies (that is, galaxies comparable in mass

---

[3]D. M. Elmegreen and B. G. Elmegreen, *Monthly Notices of the Royal Astronomical Society*, **201**, 1021 (1982).

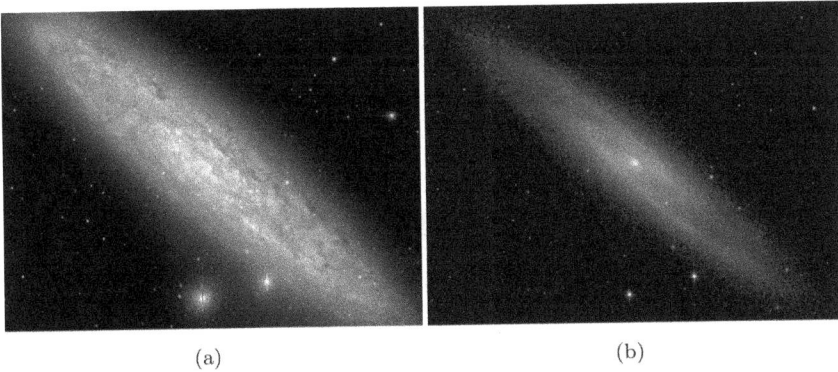

(a)   (b)

Fig. 2.   Comparison of a blue-light (*B*-band) image (a) with a near-infrared *K*-band (2.2 micron) image (b) of NGC 253, a major nearby Sc spiral lying 10.5 million light years away in the constellation Sculptor. The *B*-band image shows a flocculent pattern of scattered dust patches and ill-defined pieces of spiral structure. The *K*-band image shows a stunning two-armed grand design spiral with a highly-foreshortened bar and a trace of an inner pseudoring. Images source: The de Vaucouleurs Atlas of Galaxies published by Cambridge University Press, 2007; reprinted with permission.

and size to the Milky Way). Spirals are very rare among low luminosity galaxies, like dwarfs.

(2) As described in Chapter 8, the arms of many grand design and multi-armed spirals are traced by specific kinds of objects: (a) photoionized clouds (HII regions) and their associated young star clusters and stellar associations and (b) cold atomic hydrogen gas. Microwave emission from carbon monoxide in giant molecular clouds can also trace spiral structure, but may not be exactly coincident with the arms traced by HII regions or atomic hydrogen.

(3) Some spirals are red in color, meaning they are made mostly of old stars. Such spirals may be in barred or nonbarred galaxies.

(4) The shapes of the arms of some grand design spirals are logarithmic, meaning the arms open outward as the natural logarithm of the galactocentric distance. In contrast, some spiral patterns close into pseudorings after winding 180° or 270°, and are therefore not logarithmic.

(5) Most spirals *trail* the direction of galactic rotation. Leading spiral arms do exist, but are extremely rare compared to trailing spirals.

(6) Both stars and interstellar gas and dust play a role in the dynamics and evolution of spiral arms.

16.1. Spirals as Density Waves

We have seen that disk-shaped galaxies do not rotate like solid objects. Instead, the rotation of spirals is such that angular speeds decrease and orbital periods increase with increasing distance from the center. If one were to imagine placing two sets of objects along opposing lines from the galactic center, and launching them at the circular speeds appropriate to their distances from the center, eventually these objects would trace a two-armed spiral pattern. This is shown in Figure 3. The system is assumed to be a disk galaxy where the circular speed everywhere is 200 km/s (that is, the rotation curve is assumed to be flat). Because the circular speed is the same for all the clouds, but the distances from the center are different, the clouds closer to the center get ahead of those farther out, which leads to the apparent spiral. However, these spiral arms, which would be called "material arms" since they would always be made of the same objects, would rapidly evolve. Due to the much shorter rotation period in the inner regions compared to the outer regions, the arms would wind up until they are no longer distinct. For the material arms

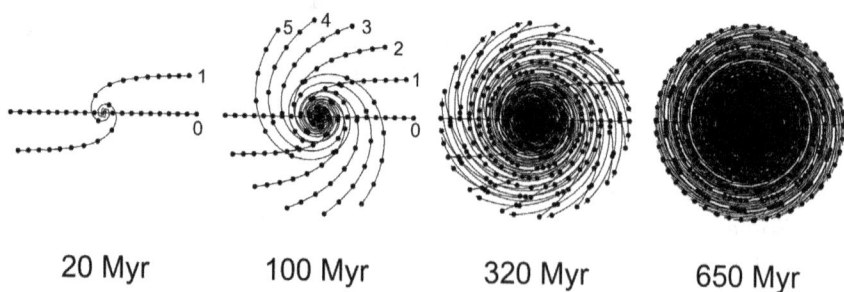

| 20 Myr | 100 Myr | 320 Myr | 650 Myr |

Fig. 3.   Illustration of the evolution of material spiral arms in a disk-shaped galaxy. A set of clouds is launched along a line on either side of the galaxy at the same circular speed, 200 km/s The four frames show the shape of the material spiral as a function of time in multiples of 20 million years (Myr). The pattern is still distinct after 100 Myr, but by 320 Myr, is so tightly wound that it is no longer recognizable. The numbered patterns are for 0 Myr (0); 20 Myr (1), 40 Myr (2), 60 Myr (3), 80 Myr (4), and 100 Myr (5).

in Figure 3, the wind up happens in only a few hundred million years, meaning that, if the arms of a grand design spiral are in fact material arms, then spirals should be very short-lived and as a consequence much rarer than they actually are.

The Swedish astronomer Bertil Lindblad (1895–1965) proposed that grand design spirals are not defined by material arms but are actually *density waves*, spiral-shaped perturbations that propogate through a disk of stars and gas clouds as a uniformly rotating *pattern*. A galaxy density wave is analogous to a slow-moving traffic jam where cars initially traveling at normal speed on a highway are forced to slow down due to a perturbation (like a blocked lane), then for a while move at a reduced speed along the road (as cars in the blocked lane try to merge with the other lanes), and then, after passing the perturbation (when, for example, the blocked lane re-opens), resume a normal speed. The car density wave is not a "material object" because it is not always made of the same cars.

In the case of a spiral galaxy, the perturbation is a spiral gravitational field. Stars and gas clouds enter the perturbation region and feel the extra gravitational pull from the pattern. A temporary slowdown occurs where an object spends extra time in the perturbed region. Eventually, stars emerge from the congested area to resume a normal speed. Gas cloud collisions in the congested region can lead to star formation in the spiral arms.

This is all best seen considering simple orbits in a galaxy with no non-circular structures. In this circumstance, orbits are just a combination of a circular angular speed of rotation and a small epicyclic motion that represents the first-order deviation from pure circular rotation (the radial excursion described in Chapter 15 and shown in Figure 12 of Chapter 15). In any disk-shaped system, and at any particular location, one can find a frame of reference where a star or gas cloud executes exactly two radial oscillations for every orbit in that frame, whose angular speed of rotation is called the "pattern speed." This means that if you were turning at the rate of the pattern speed, a star at such a location would follow a closed, slightly oval orbit whose long axis would remain fixed in relative direction. If we next imagine that the same pattern speed applies

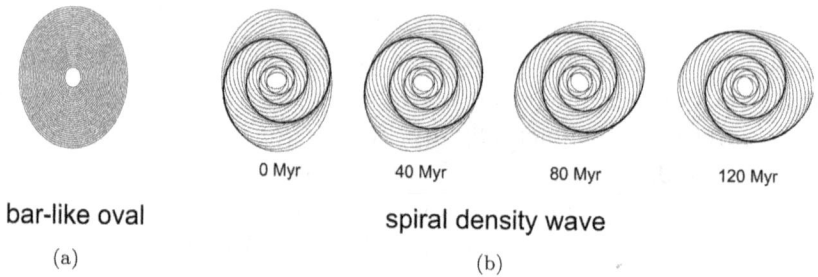

bar-like oval   spiral density wave

(a)   (b)

Fig. 4. Illustration of bar-like (a) and spiral (b) density waves in a galaxy. The spiral frames show a similar bar-like oval with a strong spiral pattern rotating with a period of 480 Myr. As the spiral rotates, it maintains its shape and does not wind up like a set of material arms would.

over a wide range of distances from the galactic center, and that the long axes of the apparently closed orbits are aligned exactly, then we would get a bar-like oval in the mass distribution like that shown in Figure 4(a). If instead we systematically misalign the orbits, we can turn the oval into a spiral pattern as shown in the other frames in Figure 4. Provided that the systematic misalignment of the orbits persists, the pattern would essentially last indefinitely. An observer rotating with the pattern speed would see a fixed spiral pattern, while an observer in our reference frame would see a rotating spiral pattern as shown by the sequence of four frames in Figure 4(b). All of this would happen even though the rotation of the galaxy is not solid-body. In the inner parts of a galaxy, the pattern will rotate more slowly than the stars.

The key issue in spiral density wave theory is finding a way to get the orbits to maintain their systematic misalignments so that a long-lived spiral pattern is achieved. Astronomer C. C. Lin and his student, Frank Shu, in 1964 proposed that grand design spirals are quasi-stationary (almost steady) patterns that are maintained by their own self-gravity. That is, to achieve such a pattern, one had to add to the gravitational field of the galaxy not only the gravity of the background circularly symmetric disk of stars and gas clouds, but also a spiral-shaped perturbation contributing less than 10% of the strength of the background gravity.[4]

---

[4]F. H. Shu, *Annual Reviews of Astronomy and Astrophysics*, **54**, 667 (2016).

In a 1985 paper, University of Athens professor George Contopoulos states that "the most important role in the dynamics of spiral and barred galaxies is played by the resonances."[5] Because the spiral constitutes a uniformly rotating pattern in a disk where the angular speed of rotation decreases with increasing distance from the center, orbital resonances will occur wherever stars and gas clouds move in step with the perturbation. Just as in the case of a barred galaxy, there will be a corotation resonance, where the angular speed of stars and gas clouds equals the pattern speed of the spiral; inner and outer Lindblad resonances (two radial oscillations per orbit in the rotating frame); and inner and outer 4:1 resonances (4 radial oscillations per orbit in the rotating frame). Unlike material spiral arms, the stars involved in a density wave pattern are not always the same. Inside corotation, stars and gas clouds first encounter the inner (concave) side of the arms. While in an arm, these objects slow down a little and, temporarily at least, stay along the arm, moving inward. Eventually, these objects leave the arm and move into the interarm region where they stay until they encounter the other arm. Outside corotation, everything happens in reverse. The spiral pattern moves faster than the stars and gas clouds, which first encounter the arms on their convex (or outer) sides.

Although individual stars can pass through the density wave without colliding with any other star, the same is not necessarily true of interstellar clouds of dust and gas. These are much bigger than stars and can suffer a shock (or sudden increase in density) as they move into the arm region and collide with other clouds. Possible evidence for this shock is how in grand design spirals, strong dust lanes are often seen on the concave (inner side) of the arms (see the color index map of M51 in Figure 3(b) of Chapter 11 and also Figure 5 of Chapter 12). The theory predicts a time sequence across a density wave arm: first, gas compression on the concave side of the arm as clouds move into the region and feel a shock from encountering ("rear-ending"?) other clouds already entering or in the arm; second, gravitational collapse of compressed clouds, followed by formation of

---

[5]G. Contopoulos, *Comments on Astrophysics*, **11**, 1 (1985).

clusters of young stars, including stars massive enough to photoionize HII regions. These will last on the order of 10 million years and will likely evolve before fully leaving the arm area; finally lower mass stars leave the arms and populate the inter-arm regions, the more massive stars not surviving long enough to reach these points. We would see a gradient across the arms from blue to reddish inside corotation and from reddish to blue outside corotation.[6]

On the whole, observations of grand design spirals agree with these predictions, but there is a caveat. If spiral structure actually enhances star formation, then we would expect lower star formation rates per unit area in galaxies that lack spiral arms. In fact, star formation rates per unit area are independent of arm class. Density wave spirals (arm classes G and M) do not have enhanced star formation rates per unit area compared to flocculent (F) spirals.[7] A study of the arm classes of supernova hosts and galaxies of different chemical abundances led to a similar conclusion.[8] All this appears to imply that density waves are not required to trigger star formation, but if they are not, why are spiral arms so often traced by star formation? It has been argued that density waves do not induce star formation, but instead through their orbit-crowding act to collect star-forming regions into the pattern.

Spiral density wave theory has natural limits for the extent of spiral patterns. The arms of a strong two-armed spiral should not extend either far beyond the OLR or within the ILR. That is, the pattern is generally restricted between these two resonances. The restriction at the ILR occurs because random stellar motions are increased at the ILR due to the resonant interaction between the wave and the stars. However, in the absence of an ILR, the density wave can propogate through the center and not be absorbed in that area.

A spiral pattern can also be characterized by its *pitch angle*. The pitch angle is a measure of how open the spiral arms are (Figure 5).

---

[6]F. Shu, *Annual Reviews of Astronomy and Astrophysics*, **54**, 667 (2016).

[7]B. G. Elmegreen and D. M. Elmegreen, *Astrophysical Journal*, **311**, 554 (1986).

[8]M. L. McCall, *Publications of the Astronomical Society of the Pacific*, **98**, 992 (1986).

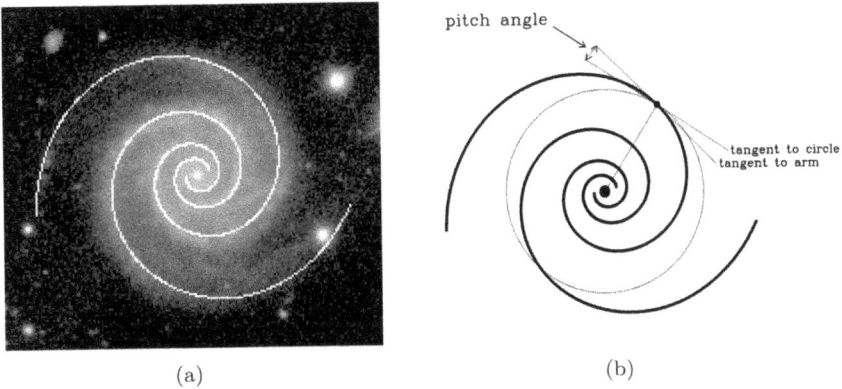

(a)   (b)

Fig. 5. (a) Blue-green (*g*-band) image of NGC 2857, a face-on spiral with a strong two-armed pattern. The white curves show a logarithmic spiral that fits the observed pattern well. (b) A schematic showing how the pitch angle of a spiral is defined. At any given point, the pitch angle is the angle between the tangent to a circle at that point and the tangent to the arm at the same point. The arms of NGC 2857 are well-represented by a logarithmic spiral having a pitch angle of 12°.

If the arms are tightly-wrapped, we say the pitch angle is low (close to 0 degrees), but if the arms are very open, we say the pitch angle is high. The most open arms have a pitch angle close to 30°.[9] Pitch angle is actually one of the factors that Hubble used for typing spirals: the more tightly wrapped the arms, the less well-resolved the arms would appear to be and a galaxy would be classified as type Sa or SBa. Going from Sa to Sc (or SBa to SBc), the arm pitch angle increases and the degree of knottiness of the arms also increases, on average.

The mathematical forms of arms, that is, how they open outward as a function of galactocentric distance, is complicated by a number of factors, not the least of which is the existence of resonances. As shown in Figure 5, some spirals do open outward in a logarithmic manner. An ideal logarithmic spiral would be characterized by a single value of the pitch angle (as for NGC 2857), but not all spirals can be so characterized. In barred galaxies, as shown in Chapter 15, the outer spiral arms can open outwards and then close inwards, forming a

---

[9]R. C. Kennicutt, *Astronomical Journal*, **86**, 1847 (1981).

pseudoring. In these cases, the pitch angle varies around the pattern and the arms are distinctly non-logarithmic.

16.2. Spirals and the Inner 4:1 Resonance

George Contopoulos examined orbits in a model galaxy having a spiral density wave, and concluded that strong spirals can extend no further than the inner 4:1 resonance (I4R) with the pattern speed of the spiral.[10] This is the same resonance that, in Chapter 15, was identified with *inner rings* in barred galaxies. Can we "see" this resonance in a pure spiral like we can see an I4R ring? The answer is perhaps. Contopoulos presented NGC 5247 (shown in Figure 13 in Chapter 18) as an example of a real spiral likely limited in extent by the I4R. The issue is that the orbits in a galaxy must support the pattern in order to keep it maintained. In Contopoulos's model, this did not happen past the I4R.

Figure 6 shows another galaxy where the I4R might be identifiable. NGC 3433 is a grand design spiral lying 117 million light years away in Leo. The image, taken in blue-green light (the *g*-band) has been cleaned of foreground and background objects and computer-deprojected to approximate a face-on view. The morphology of NGC 3433 shows a strong, inner two-armed spiral, beyond which the structure breaks up into a weaker multi-armed pattern. This is a fairly typical spiral morphology and other examples are included in Chapters 17 and 18. In the case of NGC 3433, we can see the ends of the inner, bright spiral; it is these ends which can be interpreted as extending to the I4R (white circle). Outside this circle, there are at least six fainter spiral arms.

In contrast to NGC 3433, Figure 7 shows the computer-deprojected *g*-band image of the remarkable multi-armed barred spiral NGC 5375. The image has been rotated so that the bar is oriented horizontally. This one is interesting because of the pure oval shape of its bar, the spot-like ansae at the ends of this bar, and the presence of the two dark regions (arrows) close to the minor axis

---

[10]G. Contopoulos, *Comments on Astrophysics*, **11**, 1 (1985).

Fig. 6. Blue-green light image of NGC 3433. The image has been cleaned of contaminating objects and computer-deprojected. The circle shows the approximate theoretical location of the inner 4:1 resonance. Sloan Digital Sky Survey Collaboration, http://www.sdss.org.

line of the bar. Although such regions are common in regular two-armed or four-armed barred spirals, as in the two cases described in Figure 31 of Chapter 15, it is unusual to see them in such a complex multi-armed spiral. As in ESO 566-24 and NGC 1433, these dark regions could trace the location of the L_4 and L_5 Lagrangian points in the galaxy's gravitational field. However, the zones are significantly rotated with respect to the bar minor axis, which is likely due to the significant influence of the spiral.[11] The galaxy also does not show a strong, inner two-armed spiral pattern. As in Figure 6, the white

---

[11]E. Athanassoula *et al.*, *Monthly Notices of the Royal Astronomical Society*, **400**, 1706 (2009).

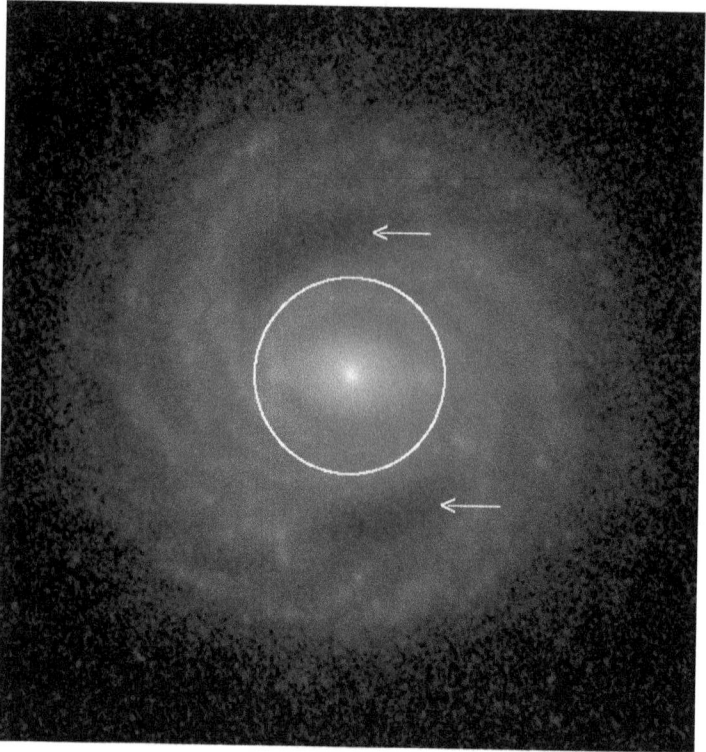

Fig. 7.   Blue-green light image of NGC 5375, a multi-armed spiral 109 million light years away in Canes Venatici. The image has been cleaned of contaminating objects, computer-deprojected, and re-oriented to make the bar horizontal. The arrows point to two peculiar dark regions that could be linked to the locations of the L_4 and L_5 Lagrangian points in the bar gravitational field. Sloan Digital Sky Survey Collaboration, http://www.sdss.org.

circle in Figure 7 is the theoretical location of the I4R; this is based on an analysis similar to that which led to the resonance diagrams for ESO 566-24 and NGC 1433 shown in Figure 31 of Chapter 15. In this case, the I4R encompasses an inner pseudoring as well as the bar.

16.3. What Causes Multi-armed and Flocculent Spirals?

Figure 31 in Chapter 15 showed that higher-order resonances than 2:1 could account for the regular multi-armed patterns seen in galaxies

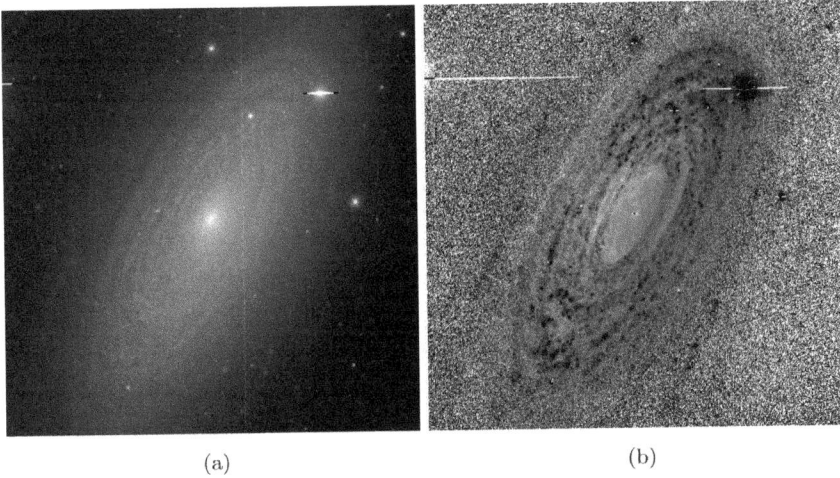

(a)   (b)

Fig. 8. (a) *B*-image and (b) *B–V* color index map of the flocculent spiral NGC 2841 in Ursa Major. Images source: The de Vaucouleurs Atlas of Galaxies published by Cambridge University Press, 2007; reprinted with permission.

like ESO 566-24. Unlike flocculent spirals, these could be density wave patterns because of their regularity and global symmetry. NGC 2841, shown with *B* and *B–V* images in Figure 8, is a nearby example of a flocculent spiral. Unlike NGC 253, its structure is not really hidden much by dust. It has clear spiral structure, but the arms are in the form of short pieces and there is no real global pattern. In fact, much of the global structure is ring-like. Thus, it is unlikely that the arms of NGC 2841 are related to density waves.

In 1978, Humberto Gerola and Philip Seiden (1934–2001), both of the IBM T. J. Watson Research Center in New York State at the time, proposed an alternative view to the density wave theory. Noting the close connection between star-forming regions and spiral arms, they proposed that spiral structure is made of individually short-lived material features that extend the lifetime of the spiral though random, self-propogating star formation.[12] The idea is that randomly created star-forming regions in a galactic disk can induce, through shock waves from supernova explosions, the creation of more massive

---

[12]H. Gerola and P. E. Seiden, *Astrophysical Journal*, **223**, 129 (1978).

stars and star-forming regions. As star formation propogates over a larger area, the non-solid body rotation of the galactic disk stretches the regions into material spirals that continually change over time. The result is a spiral pattern that, while it changes, is long-lived.

Gerola and Seiden presented models of two cosmic pinwheels, one being the famous NGC 5457 (M101) shown in Figure 9. The other is M81, shown in Figure 10. In both cases we see the prediction of a multi-armed spiral pattern that persists even though the features are constantly changing. Were it not for the self-propogating star formation aspect of the models, the material spiral features would wind up and quickly disappear. The models provide a reasonable interpretation of both multi-armed and some flocculent spirals. Many grand design spirals also have short "spurs" in interarm regions that conceivably could be related to the Gerola and Seiden process.

16.4. What Drives Spiral Density Waves?

In the random, self-propogating star formation model of spiral structure, the spiral patterns do not need a special "driving mechanism," that is, something extra to perturb the disk stars and gas clouds and make them create the spiral gravitational field that underlies the pattern. In contrast, while density wave theory can explain important aspects of observed spiral structures, it does not necessarily have built into it a mechanism for generating the spiral perturbation in the first place. The search for a natural driving mechanism for density waves became a major topic of research after the theory was proposed in 1964.

It was not until the advent of infrared imaging that it became possible to see the stellar spiral that underlies the spiral defined by massive stars and HII regions.[13] This is highlighted by the 2.2 micron (K-band)[14] and 3.6 micron (L-band)[15] images of M51 shown in Figure 11. The arms in the K-band image are smooth and

---

[13]D. L. Block *et al.*, *Astronomy and Astrophysics*, **288**, 365 (1994).

[14]E. Laurikainen *et al.*, *Monthly Notices of the Royal Astronomical Society*, **418**, 1452 (2011).

[15]R. Buta *et al.*, *Astrophysical Journal Supplement Series*, **217**, 32 (2015).

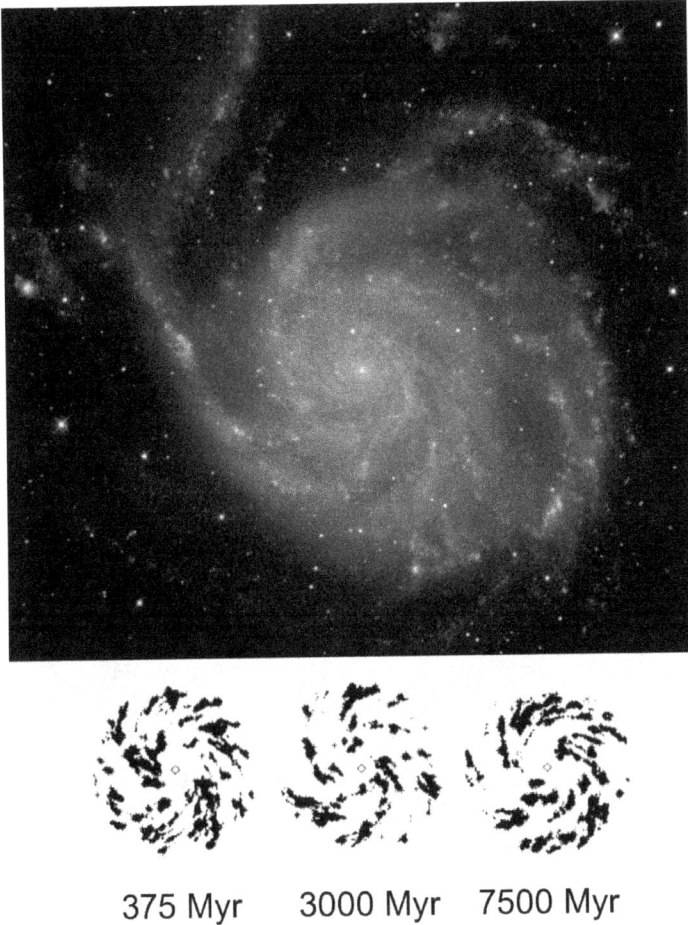

375 Myr   3000 Myr   7500 Myr

Fig. 9. A blue-light image of M101 (NGC 5457), one of the nearest and brightest cosmic pinwheels. Below the image is an evolving model of random, self-propogating star formation from Gerola and Seiden (1978) that reproduces the rather open, multi-armed spiral structure of M101 well. The times are in millions of years. From H. Gerola and P. E. Seiden, *Astrophysical Journal*, **223**, 129 (1978); © AAS; reproduced with permission. Image source: The de Vaucouleurs Atlas of Galaxies published by Cambridge University Press, 2007; reprinted with permission.

relatively symmetric, mainly because the effects of extinction by dust and the influence of star-forming regions are nearly at a minimum in that band. The 3.6 micron image appears to be more affected by star-forming regions than is the 2.2 micron image. The reason for

375 Myr   3000 Myr   7500 Myr

Fig. 10.   A blue-light image of M81 (NGC 3031), another one of the nearest and brightest cosmic pinwheels. Below the image is an evolving model of random, self-propogating star formation from Gerola and Seiden (1978) that reproduces the more tightly wrapped, more global spiral structure of M81 well. The times are in millions of years. From H. Gerola and P. E. Seiden, *Astrophysical Journal*, **223**, 129 (1978); © AAS; reproduced with permission. Image source: The de Vaucouleurs Atlas of Galaxies published by Cambridge University Press, 2007; reprinted with permission.

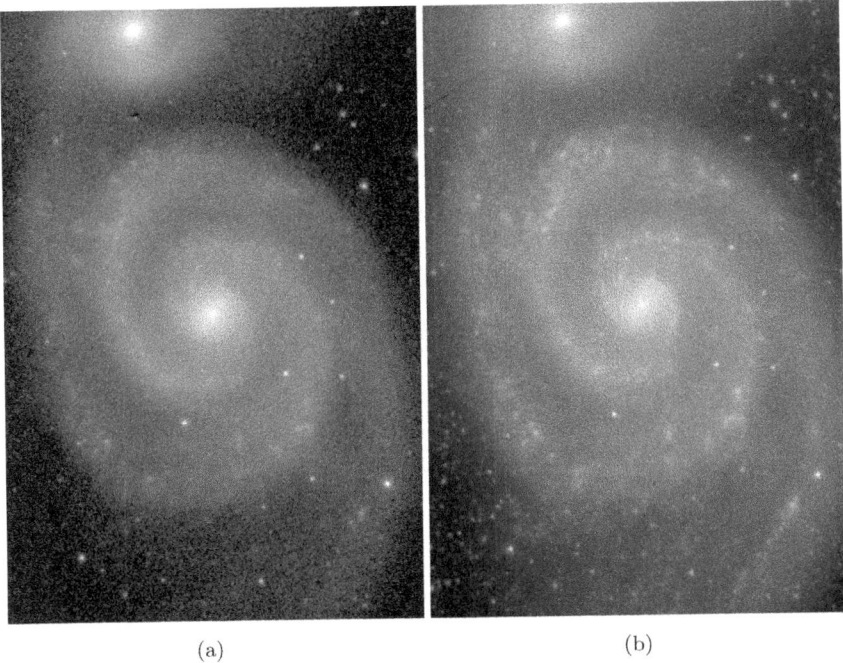

(a)   (b)

Fig. 11.   Long wavelength images of M51 (NGC 5194): (a) 2.2 microns (*K*-band, near-infrared). From E. Laurikainen *et al.*, *Monthly Notices of the Royal Astronomical Society*, **418**, 1452 (2011); Copyright © 2011, Royal Astronomical Society; (b) 3.6 microns (*L*-band, middle infrared). From R. Buta *et al.*, *Astrophysical Journal Supplement Series*, **217**, 32 (2015); © AAS; Reproduced with permission. These reveal the old stellar component underlying the bright, star-forming spiral arms seen in blue light. Vestiges of the star-forming component are nevertheless still seen in both images, including red supergiants.

this has to do with the nature of star-forming regions. These always involve a mix of interstellar dust and gas. In blue light, the hot, massive stars in these regions dominate the appearance of the arms (Figure 12(b)). These stars emit most of their light in the ultraviolet part of the electromagnetic spectrum (Figure 12(a)) and very little of their light in the infrared part of the spectrum. As a consequence, we do not see the hot, blue stars in the images in Figure 11. We are instead seeing the thermal emission from dust in the regions heated by the ultraviolet light of the hot, blue stars. In blue light, dust acts to obscure the light of stars behind it, but in

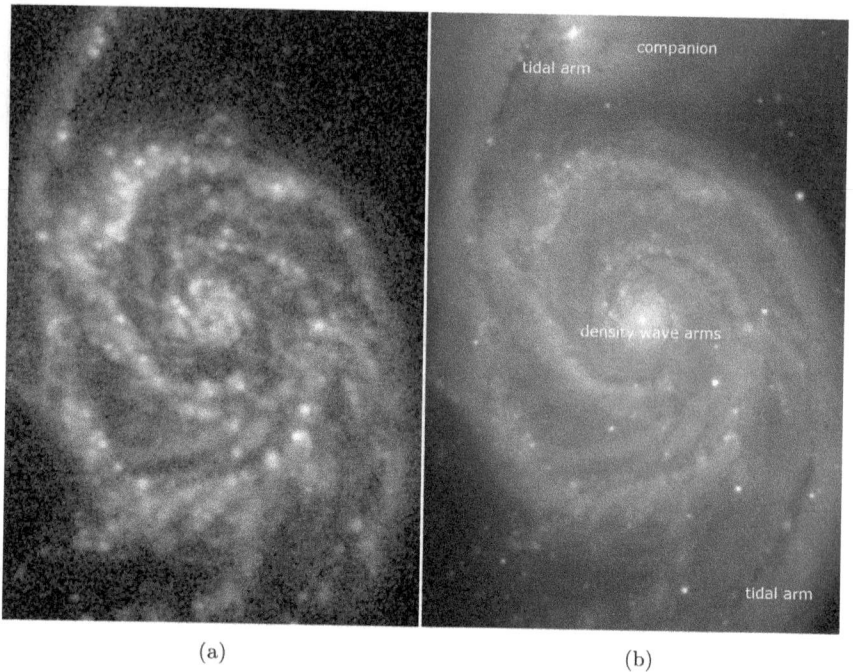

(a)   (b)

Fig. 12. Short wavelength images of M51 (NGC 5194): (a) 150 nm *Galaxy Evolution Explorer* (GALEX) image (far ultraviolet), from D. C. Martin *et al.*, *Astrophysical Journal*, **619**, L1, (2005); © AAS. Reproduced with permission. These reveal the young stellar component of the spiral arms and extinction by dust; (b) 440 nm (*B*-band, optical), from The de Vaucouleurs Atlas of Galaxies published by Cambridge University Press, 2007; reprinted with permission. The labeling indicates that the outer arms are likely tidal in origin, while the bright inner arms are likely density wave arms.

the infrared, dust glows. This heated dust becomes more prominent with increasing wavelength, until by 8.0 microns, the brightly shining dust overwhelms the light of stars so that all one sees mostly is dust.

In 1979, John Kormendy (then at the Institute of Astronomy, Cambridge, England) and Colin Norman (then at the Huygens Laboratory, The Netherlands) concluded that two major mechanisms might drive spiral density waves.[16] First, the fact that M51 has a prominent companion suggests that at least some spirals are triggered

---

[16]J. Kormendy and C. A. Norman, *Astrophysical Journal*, **233**, 539 (1979).

by a gravitational interaction. NGC 5195 is not as massive as the spiral NGC 5194 and seems to lie a little behind the spiral (as implied by the heavy foreground extinction seen in Figure 3 of Chapter 11). NGC 5195 has no star formation of its own, and as a consequence it nearly disappears in the far ultraviolet[17] image in Figure 12(a).[18]

There are many systems like M51, which suggests that an interaction is at the heart of some spirals. Spiral arms that result solely from a gravitational interaction are known as tidal arms. Such arms are caused by the great difference in the force that the side of a galaxy closest to a companion feels compared to the side farthest from the companion. Tidal forces are common among galaxies because galaxies are large compared to their typical separations. Tidal arms are material arms, but as long as a perturbation lasts, these arms can persist for a long time.

How tidal arms develop in disk-shaped galaxies was first shown by Swedish astronomer Erik Holmberg (1908–2000) in 1941.[19] At that time, computers capable of doing numerical simulations of galaxies did not exist, but Holmberg nevertheless was able to do something like simulations with an elaborate mechanical device that used light bulbs, photocells, galvanometers, and the ability to trace out orbits and rotations to simulate mass and gravity. Later, MIT astronomer Alar Toomre and his brother Juri Toomre (then at New York University) used simple but nevertheless very enlightening computer simulations to show that the outer arms of M51 are likely tidal in origin and that the inner arms could be density wave arms, possibly connected to a pre-existing pattern.[20] The inner part of M51 they were referring to is shown in Figure 5 of Chapter 12.

In 1992, Gene Byrd and Sethanne Howard used a more advanced computer code to show that a significant grand design spiral pattern can be generated in the disk of a nonbarred galaxy having a flat

---

[17]The far ultraviolet refers to the short wavelength part of the ultraviolet domain of the electromagnetic spectrum.
[18]D. C. Martin *et al.*, *Astrophysical Journal*, **619**, L1, (2005).
[19]E. Holmberg, *Astrophysical Journal*, **94**, 385 (1941).
[20]A. Toomre and J. Toomre, *Astrophysical Journal*, **178**, 623 (1972).

rotation curve by a companion having a mass of as little as 1% the mass of the disk.[21] The fact that even a very small companion can trigger a grand design spiral in a nonbarred galaxy suggests that some of the global spiral patterns seen in the isolated galaxies shown in Figure 8 of Chapter 9 could still be tidal in origin due to unrecognized small companions.

To account for the full spiral pattern in M51, Byrd and Howard found that the mass and gravity of strong, outer tidal arms could excite a spiral density wave in the inner regions that would blend together to give a continuous pattern. M51's outer tidal arms and inner density wave arms are indicated by the labeling in Figure 12(b). In this interpretation, the density wave arms represent a long-term evolutionary state that started with the development of the tidal material arms. Such a combination is likely to exist in many typical nonbarred spirals, another likely example being UGC 10330 shown in Figure 9 of Chapter 17.

The most sophisticated study of the spiral structure of M51 was made by Finnish astronomers Heikki Salo and Eija Laurikainen in a 2000 study.[22] These authors showed that the tidal arms generated by the companion in the outer regions travel like waves inward, where they interplay with a bright intrinsic central spiral (meaning a spiral that would have been there independent of the companion's influence). The interplay would lead to a continuity between the inner and outer spiral structure.

The second possible driving mechanism for spiral density waves identified by Kormendy and Norman is the presence of a bar. Barred galaxies tend to have well-defined global spiral patterns; only in very rare cases is the spiral pattern in a barred galaxy not global in nature. An example is IC 5240 shown in Figure 13.

Related to bars in spiral galaxies are features known as *ovals*.[23] These are massive structures that are bar-like in the way they can

---

[21] G. Byrd and S. Howard, *Astronomical Journal*, **103**, 1089 (1992).

[22] H. Salo and E. Laurikainen, *Monthly Notices of the Royal Astronomical Society*, **319**, 393 (2000).

[23] J. Kormendy and C. A. Norman, *Astrophysical Journal*, **233**, 539 (1979).

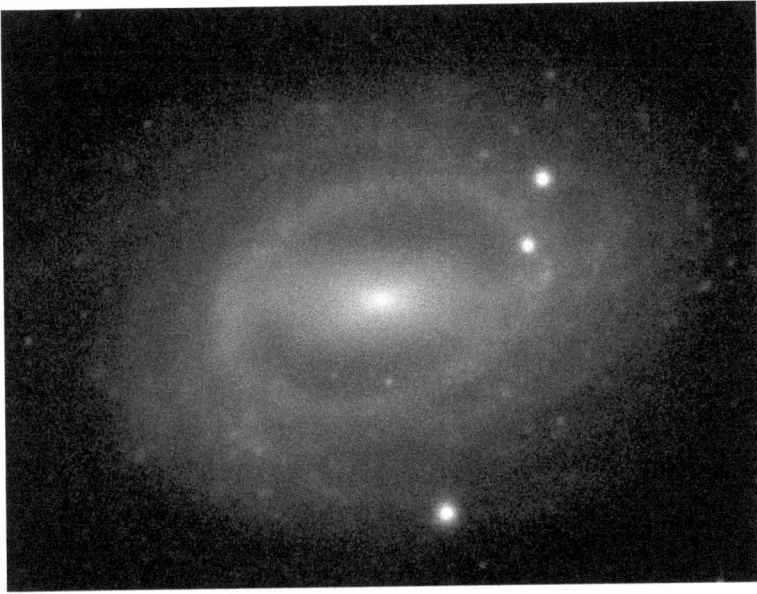

Fig. 13.   Blue light (*B*-band) image of the "Steering Wheel Galaxy" IC 5240, a ringed spiral lying 78 million light years away in Grus. The interesting features of the galaxy include a strong bar with an inner boxy zone, a bright inner ring, and a non-global outer spiral pattern, which is unusual for a barred galaxy. Image source: The de Vaucouleurs Atlas of Galaxies published by Cambridge University Press, 2007; reprinted with permission.

influence motions in a galactic disk, but are harder to recognize than bars. Examples of ovals are the intermediate zone rimmed by the inner pseudoring in ESO 565-11 (inset in Figure 20(a) of Chapter 15), and the bright, dust-lane-riddled region between the two rings of M94 (Figure 8 of Chapter 15). Ovals may not be as elongated as conventional bars, but they can be more massive than bars and therefore are still important to the evolution of disk-shaped galaxies.[24]

In the absence of either a bar or a companion, Kormendy and Norman found that a spiral density wave could exist for a long time, but only if the disk material rotated like a solid object in the area

[24]J. Kormendy, in *Secular Evolution of Galaxies*, Jesús Falcón-Barroso and J. H. Knapen, eds. (Cambridge, Cambridge University Press, 2013), p. 1.

where the pattern existed. If the disk of a non-barred galaxy did not rotate like a solid object, the galaxy would look like NGC 2841 (Figure 8) with its piecemeal (flocculent) spiral structure.

The Kormendy and Norman mechanisms probably can account for a significant fraction of spirals, but an additional mechanism that could be relevant is spontaneous development of the pattern as a *mode* in a galactic disk. Such a feature would not necessarily depend on an interaction or a bar to become manifested in the morphology of a disk galaxy. Like a natural, standing wave oscillation of a string, a galactic mode is a particular disk morphology (barred, or spiral) that self-organizes within the limits imposed by the characteristics of a particular stellar disk. These characteristics include the rotation curve, the luminosity distribution, and the degree of random motions, and how these vary with distance from the galactic center. This interpretation of bars and spirals has been advocated by George Mason University astrophysicist Xiaolei Zhang, who in her 1992 PhD thesis studied the impact of such modes on the evolution of a galactic disk.[25]

16.5. Backwards Galaxy or Backwards Astronomers?

The galaxy shown in Figure 14 is NGC 4622, a beautiful face-on spiral located about 140 million light years away in the constellation Centaurus. The galaxy is not isolated, but is a member of the Centaurus Cluster and has several nearby major companions.

In 1988, my colleague, Gene Byrd, noticed something about this galaxy that had been overlooked by even the most stalwart of spiral structure theorists: in addition to the two strong, outer spiral arms, there is a single, inner spiral arm that winds outward in the opposite sense to those arms. That is, the galaxy has *counter-winding spiral structure*. If the three spiral arms are in the same plane, then the galaxy has at least one *leading spiral arm*, the first time such a feature has been identified unambiguously in any galaxy. The irony of this

---

[25]X. Zhang, *Dynamical Evolution of Galaxies* (de Gruyter, 2017); see also G. Bertin *et al.*, *Astrophysical Journal*, **338**, 78 (1989).

Fig. 14.   *Hubble Space Telescope* color image of the spiral galaxy NGC 4622 in Centaurus. Image credit: NASA and the Hubble Heritage Team.

discovery is that Gene noticed the extra arm in a textbook written by Frank Shu, one of the authors of the famous density wave theory of spiral structure that had been proposed with C. C. Lin in 1964. Shu used a publicly-available CTIO 4-m prime focus photo of NGC 4622 to show what he considered to be a typical spiral.

In 1989, at the University of Alabama, Gene showed me the picture of NGC 4622 in Shu's book and an article he and other colleagues had written that suggested that the inner arm had the leading sense and may have been the result of a gravitational encounter where a small companion galaxy flew by in a sense opposite to the rotation of NGC 4622 (that is, a "retrograde" encounter). What intrigued me about all this is how the leading and trailing arms in NGC 4622 occupy different radial zones and blend together into a nearly circular, off-centered ring. Most rings are related to bars, but NGC 4622 appeared to be a clear case of a non-barred, ringed spiral galaxy. This made the nature of the ring uncertain and the galaxy even more mysterious. Gene and I decided to apply for telescope time at CTIO to get multi-wavelength digital images of NGC 4622 and Fabry–Perot interferometry. Our goal was to determine which of the three spiral arms of NGC 4622 are actually leading.

Figure 15(a) shows an infrared (I-band) image of NGC 4622 that I use to separate the inner and outer arms more clearly. This is done using Fourier series analysis, where the light distribution of a galaxy is decomposed into sinusoidal components of different multiplicity m. Once the strength of each m component is determined, an image of the component can be made. This is shown for $m = 0$, 1, and 2 in Figures 15(b)–15(d). The $m = 1$ image shows the inner single arm winding outwards in a counter-clockwise sense, while the $m = 2$ image shows the two outer arms winding outward in the opposite sense. The $m = 0$ component only shows the inner ring. These low-order terms are the most significant components of the galaxy's structure.

To determine which set of arms leads, we need two pieces of information: which half of the galaxy major axis is receding from us, and which half of the galaxy minor axis is the near side. Figure 16 shows how these were determined. First, the color-coded Fabry–Perot velocity field shows that the north half of the galaxy is receding from us. The spectral emission lines on that side of the galaxy are redshifted to slightly longer wavelengths relative to the center of the galaxy, and this is why those points are color-coded red. The south side is color-coded blue/violet because the spectral emission lines at those points are slightly blue-shifted relative to the center, and hence those points are approaching us. Analysis of the velocity field tells us that the line of nodes is at the angle shown in Figure 16(a). This is what we need to determine the near side. The time-honored way of telling this is due to Vesto Slipher (Chapter 5) and uses

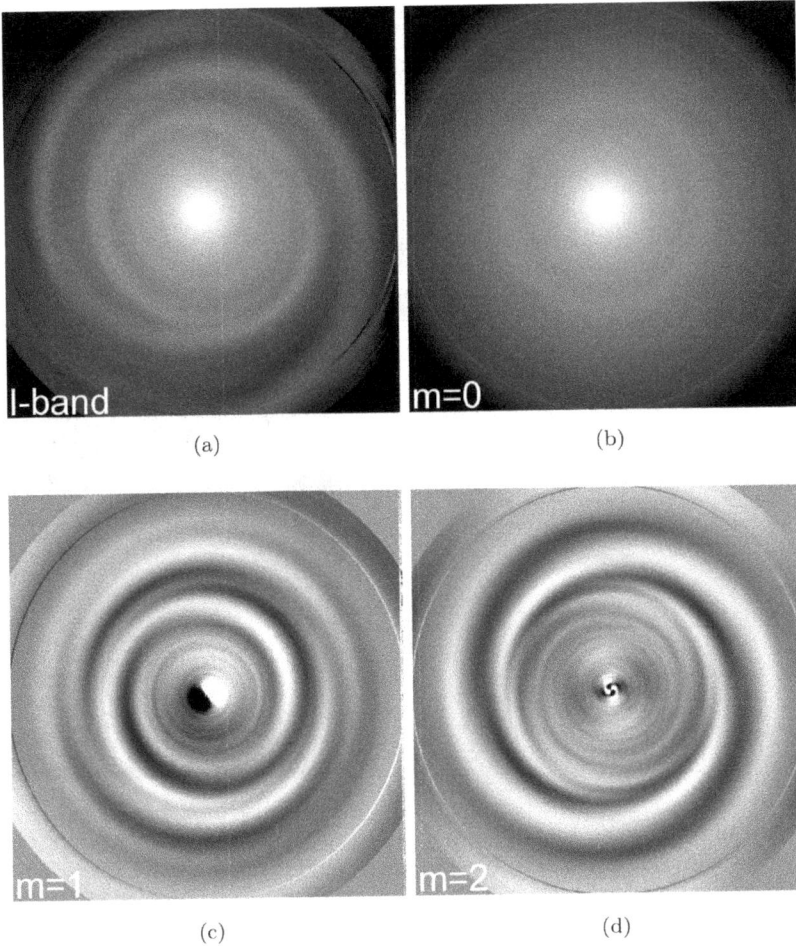

Fig. 15. An *I*-band image of NGC 4622 is separated into its *m* components, with the three lowest order terms shown. From R. Buta *et al.*, *Astronomical Journal*, **125**, 634 (2003). © AAS. Reproduced with permission.

the asymmetry in the apparent dust distribution often seen in tilted spirals. The dust is mostly confined to the disk plane, but in a galaxy with a significant bulge, the bulge light is viewed through the dust layer on the near side, while the dust layer is viewed through the bulge on the far side (Figure 17). This causes a reddening and extinction asymmetry across the galaxy's minor axis. For a sufficiently inclined, not too distant galaxy, this asymmetry can usually be detected with

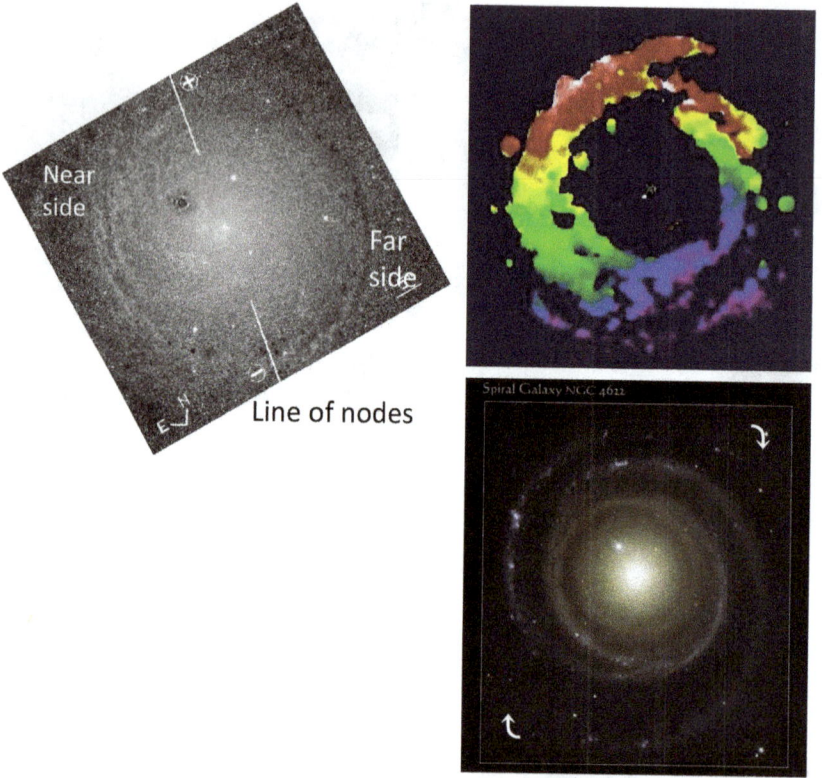

Fig. 16.   Images used to determine the sense (clockwise or counterclockwise) of rotation of NGC 4622. From R. Buta, in *Secular Evolution of Galaxies*, J. Falc-n-Baroso and J. Knapen, eds. (Cambridge, Cambridge University Press, 2013), p. 155. Reproduced with permission.

ground-based multi-color images. However, this was not possible for NGC 4622 due to its almost face-on orientation. We needed higher resolution to see this asymmetry in NGC 4622, and our best bet was multi-color imaging with the *Hubble Space Telescope*.

In May 2001, we obtained *BVI* images of NGC 4622 with the HST, and from these were able to deduce that the east side of the galaxy was the near side. *This meant that the galaxy is rotating clockwise, and that the two outer arms have the leading sense*, which led to NGC 4622 being christened as the "Backwards Galaxy." This result was so unexpected that when brought to the attention of a

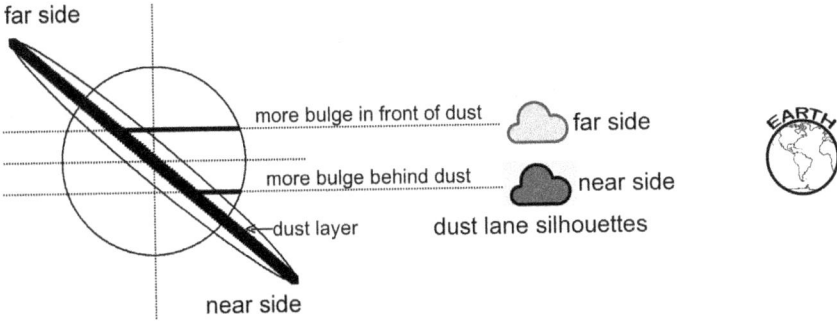

Fig. 17. A schematic showing how the reddening-extinction asymmetry across the minor axis of a tilted spiral galaxy comes about. On the near side, the bulge is seen through the dust layer, while on the far side the dust layer is seen through the bulge. Thus, more light is extinguished on the near side than on the far side.

well-known spiral structure theorist, he skeptically declared that it was the astronomers, not the galaxy, that were backwards!

The main argument against the two outer arms being leading is the "swing amplification" theory of Alar Toomre. In this theory,[26] a two-armed leading spiral density wave spontaneously develops and propogates outward in the disk of a galaxy. As these arms open outward in the direction of rotation, they are swung by the non-solid-body rotation into trailing arms that get significantly amplified in strength and which last a much longer period of time than did the leading spiral arms. The theory predicts that density wave spirals should in general be trailing, especially the strongest apparent spirals. It thus would seem unlikely that the outer arms of NGC 4622 are leading, in spite of what the dust asymmetry method implies.

Even if the dust asymmetry method is misleading in NGC 4622, as Alar Toomre believes,[27] the galaxy still has leading spiral structure. Furthermore, the peculiar combination of a single inner spiral arm and multiple outer arms is not unique and could be significant. Another example is ESO 297-27, located 280 million light years away in the constellation Phoenix. ESO 297-27 is a tilted,

---

[26] A. Toomre, in *The Structure and Evolution of Normal Galaxies*, S. M. Fall, ed. (Cambridge, Cambridge University Press, 1981), p. 111.
[27] J. Lucentini, *Sky and Telescope*, **September**, p. 37 (2002).

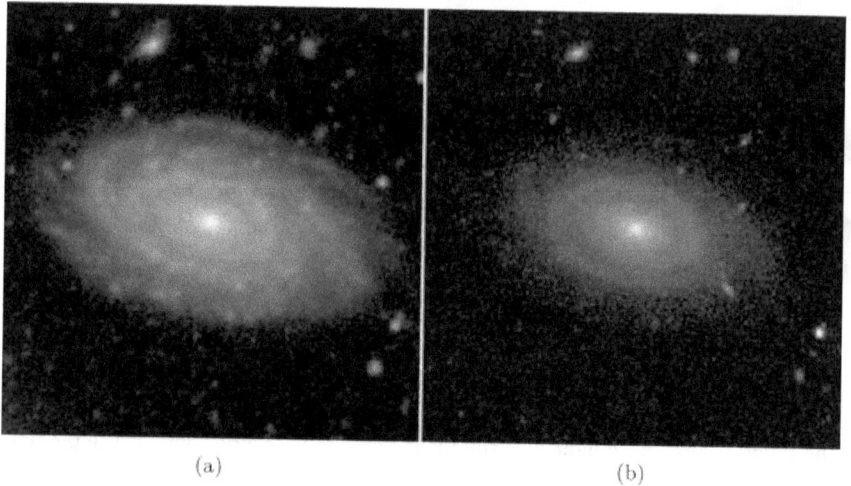

(a)   (b)

Fig. 18.   Blue (a) and infrared (b) images of ESO 297-27. A small companion galaxy of unknown distance lies to the upper left. From R. D. Grouchy *et al.*, *Astronomical Journal*, **136**, 980 (2008). © AAS. Reproduced with permission.

ordinary spiral galaxy having a fairly strong, single inner arm that is counter-winding with respect to two or three weaker outer arms (Figure 18). As for NGC 4622, the spiral arms in ESO 297-27 are separated using *m*-images in Figure 19. Using the dust asymmetry method, Rebecca Grouchy and coworkers[28] determined that the inner single arm of ESO 297-27 is leading while the two or three outer arms are trailing, the exact opposite of NGC 4622. In this case also, the galaxy lies in a sparser environment than NGC 4622, which argues that the peculiar structure may have arisen naturally rather than in an encounter with another galaxy. Nevertheless, a small companion of unknown distance, also seen in Figure 18, could be important.

The sense of winding of the spiral structure of galaxies is a topic that has not been extensively revisited since a paper written by Gérard de Vaucouleurs in 1958.[29] de Vaucouleurs' emphasis in this paper was on highly inclined galaxies, for which the approaching and receding halves of the major axis were easier to determine using the

---

[28]R. D. Grouchy *et al.*, *Astronomical Journal*, **136**, 980 (2008).
[29]G. de Vaucouleurs, *Astrophysical Journal*, **127**, 487 (1958).

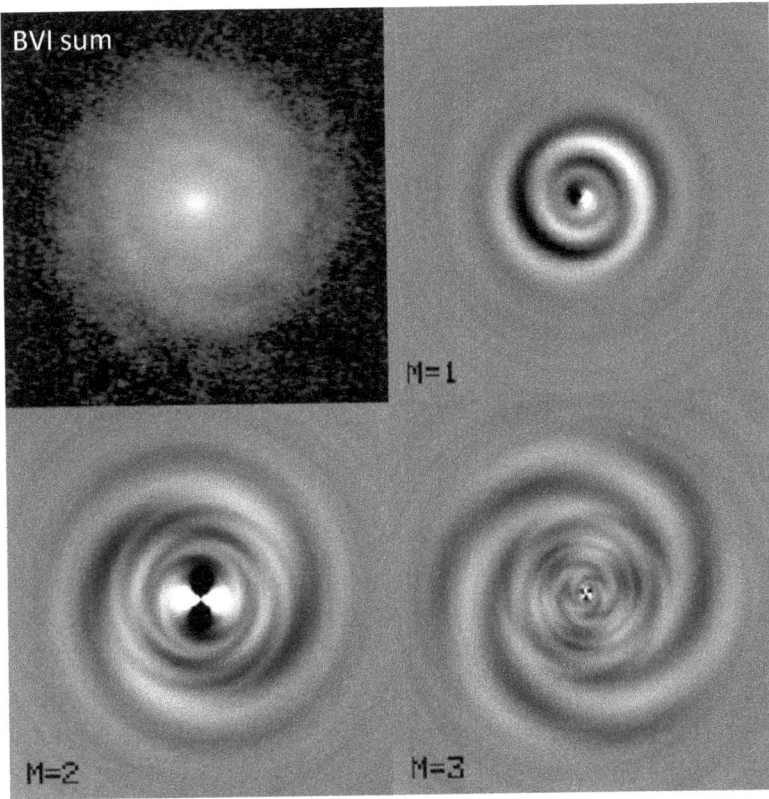

Fig. 19. *m*-images of ESO 297-27. From R. D. Grouchy *et al.*, *Astronomical Journal*, **136**, 980 (2008). © AAS. Reproduced with permission.

spectroscopic equipment of his day, and for which the near side was clear even on photographic plates. In our time, it would be possible to do a much more sophisticated study of this problem for hundreds of nearby galaxies having measured velocity fields and high quality digital imaging. Given the case of NGC 4622 and the fact that it is not necessarily unique, I would argue that the last word has not been said on this problem of leading versus trailing spiral arms.

Even so, the mystery of NGC 4622 has not yet been resolved. Because the galaxy lies in a cluster environment, perhaps a gravitational encounter with another galaxy, or a merger with a small companion, or both have contributed to the galaxy's unusual structure. A tell-tale signature of a minor merger is a small dust lane crossing the central nucleus of the galaxy that was detected only in the HST images. The dust lane is unusual in that it appears to be edge-on even though the disk of NGC 4622 is face-on.

16.6. Spiral Structure and Bars — More Recent Views

The complexity of spiral structure in galaxies has left some theorists wondering if any progress on understanding spiral patterns has actually been made since density wave theory was first proposed in the 1960s. For example, former Rutgers University astronomer Jerry Sellwood commented in a 1999 paper[30]: "It is about a century and a half since Lord Rosse first noted the spiral appearance of M51, but a satisfying and robust theory for the general spiral phenomenon in disc galaxies still eludes us." He argues that "most researchers are now convinced that spirals are driven by the stellar disk through some kind of collective gravitational process" because we see smooth stellar spirals in infrared images (as in Figure 11). He argues that the lifetime of any given global spiral may not be as long as the natural abundance of spirals may imply. This has led to the idea that spiral patterns, even density wave arms, are not necessarily long-lived. In a recent review, astronomers Claire Dobbs and Junichi Baba conclude that "With the possible exception of barred galaxies, spiral arms are transient, recurrent, and initiated by [naturally]-amplified instabilities in the disk."[31] The transient nature arises from the fact that the presence of a spiral can increase the degree of random star motions, which serves to weaken the spiral. By recurrent, it is meant that a spiral can reappear, or recur, in a disk-galaxy. One way to do this would be for a galaxy to produce an occasional weak, leading spiral that the non-solid-body rotation of the disk can swing into an amplified, longer-lived trailing spiral.

---

[30] J. Sellwood, in *Astrophysical Dynamics — in Commemoration of F. D. Kahn*, D. Berry *et al.*, eds. (Dordrecht, Kluwer, 1999).
[31] C. Dobbs and J. Baba, *Publications of the Astronomical Society of Australia*, **31**, 35 (2014).

Similar problems arise when it comes to understanding the origin of bars. Theorists have for years used numerical simulations to understand how bars form. Based on early studies, the general idea was that a cold (i.e. low random motion) rotating disk of stars with a relatively low central mass concentration is naturally unstable to the formation of a bar, via a process known as the "bar instability."[32] Some galaxies do in fact have bars and little or no central concentration, and so in principle, the bars in these cases could form by the bar instability. However, many barred galaxies have a significant central mass concentration and at the same time have strong bars (for example, NGC 1433). These are the ones that the bar instability mechanism has trouble explaining. Bar growth through the trapping of stars on highly-elongated orbits is the mechanism Sellwood eventually preferred.[33]

Studies have also shown that some bars may be generated by an interaction with a companion galaxy. However, it can likely be ruled out that this is a general formation mechanism because bars are much rarer among very distant galaxies, when interactions were more common, than they are in nearby galaxies. For example, Kartik Sheth and coworkers showed that while almost 70% of luminous nearby galaxies are barred, the fraction drops drastically (to about 20%) among very distant galaxies.[34]

There are other points of view. In the late 1980s, several theoretical studies showed that bars do not necessarily drive spirals, but that bars and spirals can be separate modes that develop in the same disk but with different pattern speeds.[35] For example, based on numerical simulations, Linda Sparke and Jerry Sellwood found that,

---

[32] F. Hohl, *Astrophysical Journal*, **168**, 343 (1971).

[33] J. Sellwood, in Dynamics of Galaxies: From the early Universe to the Present, F. Combes *et al.*, *Astronomical Society of the Pacific Conference Series*, **197**, 3 (2000).

[34] K. Sheth *et al.*, *Astrophysical Journal*, **675**, 1141 (2008).

[35] M. Tagger *et al.*, *Astrophysical Journal*, **318**, L43 (1987); J. F. Sygnet *et al.*, *Monthly Notices of the Royal Astronomical Society*, **232**, 733 (1988); L. Sparke and J. Sellwood, *Monthly Notices of the Royal Astronomical Society*, **231**, 25 (1988).

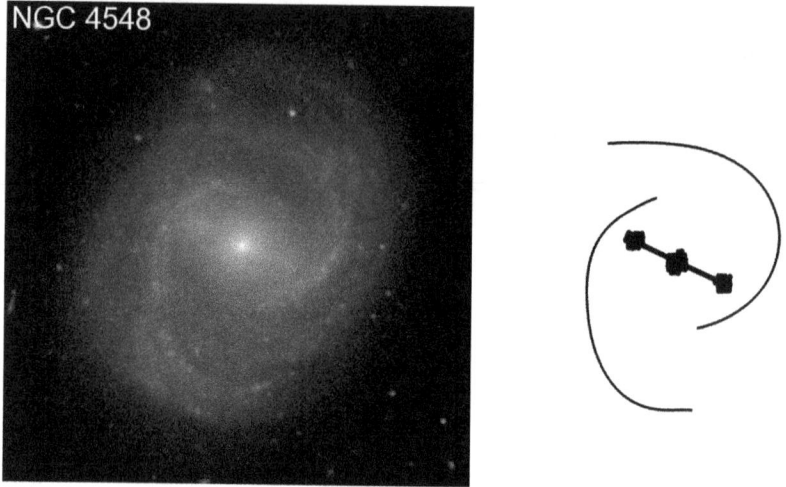

Fig. 20. NGC 4548 is a barred spiral whose bar and spiral may have different pattern speeds, which is indicated by how the ends of the bar are significantly offset from the beginnings of the spiral. Image source: The de Vaucouleurs Atlas of Galaxies published by Cambridge University Press, 2007; reprinted with permission.

in a disk where both types of features develop naturally, the spiral has a lower pattern speed than the bar. This generally means that the corotation resonance of the spiral is farther from the center than is the corotation resonance of the bar.[36] It also means that instead of the spiral breaking directly from the ends of a bar, as it likely would if the bar and the spiral had the same pattern speed, we should see mismatches between the beginning of the spiral and the ends of the bar. Such mismatches do exist, an excellent example being NGC 4548 shown in Figure 20. Other cases are described in Chapter 17.

Another point of view is known as the "manifold" theory of spiral structure, which has been developed by Marseille Observatory

---

[36]In fact, M. Tagger and coworkers in *Astrophysical Journal*, **318**, L43 (1987) noted in simulations that, from the point of view of galaxy evolution, a favorable situation for the longevity of the combined pattern is for the pattern speeds to be coupled such that the ILR of the spiral is at about the same location as the CR of the bar.

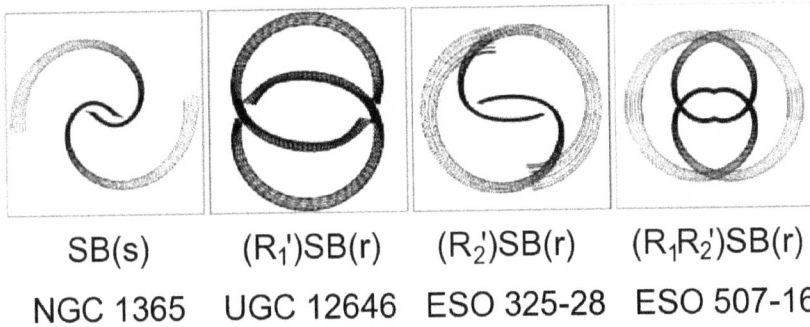

| SB(s) | (R$_1'$)SB(r) | (R$_2'$)SB(r) | (R$_1$R$_2'$)SB(r) |
| NGC 1365 | UGC 12646 | ESO 325-28 | ESO 507-16 |

Fig. 21.   Schematics showing the main types of orbits in the manifold theory of spirals and rings. Types are (left to right) escaping, heteroclinic, escaping, homoclinic (diagrams are from E. Athanassoula *et al.*, *Monthly Notices of the Royal Astronomical Society*, **400**, 1706 (2009). Copyright © 2009, Royal Astronomical Society. Reproduced with permission). Only the trajectories are shown in these plots; the bar is not shown. The bar in each case is a horizontally-oriented feature extending between the L$_1$ and L$_2$ points as shown in Figure 13 of Chapter 15. Images for reference: NGC 1365 (Figure 1 of Chapter 1); UGC 12646 (Figure 15 of Chapter 15); ESO 325-28 (Figures 14 and 15 of Chapter 15); ESO 507-16 (Figure 14 of Chapter 15).

theoretical astronomer Lia Athanassoula and her coworkers.[37] In this theory, grand design spirals are not necessarily density wave structures, nor are they companion-driven or modal structures. Instead, they are structures that originate in the dynamics of the L$_1$ and L$_2$ Lagrangian points in the gravitational field of a bar. These points lie at or near the ends of the bar (see Figure 13 of Chapter 15), and are generally unstable in the sense that quasi-periodic motions are not possible around them (as they can be around the L$_4$ and L$_5$ points). Instead, orbital trajectories will generally lead to escape from the vicinity of the points. Manifolds are like "tubes" that guide escaping particles (for example, stars and gas clouds) away from the L$_1$ and L$_2$ points in an organized manner, leading to structures that strongly resemble spirals and especially the different kinds of outer rings and pseudorings (R$_1$, R$_1'$, R$_2'$, and R$_1$R$_2'$) that were described in Chapter 15. Manifolds are trajectories of flow in a dynamical system

---

[37]E. Athanassoula *et al.*, *Monthly Notices of the Royal Astronomical Society*, 407, 1433 (2010); 400, 1706 (2009).

with L_1 and L_2 unstable equilibrium points. These trajectories create patterns of excess density that strongly resemble barred galaxy rings and spirals. The theory also almost exclusively prefers two-armed, trailing grand-design spirals.

In manifold theory, the bar and its associated spiral/ring structures have the same pattern speed, which differs from the theory of Sparke and Sellwood. There are three types of manifold orbits: those that emanate from one Lagrangian point and end at the opposite Lagrangian point (called "heteroclinic" orbits), those that emanate from one Lagrangian point and end at the same point (called "homoclinic" orbits), and those that emanate from one Lagrangian point and end not near any Lagrangian point (called "escaping" orbits, meaning they escape the vicinity of the Lagrangian points, not the galaxy as a whole). These are all shown in Figure 21, where the $(R_1')SB(r)$-type corresponds to the heteroclinic type, the $(R_1 R_2')SB(r)$-type corresponds to the homoclinic type, and the open spiral and $(R_2')SB(r)$-type correspond to escaping orbits. Both inner and outer rings are covered by the manifolds.

The numerical simulations of M. P. Schwarz demonstrated that outer rings and inner rings likely form near resonances with the bar, but in manifold theory, the features are not necessarily linked to such specific resonances. From a morphological point of view, manifold theory is just as successful as resonance theory in explaining the different observed classes of rings. Distinguishing between the two ideas would require more detailed studies of individual cases.

16.7. Red Spirals and the Galaxy Zoo Project

The Galaxy Zoo Project (mentioned previously in Chapter 6) has provided a wealth of information on spiral galaxies. One particularly interesting result from the project has been a greater appreciation of the existence of red spirals. We have seen that spiral structure in many nearby galaxies is traced by recent star formation, which gives spiral arms a characteristic bluish color. Yet, some spirals lack the the usual star-forming tracers. The arms are red, instead of blue, implying they are made entirely of very old stars. Red spirals pose an interesting problem not only for galactic dynamics, but also for galactic evolution. How does a spiral pattern last long enough to become "red and dead?"

Galaxy Zoo was effective for identifying red spirals because the survey was based on color pictures made from multi-filter images obtained in the Sloan Digital Sky Survey. In a detailed study, Karen L. Masters and her Zoo coworkers used Zoo classifications to identify a sample of red spiral-armed galaxies face-on enough to be certain of their structure.[38] These authors found that red spirals are barred more than twice as frequently as blue spirals are barred, and also tend to be more massive than blue spirals. The implication is that, in addition to possible environmental effects, something in the dynamics quenches (puts out the "fire" of) star formation in massive barred galaxies. The authors suggest that red spirals could simply be older spirals that have used up all of their interstellar gas (gas depletion), or rearranged their gaseous disk enough to leave little or no remaining gas in the spiral pattern. In a later study,[39] it is argued that red late-type spirals evolve from blue late-type spirals by the latter declining in star formation rate to nearly zero over a period of only half a billion years.

The red/blue dichotomy is clearly an interesting development in the study of spiral galaxies.

---

[38]K. L. Masters *et al.*, *Monthly Notices of the Royal Astronomical Society*, **405**, 783 (2010).
[39]R. Tojeiro *et al.*, *Monthly Notices of the Royal Astronomical Society*, **432**, 359 (2013).

Chapter 17

A Spiral Menagerie

Chapter 15 focussed on ringed spirals because the resonance idea works so well for them. Real galactic rings strongly resemble model orbits near specific resonances, based on numerical simulations. This is especially true of R_1 and R_2 rings. This chapter is focussed on mostly nonringed, (s)-shaped spirals, which constitute the majority of all spiral galaxies. It is interesting to ask why this might be so.

One way to answer this question might be to look at timing. Finnish astronomers Pertti Rautiainen and Heikki Salo suggested[1] that the lack of rings in so many galaxies could be tied to the time it takes for rings to form. Non-ringed galaxies could have the same resonances as ringed galaxies, but may not have had enough time for the resonances to organize the structure into rings, especially outer rings. They argue also that the strength of the bar and the bar pattern speed may be factors that limit the features we can see. A high bar pattern speed, for example, could prevent inner and nuclear rings from forming.

In this chapter, a sampling of typical (s)-shaped and atypical ringed spirals is described in order to show that it is not necessarily as straightforward to interpret these as compared to the regular ringed spirals. All of the illustrations are black and white images selected from the EFIGI database.[2] The illustrations are all based

---

[1] P. Rautiainen and H. Salo, *Astronomy and Astrophysics*, **362**, 465 (2000).
[2] A. Baillard *et al.*, *Astronomy and Astrophysics*, **532**, 74 (2011).

on blue-green (*g*-band) images that have been converted to units of magnitudes per square arcsecond, and the galaxies are examined on an individual basis.

The first question we could ask about these galaxies is where the corotation resonance might be. In the multi-ringed barred spirals (like NGC 3081, IC 1438, NGC 6782, NGC 7098, among others), CR likely lies in the dim region between the inner and outer rings/pseudorings, around the expected location of the L_4 and L_5 Lagrangian points (Figure 31 of Chapter 15). In a density wave spiral, CR can be located by examining where the spiral appears to end. Inside CR, stars and gas clouds move faster than the spiral with a relative speed that decreases as CR is approached. This relative speed determines the strength of the spiral shock, and this in turn will determine how prominent the arms are at different locations in the disk.[3] We expect that CR will be located where the spiral appears to end. This is also where the star formation in the arms will appear to end.

IC 769 (Figure 1) and NGC 2857 (Figure 2), are excellent examples of likely density wave spirals. In both cases, the knottiness of the arms indicates that star-forming regions are seen throughout the full extent of the pattern. Except for the sharpness of the arms, IC 769 resembles NGC 5364 but without the bright inner ring. The difference in the character of the arms is likely due to a difference in luminosity. NGC 2857 is almost five times as luminous as IC 769, and twice as luminous as NGC 5364.

NGC 2857 appears to be perfectly face-on. The arms continuously open outward with increasing galactocentric distance and, as shown in Figure 5 of Chapter 16, are well-approximated by a logarithmic spiral. (That is, they spread open in angle as the natural logarithm of the galactocentric distance). The galaxy may have a small inner ring, although being four times as distant as NGC 5364 makes the inner regions poorly resolved. The interarm regions show very faint examples of "spurs," or short spiral features.

---

[3]W. W. Roberts *et al.*, *Astrophysical Journal*, **196**, 381 (1975).

Fig. 1.   Blue-green light image of IC 769, a tilted grand design (density wave) spiral 96 million light years away in Virgo. This galaxy resembles NGC 5364 (Figure 11 of Chapter 15), but lacks the inner resonance ring. The arms are also more smooth and diffuse than those in NGC 5364. Image source: Sloan Digital Sky Survey Collaboration, http://www.sdss.org.

Figure 3 shows UGC 4074, a distant but nevertheless clear three-armed spiral. The galaxy appears genuinely non-barred, and the arms do not appear to begin close to the center. Three-armed spirals tend to be much rarer than two-armed spirals.[4] In the context of density wave theory, a strong three-armed spiral should be confined between the inner and outer 3:1 resonances (that is, resonances like the ILR and the OLR but having three radial oscillations per rotation of the pattern, rather than 2).

UGC 3920, shown in Figure 4, is a nearly face-on spiral having structure similar to NGC 1433 but only a very faint trace of a bar. There is a nearly closed inner ring and two "plumes" (secondary arms) just as in NGC 1433. Most interesting is the nearly circular

---

[4]For example, R. Buta, *Astrophysical Journal Supplement Series*, **96**, 39 (1995).

Fig. 2.   Blue-green light image of NGC 2857, a face-on grand design spiral 220 million light years away in Ursa Major. The galaxy is mainly two-armed, but between these arms there are at least half a dozen "spurs" or short spiral arcs. Image source: Sloan Digital Sky Survey Collaboration, http://www.sdss.org.

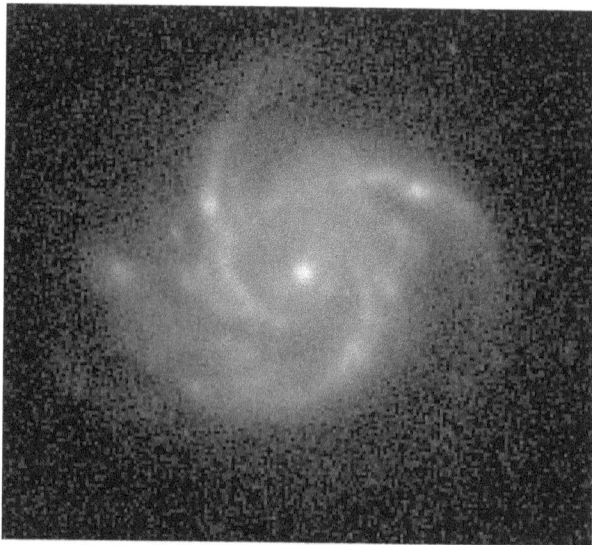

Fig. 3.   Blue-green light of image of UGC 4074, a face-on three-armed spiral 323 million light years away in Lynx. Image source: Sloan Digital Sky Survey Collaboration, http://www.sdss.org.

Fig. 4. Blue-green light image of UGC 3920, a face-on grand design spiral with a large nuclear ring, 377 million light years away in Gemini. Image source: Sloan Digital Sky Survey Collaboration, http://www.sdss.org.

nuclear ring which is so large that it is detectable as a ring even though the galaxy is ten times as distant as NGC 1433. How can this galaxy fit into conventional ideas of resonance rings if it lacks the bar needed to create such features? Perhaps the galaxy had a stronger bar at one time, but the bar mostly dissolved over time due to the development of the very unbar-like nuclear ring. There is only a weak trace of a bar in this galaxy now.

NGC 4662 (Figure 5) shows a classic multi-armed spiral pattern with at least six arms. It has a clear bar, but only a faint trace of an inner pseudoring. As is typical of such cases, the pattern is mostly two-armed only in the inner regions. Farther out, the multi-armed character is more evident. In the inner regions, the arms appear to miss breaking directly from the bar ends by substantial angles (see Figure 20 of Chapter 16). This could signify that the spiral pattern and the bar pattern are two independent features with different pattern speeds. This is significantly different from ringed galaxies

Fig. 5.   Blue-green light image of NGC 4662, a face-on multi-armed spiral 314 million light years away in Canes Venatici. There are at least six clear arms in addition to the strong bar. Image source: Sloan Digital Sky Survey Collaboration, http://www.sdss.org.

like NGC 3081, where the alignments of the rings with respect to the bar favor only a single pattern speed.

NGC 4241 (Figure 6) is a stunning example of a late-type spiral with "massive" spiral arms, one of two types of spiral arms proposed as worthy of classification recognition by John. H. Reynolds in a 1927 paper.[5] The structure of NGC 4241 is characterized by a very tiny central object, a short but clear bar, and a pair of very diffuse, open spiral arms. The arms are astonishingly different from the thin filamentary arms (the second type recognized by Reynolds) seen in, for example, NGC 5364. NGC 4241 is what de Vaucouleurs would classify as an "Scd" galaxy, a "late-type" spiral with very little bulge and very open arms. The two arms in this case appear to break almost exactly from the ends of the bar. Although well-defined as a

---

[5] J. H. Reynolds, *Observatory*, **50**, 185 (1927).

Fig. 6.   Blue-green light image of NGC 4241, a face-on grand design spiral with a broad bar, diffuse arms, and almost no bulge, 29 million light years away in Virgo. Image source: Sloan Digital Sky Survey Collaboration, http://www.sdss.org.

spiral, NGC 4241 has a blue-light luminosity of only about 2 billion suns, comparable to that of the Triangulum Spiral M33. This can account for the broadness of its spiral arms.

NGC 6116 in Figure 7 is an excellent example of a smooth-armed, nonbarred spiral (Hubble type Sa) where the arms are so broad and diffuse that they seem to blend in the inner regions. The arms appear not to be traced by star-forming regions. Even at its distance of 400 million light years, we should still be able to tell if the arms had a significant amount of recent star formation. Imaging at the resolution of the *Hubble Space Telescope* could shed more light on the star formation properties of the arms in this case. NGC 6116 has two comparable-sized companion galaxies, but neither is especially close and it seems unlikely that they are the reason for the strong spiral pattern.

NGC 3577 (Figure 8) is a face-on spiral with a clear inner pseudoring and a very strong-looking bar. The two main arms do

Fig. 7.   Blue-green light image of NGC 6116, a tilted grand design spiral with very smooth arms, 400 million light years away in Corona Borealis. Image source: Sloan Digital Sky Survey Collaboration, http://www.sdss.org.

Fig. 8.   Blue-green light image of NGC 3577, a face-on grand design barred spiral with asymmetric arms, 241 million light years away in Ursa Major. Image source: Sloan Digital Sky Survey Collaboration, http://www.sdss.org.

Fig. 9. Blue-green light image of UGC 10330, a face-on, possibly interacting grand design spiral with peculiar symmetric arms, 444 million light years away in Hercules. Image source: Sloan Digital Sky Survey Collaboration, http://www. sdss.org.

not close into an outer pseudoring feature. The arms begin a little ahead of the bar in the direction of rotation (counterclockwise in this case) which suggests that the outer arms and the bar may have different pattern speeds. NGC 3577 has a bright companion, NGC 3583, but the latter appears to be more than 100 million light years in the foreground.

UGC 10330 in Figure 9 shows a strong and fairly symmetric two-armed pattern with no rings, no bar, and little or no background disk light. The arms may not be logarithmic, at least in the outer regions, which suggests that they are tidal in origin. This is a reasonable interpretation because the galaxy has a significant nearby companion (named LEDA 214481)[6] also visible in Figure 9. The bright inner arms could be density wave features, as in M51.

---

[6]LEDA stands for Lyon Extragalactic Database.

Fig. 10.   Blue-green light image of NGC 3253, a face-on barred spiral 428 million light years away in Leo. Image source: Sloan Digital Sky Survey Collaboration, http://www.sdss.org.

The galaxy shown in Figure 10 is NGC 3253, which is a strong face-on case showing a well-defined inner two armed pattern and a complex, multi-armed pattern in the outer regions. There is only a trace of an inner pseudoring encircling a fairly well-defined bar. In this case, perhaps the inner spiral has been "driven" by the bar, while the outer structure is independent of the bar.

UGC 3822, shown in Figure 11, has four clear spiral arms, a strong bar, and half an inner ring. There is considerable asymmetry, but the galaxy has no particularly close companions.

UGC 4867 (Figure 12) is an interesting late-type spiral that de Vaucouleurs would have classified as type Sd. There is a short, stubby bar and complex outer arms defined by significant star formation. The galaxy has a luminosity of only 2.6 billion suns, a factor of two lower than for the Triangulum Spiral M33.

Fig. 11.   Blue-green light image of UGC 3822, a four-armed barred spiral 319 million light years away in Auriga. Image source: Sloan Digital Sky Survey Collaboration, http://www.sdss.org.

Fig. 12.   Blue-green light image of UGC 4867, a strong spiral with no bulge and a weak bar, located 111 million light years away in Leo Minor. Image source: Sloan Digital Sky Survey Collaboration, http://www.sdss.org.

Fig. 13. Blue-green light image of UGC 5055, a strong spiral with tightly-wrapped, ring-like, star-forming arms, located 340 million light years away in Ursa Major. There is also a prominent but broad, diffuse bar. Image source: Sloan Digital Sky Survey Collaboration, http://www.sdss.org.

Figure 13 shows UGC 5055, a spiral where two strong outer arms wind into a large pseudoring of type R'_2. The arms in this case cannot be logarithmic, because a logarithmic spiral never closes in this manner. There is a clear but not especially strong-looking bar in this case, and the arms are defined by considerable recent star formation. This is a case where the spiral and the bar likely have the same pattern speed.

Figure 14 shows the highly-inclined spiral NGC 2854. The galaxy has two fairly strong and symmetric arms, and a peculiar inner region that looks affected by dust. There is no bright central concentration, but in this case the bright center may be obscured even though the galaxy is not exactly edge-on. The structure of NGC 2854 may be influenced by gravitational interaction with a nearby comparable-sized companion, NGC 2856.

As for IC 769 and NGC 2857, NGC 4411b (Figure 15) and NGC 3362 (Figure 16) are two excellent face-on Sc spirals with very

Fig. 14. Blue-green light image of NGC 2854, a peculiar, highly-inclined smooth-armed spiral located 125 million light years away in Ursa Major. The inner regions are mottled with dust and there is a lack of any central concentration. Image source: Sloan Digital Sky Survey Collaboration, http://www.sdss.org.

Fig. 15. Blue-green light image of NGC 4411b, a face-on, Hubble-type Sc spiral lying 54 million light years away in Virgo. Image source: Sloan Digital Sky Survey Collaboration, http://www.sdss.org.

Fig. 16.   Blue-green light image of NGC 3362, a Hubble-type Sc spiral lying 365 million light years away in Leo. Image source: Sloan Digital Sky Survey Collaboration, http://www.sdss.org.

different total luminosities. In blue light, the former has a luminosity equivalent to 2.8 billion suns, while the latter has a luminosity equivalent to 95 billion suns. This difference is reflected in how well-defined the spiral arms are. Those in NGC 4411b are weak and somewhat ill-defined, while those in NGC 3362 are much sharper and well-defined. These two galaxies demonstrate how it is possible to tell a luminous spiral from a less luminous spiral just from the appearance of the spiral arms. This was first noticed and commented upon by Canadian astronomer Sidney van den Bergh in 1960, and is the basis for what he called *luminosity classification*. This is discussed further in Chapter 23.

NGC 5698 (Figure 17) is a peculiar barred spiral with asymmetric, massive-looking spiral arms. One arm has a long, faint extension on one side that is not matched by the other arm. Surprisingly, NGC 5698 has no comparable-sized nearby companions, which makes the origin of its asymmetry uncertain.

Fig. 17.   Blue-green light image of NGC 5698, a peculiar, asymmetric spiral lying 169 million light years away in Bootes. Image source: Sloan Digital Sky Survey Collaboration, http://www.sdss.org.

NGC 6004 in Figure 18 should be compared with ESO 566-24, which was described in Chapter 15. The two galaxies show a similar four-armed pattern breaking from a prominent bar. Both galaxies have an inner ring, except the feature in NGC 6004 is a more open pseudoring. Like ESO 566-24, the structure of NGC 6004 may be influenced by 4:1 resonances with the bar pattern speed.

UGCA 322 in Figure 19 is what de Vaucouleurs would call an "Sdm" galaxy, that is, an extreme late-type spiral with ill-defined arms sprinkled with star-forming regions. There is a faint trace of a bar also. The galaxy has a blue light luminosity of 3.6 billion suns.

Figure 20 shows UGC 6093, a bright spiral with a fairly strong bar, a trace of an inner pseudoring, and two well defined, relatively smooth outer arms. What is unusual about the galaxy is that the spiral appears much stronger than the bar. It is likely that the bar and the spiral are independent features having different pattern speeds. The galaxy has no major nearby companions.

Fig. 18.   Blue-green light image of NGC 6004, a four-armed spiral similar to ESO 566-24, lying 175 million light years away in Serpens Caput. Image source: Sloan Digital Sky Survey Collaboration, http://www.sdss.org.

Fig. 19.   Blue-green light image of UGCA 322, a bulge-less, complex spiral lying 57 million light years away in Virgo. Image source: Sloan Digital Sky Survey Collaboration, http://www.sdss.org.

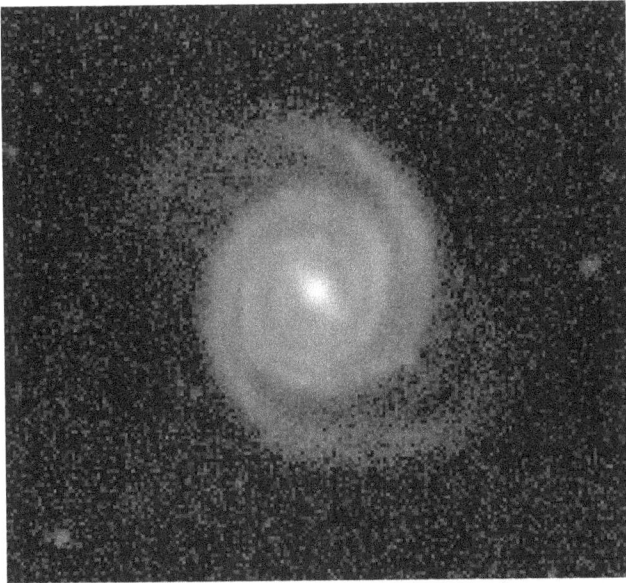

Fig. 20.   Blue-green light image of UGC 6093, a large spiral 480 million light years away in Leo. Image source: Sloan Digital Sky Survey Collaboration, http://www.sdss.org.

NGC 7738 (Figure 21) is a high luminosity (48 billion suns in blue light), two-armed, mostly (s)-shaped barred spiral with an exceptionally strong bar. The object in the center is part of a small nuclear ring (a dust ring in this case). This seems to be a general characteristic of nuclear rings in strongly-barred galaxies. As noted by University of Oulu astronomer Sebastien Comerón and coworkers, "stronger bars host smaller [nuclear] rings."[7] NGC 7738 also shows a trace of a highly-elongated inner ring. The two outer arms form an R'_1 outer pseudoring.

NGC 5406 in Figure 22 strongly resembles NGC 4662 in having a well-defined bar and an extensive outer multi-armed spiral pattern.

---

[7]S. Comerón *et al.*, *Monthly Notices of the Royal Astronomical Society*, **402**, 2462 (2010).

Fig. 21.   Blue-green light image of NGC 7738, a two-armed, mostly (s)-shaped barred spiral lying 307 million light years away in Pisces. Image source: Sloan Digital Sky Survey Collaboration, http://www.sdss.org.

Fig. 22.   Blue-green light image of NGC 5406, a multi-armed barred spiral lying 236 million light years away in Canes Venatici. Image source: Sloan Digital Sky Survey Collaboration, http://www.sdss.org.

Fig. 23. Blue-green light image of NGC 3049, a spiral with a long, narrow bar and short arms, lying 60 million light years away in Leo. Image source: Sloan Digital Sky Survey Collaboration, http://www.sdss.org.

In this case, there is a nearly closed inner pseudoring, but, like NGC 4662, the galaxy bears little resemblance to conventional ringed galaxies. Both galaxies strongly resemble NGC 5375 but lack dark areas like those seen in that galaxy (arrows in Figure 7 of Chapter 16) and that were interpreted in Chapter 16 in terms of the L_4 and L_5 Lagrangian points. The lack of the dark spaces could signify either that these points are stable, and therefore the orbits around them are not chaotic, or that the patterns in NGC 4662 and 5406 are less evolved than that in NGC 5375.

The final two cases, NGC 3049 and NGC 4389, are shown in Figures 23 and 24. Both are similar in having a very thin, knotty bar and rather stubby spiral arms. The knottiness in the bars implies recent star formation, which is very atypical of bars found in other galaxies (Chapter 11).

Fig. 24.   Blue-green light image of NGC 4389, a spiral with a long, narrow bar and short arcs, lying 35 million light years away in Canes Venatici. The presence of dust lanes oriented nearly perpendicular to the apparent bar suggest that the galaxy has experienced an interaction, possibly a minor merger with another galaxy. Image source: Sloan Digital Sky Survey Collaboration, http://www.sdss. org.

This discussion has shown how difficult it can be to interpret the structure of spiral galaxies. Spirals cover a wide range in mass and luminosity, and the impact of resonances on their structure is an important issue.

Chapter 18

Lifting the Dusty Veils

In the blue-light photographs of the old days, interstellar dust inside a galaxy could have a major impact on its appearance, especially with increasing tilt (see Figure 6 of Chapter 3). Planar dust lanes, bar dust lanes, spiral arm dust lanes, and other dusty features (like dust rings) impacted the apparent structure and even how well we could see certain features. For example, in blue light, a bar or a galactic nucleus could be partly hidden by internal dust. The standard filters that were used for galaxy photometry in the 1980s and 1990s, $UBVRI$, were all fairly sensitive to the influence of dust.

Things changed when detailed mid-infrared[1] imaging became possible. In 2003, NASA launched the *Spitzer Space Telescope* (SST), which was an orbiting observatory designed in part to obtain images at 3.6, 4.5, 5.8 and 8.0 microns. This is the wavelength domain that allows us to see the "backbone" of the stellar mass distribution, that is, the real distribution of stars unencumbered by interstellar extinction. It is also the wavelength domain that exposes the interstellar medium (i.e., the space between the stars) in a manner that no other domain can.

Throughout this book, the near-infrared I-band (effective wavelength 0.8 microns) has been used for color index maps. Groundbased near-infrared imaging to 2.2 microns (known as the "K-band"),

---

[1] The mid-infrared refers to wavelengths from 3 to 8 microns (3,000–8,000 nm) in wavelength.

was possible not long before the launch of the SST, but there is a fundamental difficulty with observing in the infrared from the ground: the very high sky background brightness due to airglow. This glow rises so rapidly with increasing wavelength in the infrared that even a full moon has little additional impact. It is as if the night sky in the infrared is a perpetual twilight. When I was doing *UBV* photometry, I could only observe at dark time (new moon), but when I became an infrared observer in the 1990s, dark time was no longer necessary. Like other former optical photometrists, I became a bright time (around full moon) observer. The high sky brightness made it difficult to get *K*-band images of the same depth in surface brightness as could be achieved with an *I*-band filter.

The SST made it possible to do infrared imaging with a much reduced background brightness. The SST orbits above the main source of the airglow, the upper atmosphere, which allowed the deepest infrared imaging that had ever been obtained. This led to the *Spitzer* Survey of Stellar Structure in Galaxies, a survey of 2400 bright galaxies using the 3.6 and 4.5 micron filters.[2] This survey is usually abbreviated to "S⁴G."

The 3.6-micron filter used in the S⁴G provides images of galaxies that are minimally affected by extinction. At this wavelength, all dust lanes largely disappear and what we see is starlight. From 3.6 to 8.0 microns, the proportion of starlight drops while the light of heated dust increases, until by 8.0 microns most of what we are seeing is the heated dust. This means 8.0 micron images reveal the dusty interstellar medium just as 21-cm radiation reveals the gaseous interstellar medium.

Figures 1–13 show a sampling of the beautiful views of spirals in the 3.6-micron filter. Figure 14 shows an 8-micron image of M81 as compared to a *B–I* color index map. The starlight that is being emphasized in the 3.6-micron filter is due largely to old red giant

---

[2]K. Sheth *et al.*, *Publications of the Astronomical Society of the Pacific*, **122**, 898 (2010).

Fig. 1. 3.6 micron S^{4}G image of the interacting barred spiral NGC 1097. This can be compared with the blue light image shown in Figure 23 of Chapter 15. Most striking is the appearance of the starburst nuclear ring: in blue light, it is broken up by the bar dust lanes, but at 3.6 microns, the ring is a nearly circular, uniform feature. The brightness of the feature likely signifies the presence of large numbers of unresolved, young red supergiants. Image source: R. Buta *et al.*, *Astrophysical Journal Supplement Series*, **217**, 32 (2015). © AAS. Reproduced with permission.

branch stars. Nevertheless, we still see in these images the knottiness that characterizes star-forming regions in the spiral arms.

One of the interesting perspectives that infrared imaging provides is how the morphology of different Hubble galaxy types translates into the infrared. The question is, would a galaxy classified as type Sc in blue light still be classified as Sc in a 3.6-micron mid-infrared image? The answer is yes in many cases and no in some, although the latter might only involve a change of classification from Sc

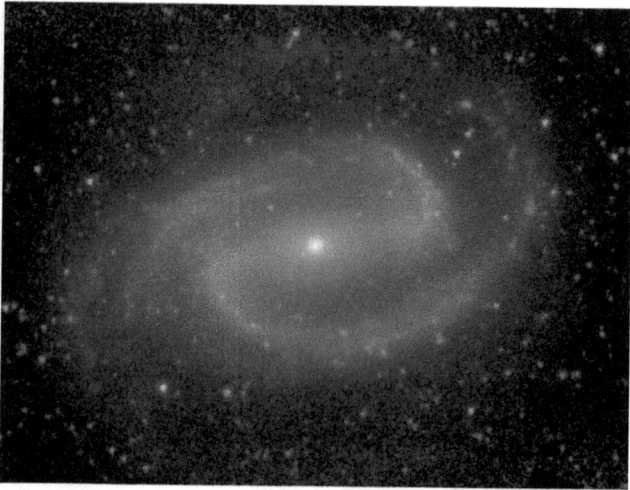

Fig. 2.   3.6 micron S^{4}G image of NGC 1300, a bright, classic (s)-shaped barred spiral lying 67 million light years away in the constellation Eridanus. This can be compared with the blue-light image shown in Figure 1 of Chapter 15. Like NGC 1365, NGC 1300 is far enough south that it was never observed at Birr Castle. The galaxy is nearly at the same distance as NGC 1365, but the latter's spiral arms are much brighter. The concentrated star formation near the ends of the bar is thought to be an "orbit-crowding" effect. Image source: R. Buta *et al.*, *Astrophysical Journal Supplement Series*, **217**, 32 (2015). © AAS. Reproduced with permission.

to Sbc.[3] The reason for this is that Hubble's three classification criteria for spirals: the degree of openness of the arms, the degree of resolution of the arms into distinct "knots," and the degree of central concentration (relative strength of the bulge) are still detectable and applicable at 3.6 microns. As shown in Figure 11 of Chapter 16, the effects of star formation on morphology are minimal at 2.2 microns, but at 3.6-microns, the knottiness of spiral structure reappears. The degree of openness of the arms hardly changes, but the bulge may be slightly more prominent because it is generally made of older stars than are found in the disk. Older stars tend to be redder than younger stars, and thus can be more prominent in a 3.6 micron image. The result is that Hubble's tuning fork does not necessarily get twisted

---

[3]R. Buta *et al.*, *Astrophysical Journal Supplement Series*, **217**, 32 (2015).

Fig. 3.   3.6 micron S^{4}G image of NGC 1365, showing the stellar backbone of the spectacular barred spiral. This can be compared with the blue light image shown in Figure 1. The most interesting aspects of the 3.6-micron image are how much clearer the bar is, and how penetrating the bar dust lanes reveals a small nuclear ring that was not evident in the blue light image. Considerable star formation is still apparent in the arms. Image source: R. Buta *et al.*, *Astrophysical Journal Supplement Series*, **217**, 32 (2015). © AAS. Reproduced with permission.

out of shape in the infrared. Instead, some galaxies merely shift into "earlier" stages along the revised tuning fork shown in Figure 10 of Chapter 6. There are no stages in the blue-light tuning fork that are not recognizable in the mid-infrared.

The SST is capable of imaging at longer wavelengths than 8.0 microns, such as 24, 70, and 160 microns. These wavelengths are considered to be in the "far-infrared." Imaging at 8.0 microns mostly shows us the location of warm dust, but imaging at these longer wavelengths is more sensitive to colder dust.

Fig. 4.   3.6 micron S^{4}G image of NGC 1512, a ringed spiral lying 33 million light years away in the constellation Horologium. This can be compared with the blue-light image in Figure 22 of Chapter 15. Like NGC 1097, NGC 1512 has a star-forming nuclear ring that is prominent in the mid-infrared due to the presence of young red supergiants. Image source: R. Buta *et al.*, *Astrophysical Journal Supplement Series*, **217**, 32 (2015). © AAS. Reproduced with permission.

Fig. 5.   3.6 micron image of the grand design spiral NGC 1566. The space-based image is deep enough to reveal the complex and possibly unrelated arms in different regions, ranging from the very bright and sharp inner arms to the more diffuse outer arms. The bright inner arms are like those seen in NGC 3433 (Figure 6 of Chapter 16) and likely extend to near the inner 4:1 resonance. The outer arms have the morphology of an R'_1 outer pseudoring with respect to the inner structure, which is unusual. Image source: R. Buta *et al.*, *Astrophysical Journal Supplement Series*, **217**, 32 (2015). © AAS. Reproduced with permission.

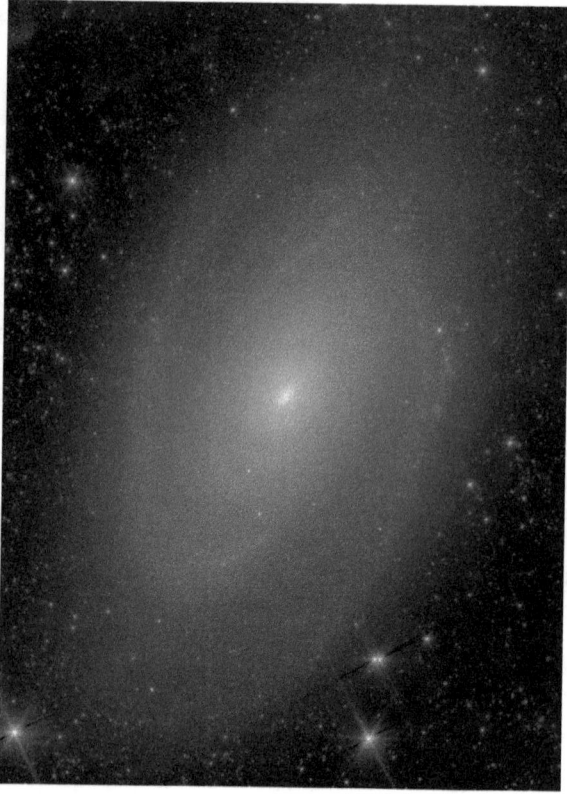

Fig. 6.   The magnificent M81, lying 12 million light years away in Ursa Major, as seen in a 3.6 micron S^{4}G image. This can be compared with the blue-light image in Figure 1 of Chapter 15. Image source: R. Buta *et al.*, *Astrophysical Journal Supplement Series*, **217**, 32 (2015). © AAS. Reproduced with permission.

Fig. 7.   3.6 micron S^{4}G image of NGC 3184, a large face-on spiral lying 27 million light years away in Ursa Major. The spiral pattern is two-armed in the inner regions and multi-armed in the outer regions, as is typical of the Sc Hubble type. If, as is likely in NGC 1566 and 3433, the inner two-armed pattern in NGC 3184 extends to the inner 4:1 resonance, then corotation in this object would be in the outer disk. Image source: R. Buta *et al.*, *Astrophysical Journal Supplement Series*, **217**, 32 (2015). © AAS. Reproduced with permission.

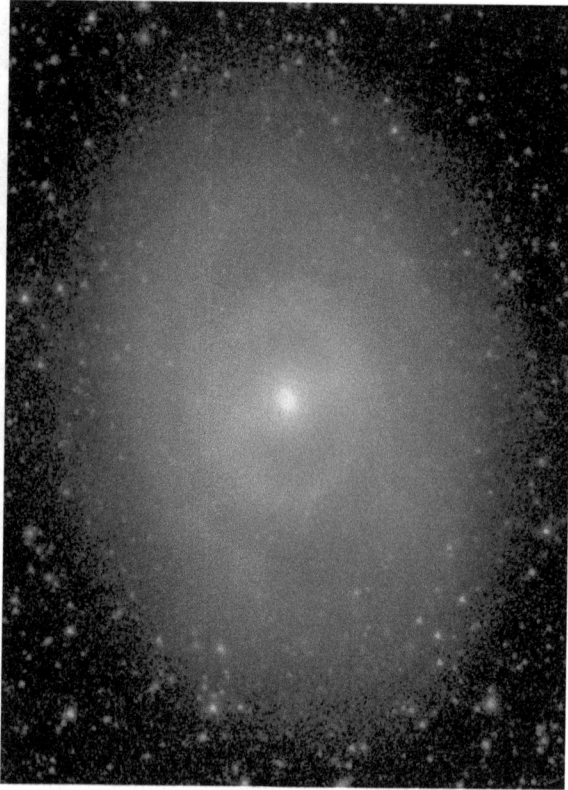

Fig. 8.   3.6 micron S^{4}G image of M95 (NGC 3351), an inclined spiral lying 30 million light years away in Leo. In blue-green light, this galaxy is classified as Hubble-type SBb (see Figure 7 of Chapter 3), but at 3.6 microns, the arms are much smoother and the Hubble type is SBa. Thus, observing in the mid-infrared shifts the galaxy to an "earlier" stage along the Hubble tuning fork. A small nuclear ring is visible in the center. Note how the inner ring and bar resemble the greek letter theta. This is only because of tilt; the ring is actually nearly circular in the plane of the galaxy. Image source: R. Buta *et al.*, *Astrophysical Journal Supplement Series*, **217**, 32 (2015). © AAS. Reproduced with permission.

Fig. 9.   3.6 micron S^{4}G image of M100 (NGC 4321), located 53 million light years away in Virgo, and one of the major spirals in the Virgo Cluster of galaxies. The galaxy is face-on and has an inner, well-defined two-armed spiral imbedded within a more diffuse two-armed pattern. The nature of the broader pattern is uncertain. The infrared image also reveals well a small nuclear ring in the center. Image source: R. Buta *et al.*, *Astrophysical Journal Supplement Series*, **217**, 32 (2015). © AAS. Reproduced with permission.

Fig. 10.   3.6 micron S⁴G image of NGC 4535, a massive, nearly face-on Virgo Cluster spiral lying 52 million light years away. Like NGC 3184, this object shows a well-defined two-armed inner spiral and a multi-armed outer spiral structure. The galaxy also has a bar, but it is not strong and is a feature de Vaucouleurs classifies as type "SAB." Image source: R. Buta *et al.*, *Astrophysical Journal Supplement Series*, **217**, 32 (2015). © AAS. Reproduced with permission.

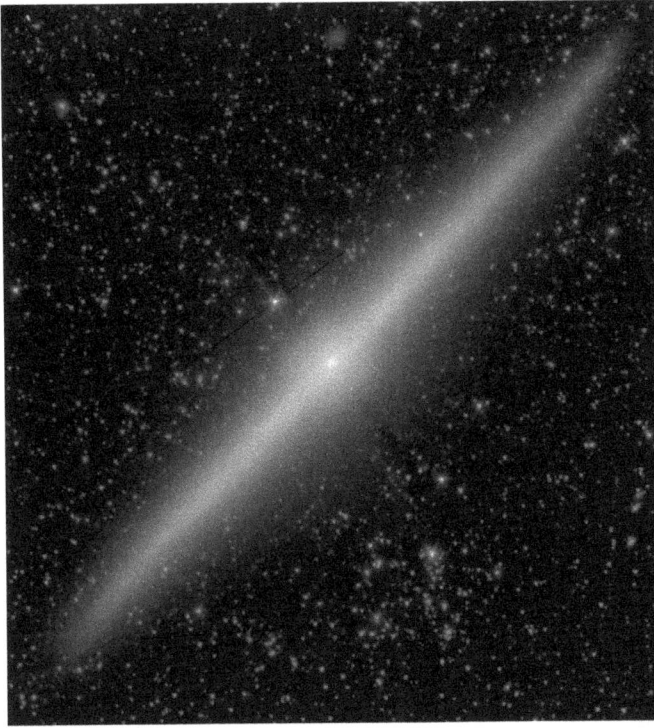

Fig. 11.   3.6 micron S^{4}G image of the superb edge-on spiral NGC 4565, located 55 million light years away in Coma Berenices. This can be compared with the blue light image shown in Figure 1 of Chapter 10. The infrared image reveals the nucleus much more brightly, and the bulge is seen as a boxy/peanut type, implying that NGC 4565 is actually a barred spiral galaxy. The image also reveals the subtle warping of the galaxy's disk: near the ends of the major axis, the light twists very slightly. Warping can naturally occur in a galaxy, or may be caused by an interaction. Image source: R. Buta *et al.*, *Astrophysical Journal Supplement Series*, **217**, 32 (2015). © AAS. Reproduced with permission.

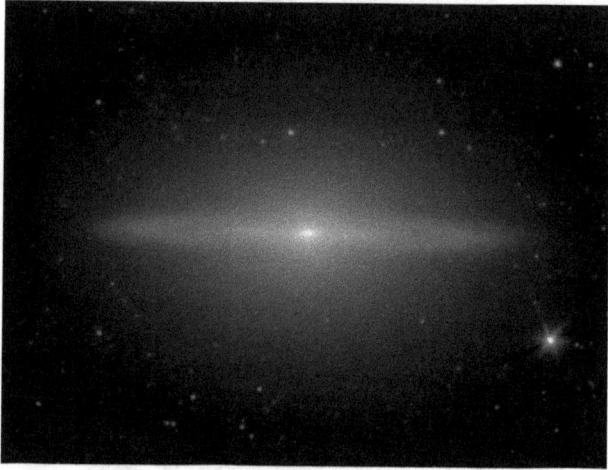

Fig. 12.   3.6 micron S⁴G image of M104 (NGC 4594). This can be compared with the *Hubble Space Telescope* image shown in Figure 1 of Chapter 4. The dusty disk of flocculent spiral structure is revealed, in the infrared, to be an outer ring. This suggests that the galaxy is a highly-inclined version of NGC 7217, which is shown in Figure 32 of Chapter 15. The bulge is not of the boxy/peanut type, implying that M104 is a non-barred galaxy. Image source: R. Buta *et al.*, *Astrophysical Journal Supplement Series*, **217**, 32 (2015). © AAS. Reproduced with permission.

Fig. 13.   3.6 micron S⁴G image of NGC 5247, a face-on Sc spiral lying 56 million light years away in Virgo. It has been suggested that the spiral pattern in this galaxy extends to the inner 4:1 resonance (G. Contopoulos, *Comments on Astrophysics*, **11**, 1 (1985)). Image source: R. Buta *et al.*, *Astrophysical Journal Supplement Series*, **217**, 32 (2015). © AAS. Reproduced with permission.

(a) (b)

Fig. 14.   Comparison between a blue minus infrared (*B–I*) color index map (a) and a *Spitzer Space Telescope* 8.0 micron image (b) of the nearby spiral M81. The color index map is as usual coded such that blue features are dark and red features are light. The 8-micron map shows how the distribution of warm dust in M81 is closely associated with the galaxy's recent star formation. A noteworthy feature in the *B–I* map is how the dust lanes in the bulge region are more conspicuous on the west (right) side of the galaxy, indicating that side is the near side (Chapter 5). Because starlight contributes little at 8 microns, this asymmetry disappears in that filter. Compare this image with the 3.6-micron image of M81 shown in Figure 6. From R. Buta, *Galaxy Morphology, in Planets, Stars, and Stellar Systems*, Vol. 6, T. Oswalt, W. C. Keel, eds. (New York, Springer, 2013), p. 1.

Chapter 19

Strange Nuclei

In most spiral galaxies, the nucleus is a faint, star-like object centered within the bulge. The spectra of typical nuclei are dominated by starlight, although diffuse ionized gas may also be present. If one could see a typical nucleus close-up, it would be packed with hundreds of thousands of stars per cubic light year. Instead of being light years apart on average, as are the stars in our neighborhood, these stars would be only a small fraction of a light year apart. If we could imagine traveling to a galactic nucleus and observing the night sky, we would see literally millions of first magnitude and brighter stars. There would be so many bright, naked eye stars that there likely would not be enough names to cover them all. Not only this, but many of these stars would move noticeably on relatively short time scales, so that any "constellations" we might imagine would be much more transient than anything we see from our current vantage point.

In about 15% of all spirals, the nucleus is much brighter than usual, and the spectrum of the nucleus shows strong emission lines. Such lines cannot come from ordinary stars, because stars typically have spectral absorption lines. Instead, the lines must come from interstellar gas. Some of the lines are also unusually broad, which indicates turbulent gas motions of thousands of kilometers per second.

A nucleus which shows such a spectrum is said to be "active." The study of active galactic nuclei has been one of the most dynamic and important subfields of extragalactic astronomy.

The discovery of active galaxies formally began with the work of Carl Seyfert (1911–1960), a student of Harlow Shapley and one of the most dynamic and productive astronomers of his time. In a 1943 paper,[1] Seyfert presented measurements of the intensities of the spectral features in active nuclei, and noted how all the same emission lines that are found in typical planetary nebulae are found in the spectra of such nuclei, especially for the two shown in Figure 1, which both became prototypes of *Seyfert galaxies*. Both M77 and NGC 4151 are Sb spirals at about 50 million light years distance. In M77, the active nucleus is not exceptionally bright, and lies within a small ring-like spiral pattern. In contrast, the active nucleus of NGC 4151 is very bright and easily seen with small telescopes.

Figure 1 shows a small portion of the spectrum of the active nuclei in M77 and NGC 4151 (from Seyfert's original paper). Only three emission lines are highlighted: the permitted hydrogen line called H-beta (or $H\beta$, at 4861 Å)[2] and the two "forbidden" lines due to doubly-ionized oxygen (at 4959 and 5007 Å). Whether spectral emission lines are "permitted" or "forbidden" has to do with our ability to see them in a laboratory spectrum. A laboratory provides a much denser environment than is normally encountered in interstellar space. All Balmer emission lines[3] of hydrogen can be seen in a laboratory spectrum, and thus are permitted. The doubly ionized oxygen lines are, however, produced only in the extremely low density environments of interstellar space; the high density of a laboratory environment "forbids" these lines from being seen on Earth.[4]

The spectrum of NGC 4151 defines what is now known as a Seyfert 1 galaxy, the hallmarks of which are broad permitted emission

---

[1]C. K. Seyfert, *Astrophysical Journal*, **97**, 28 (1943).

[2]One Angstrom equals 10^{-8} cm (0.1 nm), and is a commonly used unit for wavelength among astronomers.

[3]The Balmer series of hydrogen emission lines refers to all lines produced by quantum leaps of electrons in excited energy levels having index $n > 2$ down to the $n = 2$ energy level; see Figure 4 of Chapter 8.

[4]The forbidden oxygen lines are usually specified as [OIII] 4959, 5007, where the roman numeral indicates the degree of ionization (I = neutral, II = singly-ionized, and III = doubly-ionized, etc).

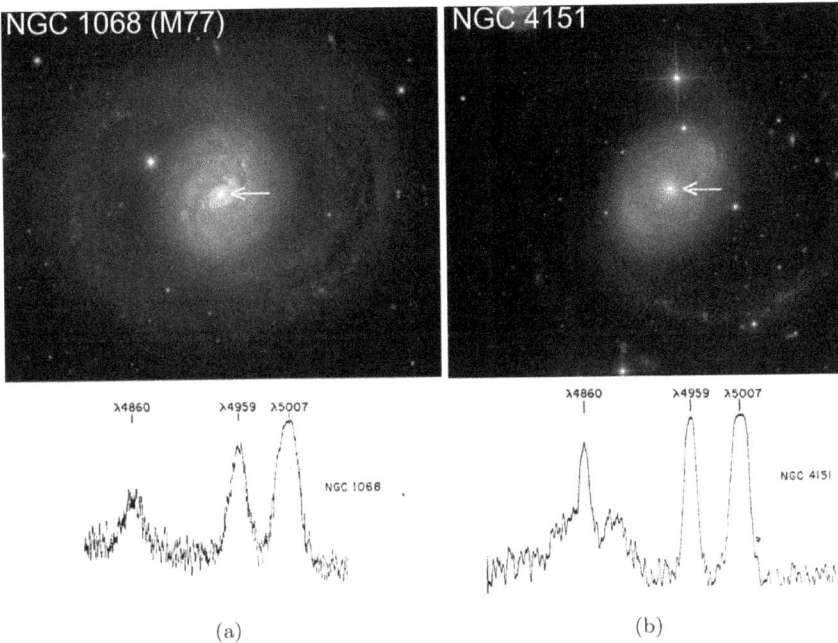

Fig. 1. The Seyfert galaxies NGC 1068 (M77) (a) and NGC 4151 (b). The images are from the de Vaucouleurs Atlas of Galaxies published by Cambridge University Press, 2007; reprinted with permission. The arrow points to the active nucleus in each case. The spectrum beneath each image shows the appearance of three major emission lines: Hβ, a bluish-green "permitted" line due to atomic hydrogen, and two "forbidden" green emission lines due to doubly ionized oxygen. From C. K. Seyfert, *Astrophysical Journal*, **97**, 28 (1943); © AAS. Reproduced with permission.

lines like the Hβ line and narrow forbidden emission lines like the doubly ionized oxygen lines. The Hβ line also shows a narrow component sitting on top of the broad line. The spectrum of M77 is different and defines a Seyfert 2 galaxy as an active galaxy having the same narrow forbidden lines as a Seyfert 1 galaxy, but lacking the broad permitted lines.

A peculiar aspect of the spectra of Seyfert galaxies is the high excitation and ionization level of the gas. For example, as a neutral atom, iron normally has 26 electrons, but in the spectra of Seyfert galaxies, emission lines due to Fe VII, a species of iron that has lost 6 of its 26 electrons, are found. In addition, emission lines due to

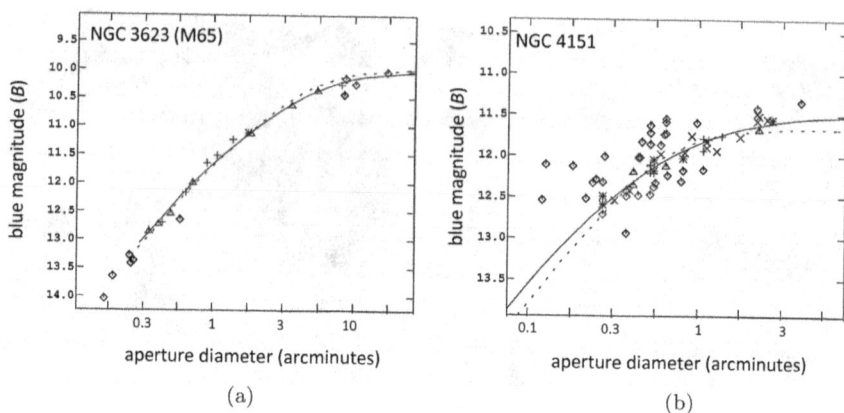

Fig. 2.   Graphs showing blue-light multi-aperture photometry of two galaxies: NGC 3623 (M65) (a) and NGC 4151 (b). M65 has a normal galactic bulge and nucleus, while NGC 4151 is the famous Seyfert 1 prototype active galaxy. The excessive scatter in the measurements of NGC 4151 compared to M65 is due to variability of the active nucleus of NGC 4151. In each case, the observations were made over a period of several decades.

three-times ionized carbon and argon, and four times ionized calcium, are seen. The high turbulence in the gas is likely responsible for such lines.

Not only do Seyfert nuclei emit a peculiar nuclear spectrum, they also emit excessive amounts of radiant energy at non-optical wavelengths. Some are bright at X-ray, infrared, and radio wavelengths. Except for the Sun, stars are not generally detected at radio wavelengths, so these non-optical wavelengths are not coming from the stars in the nucleus.

Figure 2 shows multi-aperture photoelectric photometry of NGC 4151, with its active nucleus, and another galaxy, NGC 3623 (M65), with a normal nucleus. The curves are used to extrapolate the total apparent brightness of the galaxies as was described in Chapter 7. This appears to be reasonably good for M65, but the scatter in the photometric measurements for NGC 4151 is very large and the extrapolation is clearly not reliable. The scatter cannot be attributed to noise in the photometric observations because NGC 4151 is a bright galaxy and easy to measure with multi-aperture photoelectric photometry. Instead, the scatter would seem to imply some kind of variability.[5] It cannot be variability in the brightness of the whole galaxy, because an object cannot vary on a timescale shorter than the light travel time across the object. NGC 4151 is nearly a hundred thousand light years in diameter, meaning it can't vary on a timescale less than 100,000 years. Instead, the large scatter of the data points for NGC 4151 must be due to variability in brightness of the active nucleus.

Twenty years after the publication of Seyfert's paper, the field of active galactic nuclei was thrown into a tail spin by the discovery of "radio stars." A radio star is an object that looks like a star at optical wavelengths, but which was first detected in radio waves. Radio astronomy took off as a new field of scientific research in the 1950s, and began mainly as surveys for radio sources around the sky. The surveys provided information on radio apparent brightnesses, but most importantly the surveys yielded accurate positions of the sources on the sky. Since it was not necessarily possible to tell what the radio sources were from these observations, the radio positions would be plotted on ground-based optical sky survey charts (the famous Palomar Sky Survey) to see if anything familiar was at the radio position. Many sources turned out to be obvious galaxies or nebulous objects, but some surprisingly turned out to be stars. Not only was it peculiar for these apparent stars to be bright in radio waves, but also when spectra of these radio stars were obtained, they were found to contain bright emission lines, which is also a characteristic not found for normal stars. Because these objects only looked like stars, but couldn't actually be stars, they became known as quasars, which is short for quasi-stellar objects.

At first the emission lines in quasars were hard to decipher — the wavelengths did not match up with normal spectral emission

---

[5]It is believed that Antoinette de Vaucouleurs first suspected the variability of NGC 4151 when analyzing photoelectric photmetry of the galaxy obtained by Gérard in 1957–1958. However, Gérard, believing at the time that it was impossible for a galaxy to be variable, dismissed the idea as improbable. The first report demonstrating the reality of the light variations was a paper by Fitch, Pacholcyk, and Weymann in 1967. The de Vaucouleurs presented their results on photometry of Seyfert nuclei in 1974.

lines in the several cases that had been found. The key to solving the mystery was the discovery that a radio source known as 3C 273[6] had an optical counterpart in the form of a 13th magnitude star. In the spectrum of this particular quasar, which was bright enough to be visible in a 6-inch telescope and was much brighter than any of the other radio stars, Mount Palomar astronomer Maarten Schmidt in 1963 identified the Balmer series of hydrogen in emission, redshifted by 16% of the speed of light. The emission lines of other quasars were similarly determined to be even more severely redshifted conventional lines. Most interesting was what the high redshifts implied about the distances and brightnesses of quasars. The expanding universe was already well established by the 1960s, and the high redshifts of quasars implied distances of billions of light years. For objects billions of light years away to appear as bright as quasars, they had to have luminosities hundreds of times as bright as a typical spiral galaxy. For example, modern estimates place 3C 273 at a distance of 2.1 billion light years. In order to appear as bright as 13th magnitude, 3C 273 would have to be a hundred times as luminous as the Andromeda Galaxy, one of the largest and most massive nearby spirals.

The discovery led to a flurry of research papers that lasted for decades. Quasars were found to vary erratically in brightness on short timescales, implying the source of the huge amount of energy being emitted could not be a galaxy-sized object, but something much smaller. Eventually, it was concluded that quasars are basically extremely remote Seyfert galaxies where the active nucleus is so bright that it drowns out the light of the host galaxy. It was also concluded that the only way to explain the extreme luminosities is by the presence of a supermassive black hole[7] accreting matter in the

---

[6] "3C" stands for *Third Cambridge Catalogue of Radio Sources*, which was published in 1959.

[7] All stellar and galactic black holes are distorted regions of space where a massive object or collection of material has collapsed into a singularity, or point of infinite density. Typical stellar black holes are several tens of solar masses and probably come from the evolution of stars more massive than 30 solar masses. Supermassive black holes are millions to billions of solar masses in comparison, and likely are not related to stellar evolution.

center of a galaxy. Material can collect in the center of a galaxy since this is where the gravitational well is deepest. Through interactions, interstellar gas clouds can lose orbital angular momentum and fall inward towards the black hole. Instead of falling directly into the hole, the material collects first into an accretion disk where it slowly spirals inward. In the presence of a powerful gravitational field, the material gets heated enough to emit X-rays before it goes in. Depending on the rate of accretion (that is, on the rate at which the matter falls in per unit time), material may be accelerated out of the poles of the disk and carried outward nearly at the speed of light in the form of bipolar jets. It is these jets that account for the high radio brightness of quasars.

There are other classes of active galaxies. Radio galaxies are galaxies emitting the bulk of their luminosity in radio waves; in most of these cases, the host galaxy is not a spiral but a giant elliptical galaxy that when mapped in radio waves appears as a double-lobed structure with one or two radio-emitting jets. A BL Lacertae object is an active galaxy that does not show the usual emission-line spectrum, only a continuous spectrum of light. Like quasars, BL Lacertae objects appear star-like in visible light, but the spectrum betrays the nature of the object as an active galaxy.

The standard (or unified) model of active galaxies (Figure 3) interprets the differences between quasars, Seyfert galaxies, radio elliptical galaxies, and BL Lacertae objects in terms of how the central accretion disk and its surroundings are viewed. The model interprets the broad spectral lines as coming from a region (the "broad line region") closest to the hole, in the main part of the accretion disk. The turbulence in this region, on the order of 10,000 km/s, is higher than elsewhere and accounts for the broadness of the region's spectral emission lines. Outside this region, a dusty, donut-shaped expanse of gas and dust called a torus is surrounded by a flurry of discrete ionized gas clouds that account for the narrower emission lines (turbulence on the order of several hundred kilometers per second) seen. The torus has the effect of limiting how we see the accretion disk closest to the hole. If the disk is seen at an intermediate angle (neither exactly face-on, nor edge-on), then we

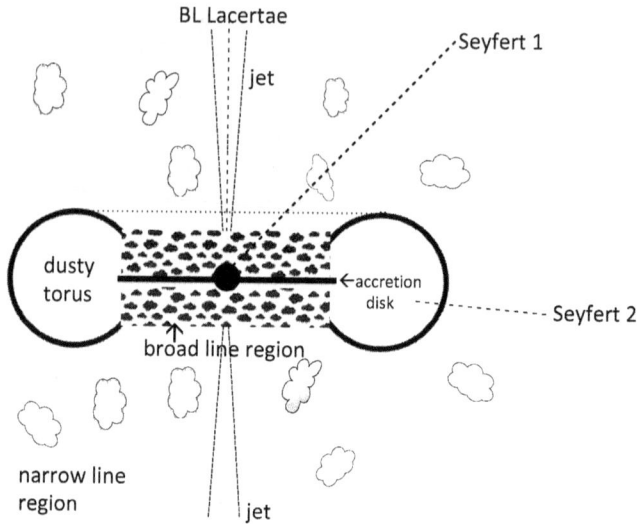

Fig. 3. The standard model of active galaxies, including Seyfert galaxies, quasars, radio elliptical galaxies, and BL Lacertae objects. Adapted from M.-P. Véron-Cetty and P. Véron, *Astronomy and Astrophysics Review*, **10**, 81 (2000).

can see into the broad-line region, and the active nucleus will be classified as Seyfert 1. Seen more nearly edge-on, the dusty torus will block the light from the broad line region, leading to an active nucleus classified as Seyfert 2. In some cases, one of the two radio jets will be pointing almost exactly towards us. In this circumstance, the continuous radiation of the jet, which involves material moving nearly at the speed of light, will swamp the light of the emission lines and we will see a BL Lacertae object.

The standard model leaves some important questions. For example, how do supermassive black holes form, and how many galaxies actually have such a thing at their center? It turns out that there are ways of determining whether a black hole is present in a galaxy even when the hole is not accreting any matter. Both random and orbital speeds of stars in the nucleus will increase substantially as the center of the galaxy is approached, and the level these speeds achieve depends on the mass of the black hole. When examined this way, it is found that most massive galaxies host a supermassive black hole, irrespective of whether the object is a disk-shaped galaxy or not.

The masses estimated range from millions to billions of solar masses and correlate strongly with the mass of a galaxy's bulge, in the sense that more massive bulges host more massive central black holes.[8] The implication of this is that supermassive black hole formation is likely tied to the process of galaxy formation itself.

19.1. Supermassive Black Holes: Up Close and Personal

Up until April, 2019, no supermassive black hole had been imaged close up. Although some have event horizons[9] as large as the solar system, these occur in galaxies tens of millions of light years away and as a consequence have very small angular sizes. Nevertheless, an image of a supermassive black hole was obtained in April 2019 using widely separated radio telescopes to synthesize an almost Earth-sized telescope.[10] The black hole imaged lies in the center of the giant elliptical galaxy M87, located in the Virgo galaxy cluster 55 million light years away. Figure 4 shows the central part of M87 with its 5,000 light year long jet of fast-moving gases. There should actually be two jets in opposite directions, but the second is invisible because of peculiar optical effects due to the near light speed of the jet gases.

The spectacular inset shows the expected "shadow" of the black hole, a dark zone about 2.5 times the event horizon radius within which photons are captured by the hole. The radius of the event horizon depends mainly on the black hole mass which has been estimated to be 6.5 billion solar masses. At the boundary of the shadow, photons are trapped in peculiar circular orbits, in a region called the "photon sphere." The nearly circular ring is related to the accretion disk and is not necessarily as face-on as it looks. The gases in the disk are rotating around the hole at near light speed, and even if the disk is highly inclined, optical effects will distort it into a more

---

[8] J. Kormendy and L. C. Ho, *Annual Reviews of Astronomy and Astrophysics*, **51**, 511 (2013).

[9] The event horizon of a black hole is the imaginary surface where the escape speed equals the speed of light.

[10] Event Horizon Telescope Collaboration (K. Akiyama *et al.*), *Astrophysical Journal*, **875**, L1 (2019).

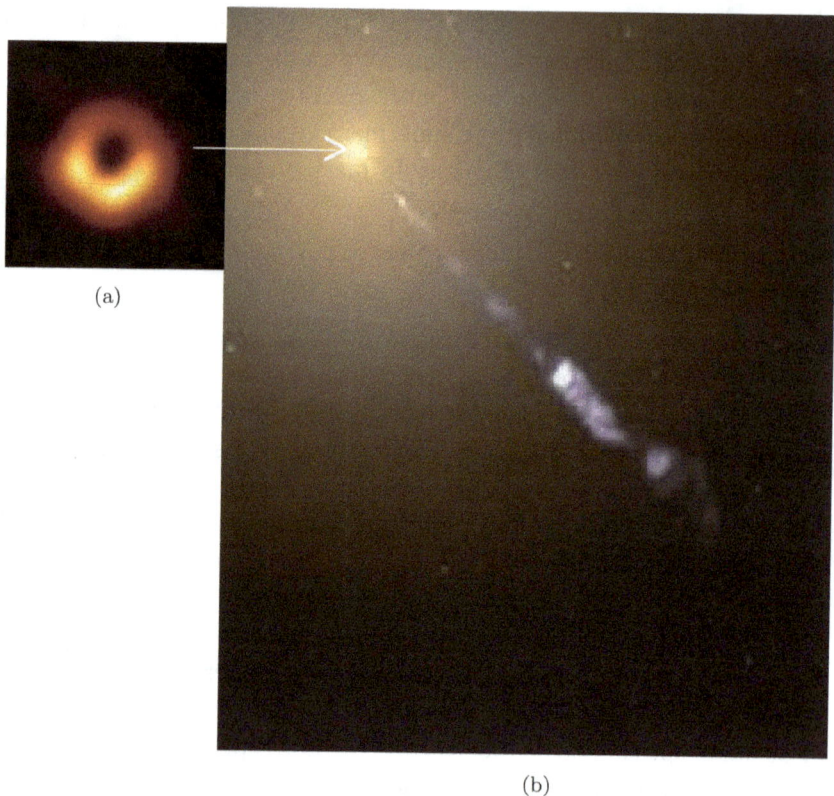

Fig. 4. (a) The supermassive black hole in M87, as imaged with the Event Horizon Telescope. The event horizon is well within the dark shadow of the hole. Image credit: The Event Horizon Telescope Collaboration, K. Akiyama *et al.*, *Astrophysical Journal*, **875**, L1 (2019); © AAS. Reproduced with permission. (b) *Hubble Space Telescope* image of the gaseous jet of M87. Image credit: J. A. Biretta *et al.*, Hubble Heritage Team (STScI/AURA), NASA.

circular shape. The asymmetric brightness of the ring is consistent with this idea. The side of the disk coming towards us gets boosted in brightness compared to the side moving away from us, due to the Doppler effect.

19.2. Monster in the Milky Way

Does our Galaxy host a supermassive black hole? The answer is yes, although it is not an especially massive example. The radio source

Sagittarius (Sgr) A* is interpreted as being the true center of the Milky Way and the location of the Milky Way's central black hole. Although its mass is only 4 million solar masses, it is by far the nearest example of this phenomenon to us, being only 27,400 light years away.[11] In fact, it is so close that the orbits of individual stars around the black hole have been mapped in detail. Six orbits are shown in Figure 5.[12] The orbits are compared in size with those of Sedna and Pluto, the former being one of the most distant solar system bodies known. Sedna orbits so far from the Sun that it has an orbital period of 11,400 years, but the stars orbiting the central black hole have much shorter periods, ranging from 16 years for star S2 to 166 years for star S1. The reason for the difference is that Sedna is orbiting a one solar mass object (the Sun) while S1 and S2 are orbiting an object having a mass of 4 million solar masses. To maintain their large orbits, the stars orbiting the black hole have to move much faster than any solar system object has to move. When closest to the black hole, star S1 is moving at a speed of 1,650 km/s while star S2 is moving at a speed of 7,680 km/s,[13] compared to 0.4 km/s for Sedna. Two of the illustrated cases get very close to the black hole. S2 gets as close as 118 AU[14] while S14 gets as close as 56 AU; the latter is just a little larger than Pluto's average distance, 40 AU, from the Sun. These distances are still very large compared to the event horizon radius, 0.08 AU, of the black hole.

The shadow of the Milky Way's black hole has not yet been published, but the astronomers who released the image of M87's black hole are working to do the same for the Milky Way's hole. As it turns out, the event horizon of the Milky Way's black hole is 1625 times smaller than that in M87. By the same token, the Milky Way's hole is 2007 times closer to us. This means the apparent angular size of the two holes will be very similar.

---

[11] A. M. Ghez *et al.*, *Astrophysical Journal*, **689**, 1044 (2008).

[12] F. Eisenhauer *et al.*, *Astrophysical Journal*, **628**, 246 (2005).

[13] https://en.wikipedia.org/wiki/Sagittarius_A*.

[14] AU stands for astronomical unit; 1 AU equals the average Earth-Sun distance of 92,957,000 miles.

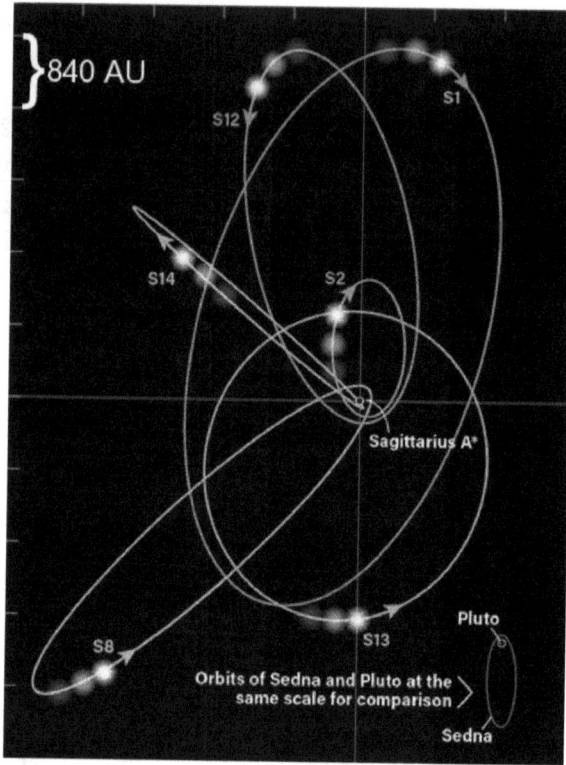

Fig. 5.   Orbits of six stars around the central black hole of the Milky Way. The black hole is labeled "Sagittarius A*" and was originally discovered as a radio source. The orbits are compared to those of Pluto and Sedna in the Solar System at lower right, while the scale of the map is 840 astronomical units (AU) per 0.1 arcsec (upper left). An AU is the average distance between the Earth and the Sun. Illustration credit: Roen Kelly, after Cmnglee/Wikimedia Commons/.

It is interesting to contemplate what it would be like if the Milky Way harbored a quasar as luminous as the one in 3C 273 which, from a great distance, swamps the light of its parent galaxy. The 3C 273 quasar has an absolute visual magnitude of −26.3, meaning that from a distance of 10 pc, it would appear visually as bright as the Sun (visual magnitude −26.7). If the quasar were instead placed at the center of the Milky Way, at a distance of 27,400 light years, it would appear 75 times as bright as Venus and would be easily visible during the day. (This would be in the absence of dust.) While the visible light may not be harmful, the non-optical wavelengths, like X-rays, emitted from the quasar region could be a problem, although the Earth's atmosphere prevents X-rays from space from reaching the ground. If we were in the path of one of the bipolar jets, the fast-moving gases could interact with our atmosphere, and may have a significant impact.

19.3. The Quasar Time Machine

One of the most interesting aspects of quasars is how none can be considered to be "nearby." Of hundreds of thousands of known examples, none is closer to us than 500 million light years, and most are more than a billion light years away. The implication of this is that quasars are a phenomenon of the remote past. Although most massive galaxies have a central black hole, the Universe seems to have passed the phase of high accretion rates onto these holes, leaving nearby Seyfert galaxies like M77 and NGC 4151 as relics of a bygone era. NGC 4151 is almost a mini-quasar in the brightness of its active nucleus, implying there is possibly an evolutionary connection between nearby active galaxies and quasars.

Chapter 20

Distant Spiral Galaxies

An important question is: how have spiral galaxies evolved over time? Were some galaxies merely born as spirals, and then maintained their shapes indefinitely, or were they born with some other morphology, and then developed a spiral shape later? Fortunately, there is a way to answer this kind of question: examine the morphology of extremely distant galaxies, known as *high redshift* galaxies. The redshift z of a galaxy is the ratio of the increase in wavelength of its spectral emission or absorption lines (due to the expansion of the universe) to the laboratory wavelengths of those lines. A nearby galaxy generally has $z < 0.05$, while a high redshift galaxy has $z > 0.5$. z correlates with something astronomers call the *lookback time*: the higher the redshift, the greater the distance to a galaxy, and the farther back in time we are looking.[1]

While galactic redshifts can be easy to measure, lookback times are not because high redshifts correspond to recession speeds approaching the speed of light. The linear relationship between recession speed and distance that Hubble discovered can only be used to get distances to nearby galaxies. Lookback times depend on the general curvature of space, which is tied to the average density of matter and energy in the universe, including the relative

---

[1]Here, lookback times are estimated using the cosmology calculator of N. Wright, *Publications of the Astronomical Society of the Pacific*, **118**, 1711 (2006) and http://www.astro.uda.edu/~wright/CosmoCalc.html.

contributions of both ordinary and dark matter and ordinary and dark energy.[2]

Interpreting high redshift galaxy morphology is complicated by a number of effects. One problem is that high redshift galaxies are generally not as well resolved as nearby galaxies. This is a problem that can be addressed by building larger space observatories, such as the James Webb Space Telescope scheduled to be launched in 2021. Another problem is "band-shifting." The spectrum of a high redshift galaxy is so stretched and shifted that filters such as $UBVRI$ do not sample the parts of the spectrum for which they were designed. For example, the B-band filter is designed to transmit light between 400 and 500 nm in wavelength, and we have seen that such light from a nearby galaxy is strongly affected by young stars and dust. For a $z = 1$ high redshift galaxy, the same light is redshifted to the range 800–1000 nm, in the near-infrared wavelength domain. The B-band filter transmits instead light that was originally emitted in the far- to mid-ultraviolet part of the spectrum, from 200 to 250 nm, a domain which will emphasize only very hot stars. This leads to another complication: selection effects. At high redshift, the B-band filter is most sensitive to galaxies with a high degree of star formation. Also, any sample of high redshift galaxies would have a natural bias towards the largest and most luminous cases.

The study of high redshift galaxy morphology has benefitted greatly from deep field imaging with the *Hubble Space Telescope*. "Deep field" means that the total exposure times are so long that the final images reveal the faintest galaxies ever seen. Deep field imaging is also usually multi-color, spanning from ultraviolet to near-infrared wavelengths, in order to deal with the band-shifting effect. One of the deepest fields thus far obtained is called the Hubble Ultra-Deep Field (HUDF), an image of a very small piece of sky in the constellation Fornax that represents approximately 400 orbits worth of exposure time (Figure 1).[3] The image reveals thousands of faint galaxies, including some very distant spirals.

---

[2]Dark energy refers to an energy that permeates the vacuum of space and acts as a repulsive force that is accelerating the universal expansion.

[3]S. Beckwith *et al.*, *Astronomical Journal*, **132**, 1729 (2006).

Fig. 1.   The Hubble Ultra-Deep Field (HUDF) in the constellation Fornax, showing galaxies spread over more than 10 billion light years of distance from us and seen from lookback times spanning almost the entire age of the universe. Credit: NASA, ESA, and S. Beckwith (STScI) and the HUDF Team.

Figure 2 compares one of these spirals, named HUDF 423, with the nearby multi-armed spiral NGC 4897 in Virgo. Despite the great difference in lookback times (3.2 billion years versus 128 million years, a factor of 25 difference), the two galaxies look astonishingly similar. Both have a bright inner ring, a weak bar, and extensive spiral structure. The universe was already 10 billion years old when the light we now receive from HUDF-423 was emitted. Even a lookback time of 3 billion years is close enough to our time that there is little difference in galaxy morphologies due to evolutionary effects. Figure 3 shows a sampling of more distant galaxies with higher lookback times. In these we can see possible evidence for galaxy evolution. The lowest lookback time is 4.3 billion years ($z = 0.42$ case), while the highest is 11.4 billion years ($z = 3.35$ case), the latter being seen as it appeared

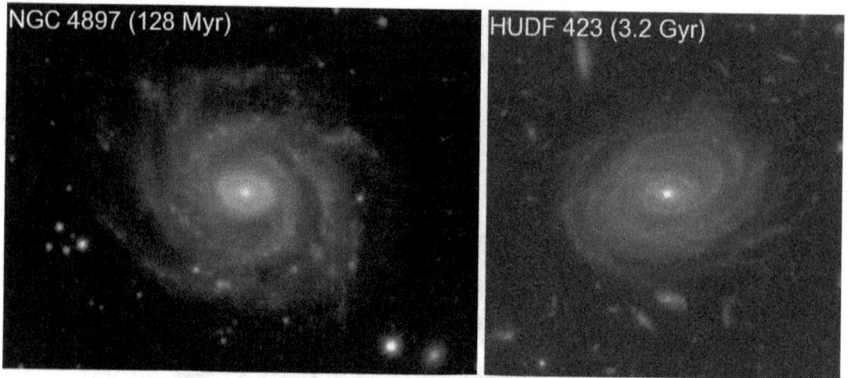

Fig. 2.   HUDF 423, a redshift 0.29 galaxy in the Hubble Ultra-Deep Field, bears a strong resemblance to the nearby galaxy NGC 4897, redshift 0.0085. The lookback times are indicated. Image credits: NGC 4897, The de Vaucouleurs Atlas of Galaxies published by Cambridge University Press, 2007; reprinted with permission; HUDF 423, NASA, ESA, and S. Beckwith (STScI) and the HUDF Team.

only 1.9 billion years after the Big Bang. While there are normal-looking galaxies in this range (the $z = 0.59$ system is a Hubble type Sb–Sc spiral while the $z = 0.66$ case is a Hubble type E galaxy), most appear unusual. Clump clusters, tadpoles, chains, and bent chains are peculiar morphologies that do not fit neatly within the Hubble tuning fork.[4] Irregular shapes become much more common at $z > 1$, and not only this, but galaxies at high redshift are generally significantly smaller than nearby galaxies.

The likely reason for the irregular shapes is that $z > 1$ galaxies are being seen when the universe was less than 6 billion years old. At these early times, some galaxies are still in their formative phase, meaning they are still accreting matter and building up their mass. During this time, they can be unstable to forming massive star-forming regions called clumps, and also may be subjected to frequent mergers with other forming galaxies or groups of clumps. At some point, mergers become less frequent and the formative phase by and large ends, allowing the formed galaxies to evolve more slowly into the shapes we see today. We say that this is the time when

---

[4]D. M. Elmegreen *et al.*, *Astrophysical Journal*, **631**, 85 (2005).

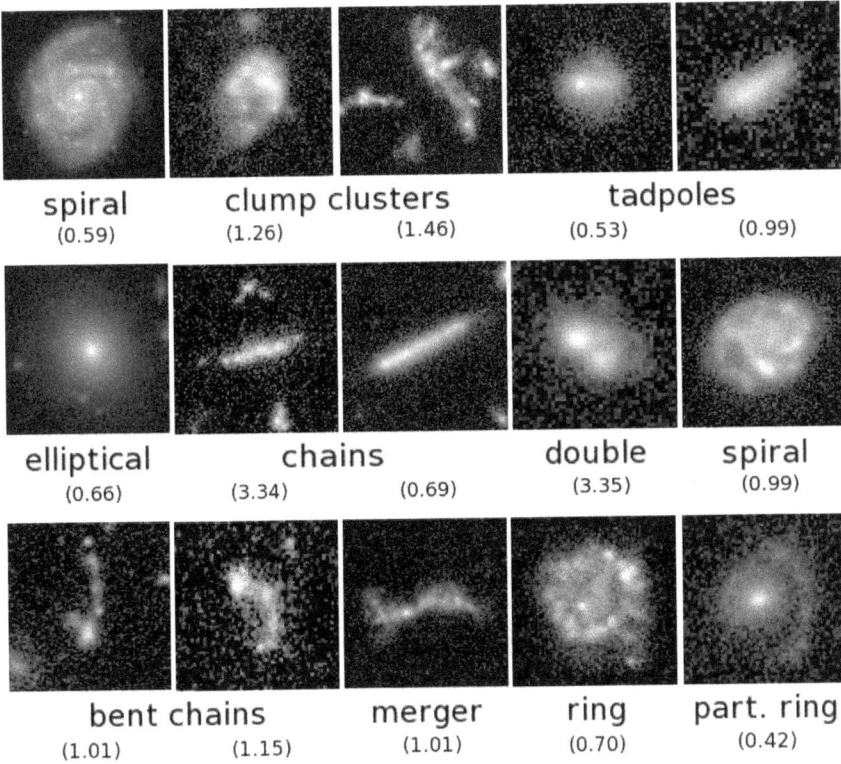

Fig. 3. High redshift galaxy morphology. The number in parentheses beneath each image is the redshift z of the galaxy. From R. Buta, *Galaxy Morphology, Planets, Stars, and Stellar Systems*, W. C. Keel, ed. (New York, Springer, 2013), p. 1; reproduced with permission.

secular evolution takes over as the dominant mechanism of change.[5] The idea is that the formative phase of galaxy evolution is fairly rapid compared to the secular evolutionary phase. To interpret the morphology of some nearby galaxies, we should appeal to secular evolutionary processes.

Secular evolution can be driven entirely by internal processes or by external processes. When it comes to an object like a galaxy, secular evolution has a natural course: to enhance the central

---

[5] J. Kormendy, R. C. Kennicutt, *Annual Reviews of Astronomy and Astrophysics*, **42**, 603 (2004).

concentration, and extend the outer disk, a process that can lead to a more tightly bound gravitational system.[6] This doesn't necessarily just happen — the process needs an "engine," something to drive it. Bars can be thought of as internal engines of secular evolution in disk-shaped galaxies, because they act from within to rearrange the material, particularly gas and dust clouds, in a galactic disk. Bars can drive interstellar gas into rings, as was described in Chapter 15. Rings are therefore products of secular evolution.[7] Although the effects of secular evolution are most evident in the gaseous disk of a galaxy, the stars can also participate in such evolution. In a 1996 paper, George Mason University astronomer Xiaolei Zhang showed that a subtle mismatch between a bar or a spiral and the orientation of the gravitational fields that they generate can drive the slow redistribution of both stars and gas clouds across an entire disk.[8] Movement of a galaxy in a cluster can also effect secular evolutionary change.[9]

The process of galaxy formation never really ends. The Earth, for example, is believed to have formed by accretion of small objects (called planetesimals) early in the history of the solar system. Any time a meteor flashes through the Earth's atmosphere, or a meteorite hits the ground, the Earth gains a little more mass. The Earth is therefore in a way still forming, although at a much slower rate than in the early days. By the same token, a galaxy today can still accrete smaller objects, like dwarf galaxies, and be considered to be still forming at a very low rate.

Studies have shown that the first spirals appear in the HUDF at $z \approx 2$,[10] corresponding approximately to 3 billion years after

---

[6] J. Kormendy, in *Secular Evolution of Galaxies*, J. Falcon-Barroso, J. Knapen, eds. (Cambridge, Cambridge University Press, 2012), p. 1.

[7] For example, J. Knapen, in *Galaxies and their Masks*, D. L. Block *et al.* (Springer Science, 2010), p. 201.

[8] X. Zhang, *Astrophysical Journal*, **457**, 125 (1996).

[9] J. Kormendy and R. Bender, *Astrophysical Journal Supplement Series*, **198**, 2 (2012).

[10] D. M. Elmegreen and B. G. Elmegreen, *Astrophysical Journal*, **781**, article 11 (2014).

the Big Bang. Grand design spirals are the first spirals that appeared, possibly because spiral density waves can be induced by a gravitational interaction with a nearby companion, a phenomenon which would have been more frequent at higher redshifts.

20.1. The Hubble Tuning Fork: Yesterday versus Today

The modified Hubble tuning fork classification of galaxies (Figure 10 of Chapter 6) has cosmological significance. At the present time, spirals are the dominant morphology of massive galaxies. Yet, this could not have always been true because few spirals are found at redshifts $z > 1$. One could ask the question: when did galaxies "settle" into their current tuning fork positions? The key to answering this question is a careful comparison of classifications of galaxies from low to high redshift using images of comparable depth and quality. One of the most informative studies of the evolution of the tuning fork was made by R. Delgado-Serrano and coworkers.[11] These authors derived the relative proportion of ellipticals, S0s, spirals, and irregular/peculiar galaxies for two galaxy samples: a nearby ($z \approx 0$) sample, and a more distant ($z = 0.65$) sample having a lookback time of 6 billion years. They found no change in the relative proportion ($\approx 20\%$) of ellipticals and S0s between these two samples, but found spirals to be more than a factor of two less frequent in the high z sample (72% nearby versus 31% at $z = 0.65$). Instead of spirals, the most common morphology at a lookback time of 6 billion years was the irregular/peculiar type (10% nearby versus 52% at 6 billion years). This suggests that the irregular, peculiar systems that are so common at high redshift are the *precursors* of spiral galaxies, one of the most interesting results to have come out of studies of high redshift galaxies.

---

[11]R. Delgado-Serrano *et al.*, *Astronomy and Astrophysics*, **509**, 78 (2010).

Chapter 21

The Nature of S0 Galaxies

From Chapter 6, we saw that S0 galaxies are disk-shaped systems just like spirals, only they lack spiral arms; they also seem to be almost completely free of interstellar gas and dust, the raw materials needed for forming new stars. How did such galaxies come about?

The interstellar gas and dust in a spiral galaxy are confined mainly to the plane of the disk. Neither the halo nor the bulge are places where such materials are abundant. The rate at which a whole galaxy forms stars, or the rate at which a specific galactic feature (such as a ring) forms stars, can be estimated by how much light a galaxy emits in its Hα emission line. For example, the galaxy NGC 210, which is seen in Figure 8 of Chapter 11 to have very bright spiral arms in the light of Hα, has been estimated to be converting 1.5 solar masses of interstellar gas into stars *per year*.[1] This does not sound like much in a galaxy likely to have several hundred billion stars, but it can add up over time. If a typical spiral galaxy has a mass of 100 billion times the mass of the Sun, 10% of which is in the form of interstellar gas, then at a rate of 1.5 solar masses per year, such a galaxy would use up all of its interstellar gas in less than half the age of the universe. In view of this, one could imagine a spiral galaxy using up all of its raw star-forming material well before the present time, thereby losing its ability to be seen as a conspicuous spiral.

---

[1] D. A. Crocker *et al.*, *Astrophysical Journal Supplement Series*, **105**, 353 (1996).

One might also question whether taking away all of the interstellar gas and dust from the M51 system would cause the spiral arms of the main component to disappear. As shown in Figure 11 of Chapter 16, the spiral pattern is very prominent in the light of very old stars, indicating that the gravitational field has this shape. If the field is being triggered by the companion, then it might seem that to convert NGC 5194 into an S0 would also require removing the companion from the system. This is not necessarily the case because, in general, spirals are expected to fade and eventually disappear in a disk-shaped galaxy if no interstellar gas is present. Stellar spirals will tend to fluctuate in strength due to collective effects, which increases random stellar motions over time. Such motions will only serve to weaken a stellar spiral pattern until it disappears.[2] Thus, if some mechanism removes the interstellar gas from a spiral galaxy, the end result should be an armless disk-shaped galaxy.

The trouble with the "running out of gas" idea is that it was noticed in the 1930s that S0 galaxies are remarkably abundant in the dense parts of rich galaxy clusters, like the great cluster Abell 1689 in Virgo (Figure 1). Spirals are also found in such clusters, but tend to avoid the denser areas. Some clusters are actually spiral-rich as in the Hercules Cluster Abell 2151 (Figure 2).

In 1951, Lyman Spitzer (1914–1997) and Walter Baade (1893–1960) proposed that S0 galaxies are spiral galaxies that have been "stripped" of their interstellar gas and dust through an interaction with their environment.[3] They proposed that multiple galaxy collisions could account for the high number of S0s in clusters. Then in 1971, astronomers discovered that rich galaxy clusters include a huge mass of gas so hot that it emits X-rays. X-rays are electromagnetic waves of very high energy, intermediate between ultraviolet light and gamma rays. In astronomy, anything emitting X-rays would have a temperature close to 100 million degrees Kelvin, much hotter than the surface of any star. The X-ray-emitting gas in clusters fills the space between the galaxies, such that the galaxies move within this

---

[2] J. Sellwood and R. Carlberg, *Astrophysical Journal*, **282**, 61 (1984).
[3] L. Spitzer and W. Baade, *Astrophysical Journal*, **113**, 413 (1951).

Fig. 1. *Hubble Space Telescope* color image of the rich galaxy cluster Abell 1689, where the majority of the galaxies are elliptical and S0 types. Located 2.5 billion light years away in the constellation Virgo. Image source: John P. Blakeslee and Holland C. Ford, used with permission.

gas, which defines what astronomers call the "intra-cluster medium." This led to the idea that a process called "ram pressure stripping" could sweep cluster galaxies of their interstellar matter, and thereby account for the high population of S0 galaxies in such clusters.[4] The "ram pressure" in ram pressure stripping is the pressure of the hot gas against the interstellar matter that has to overcome the pull of gravity within a galaxy to strip the galaxy clean, like a cosmic "broom."

Although there is likely to be more than one path to becoming an S0 galaxy, the general conclusion of many studies is that S0 galaxies are not formed as such (nature), but are *transformed* from spirals (nurture). There is much to support this idea, given the high diversity

---

[4] J. E. Gunn and J. R. Gott, III, *Astrophysical Journal*, **176**, 1 (1972).

Fig. 2. Color image of the spiral-rich galaxy cluster Abell 2151, based on observations with a 20-inch telescope. Located 495 million light years away in the constellation Hercules. Image source: Ken Crawford — imagingdeepsky.com; used with permission.

in morphologies of S0 galaxies. Figure 3 shows the three features that Hubble considered to make a non-barred S0 galaxy: the nucleus, the lens, and the envelope. The nucleus is the bright center, but the *lens* is a type of galactic morpohological feature that only recently has received attention. A lens is a feature with a shallow brightness gradient interior to a sharp edge. If S0 galaxies are in general former spirals, then it is not clear exactly what a lens represents. In 1979, John Kormendy proposed that lenses are disintegrated bars, that is, bars that lost their elongated shape through an interaction within the galaxy, possibly with the bulge.[5] However, the facts that (a) lenses are found in the same location where inner rings are found, and

---

[5] J. Kormendy, *Astrophysical Journal*, **227**, 716 (1979).

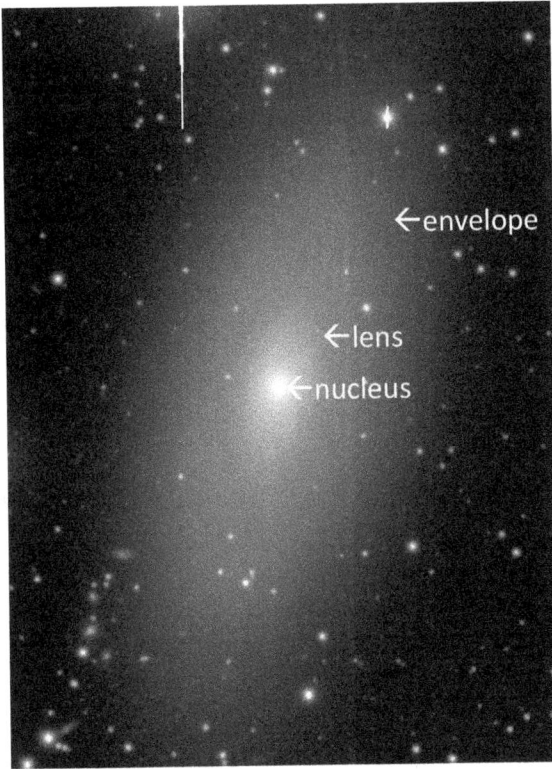

Fig. 3. A standard S0 galaxy, NGC 2784, showing the three main features that define such galaxies. Image source: The de Vaucouleurs Atlas of Galaxies published by Cambridge University Press, 2007; reprinted with permission.

(b) there are also features called nuclear lenses and outer lenses that lie in the same locations where we would see nuclear rings and outer rings, could imply that the origin of some lenses may be tied to ring evolution.

Studies of the colors of galaxies show that many S0s have the same color as elliptical galaxies, implying a similar stellar content to ellipticals. Some S0s do show evidence of recent star formation, as in the galaxy NGC 7187 (Figure 4), located 121 million light years away in the constellation Piscis Austrinus. In this case, the most recent star formation is confined to an inner ring.[6]

---

[6]R. Buta, *Astrophysical Journal*, **354**, 428 (1990).

Fig. 4.   NGC 7187 in blue light. The bright circular inner ring is a zone of active star formation. The inset shows there is also a faint, elongated outer ring. Image source: The de Vaucouleurs Atlas of Galaxies.

It is interesting to imagine what the night sky would be like in an S0 galaxy. If the Milky Way were an S0 like NGC 2784, but with a bar, and if we were located the same distance from the Galactic center, we would see a very different night sky. First, the brightest stars in our night sky would be red giants. All of the stars would likely be very old. Stars like Rigel, Sirius, Altair, Vega, Betelgeuse, and possibly even Alpha Centauri and the Sun would simply not be seen. There would still be a "milky way," that is, a band of mostly unresolved starlight going all around the sky, except that band would be very regular-looking compared to what we see within our Galaxy. We would likely see that our S0 galaxy has a disk shape in two parts: a thin disk and a thick disk (see Chapter 10). The Galactic thin disk would be the brightest feature of the disk. In addition, the Galactic center, instead of being completely invisible, would be very bright on warm summer nights in the northern hemisphere. The absence of interstellar dust would guarantee this. To the naked eye, the Galactic center would be a diffuse glow with a bright nucleus, but even a modest-sized telescope would probably resolve the mass of faint stars at the center. If the Galactic bar in our S0 version of the Milky Way was similar to what we see in the actual Milky Way, we would likely notice it more clearly as an asymmetry in the star distribution of the Galactic bulge as was discussed in Chapter 8.

If instead of being in a barred version of NGC 2784, we were in NGC 7187, then we would see some younger, massive stars. Not only would these be seen from almost anywhere in the galaxy, it is likely that we would be able to tell the star formation in our night sky was entirely confined to a ring around the center. Just as for spiral arms, our perspective on an NGC 7187-type of galaxy would be an edge-on view, meaning the two rings of the galaxy would not be seen as rings. Unless the rings are not in the same plane, which is possible, the two features should project into linear brightness enhancements as seen from within the galaxy. The inner ring would project into a straight line of star formation, and it would likely be easy to figure out its face-on shape as being circular. The outer ring has little or no star formation, but it is likely that detailed star counts would reveal the ring's presence.

I have already mentioned NGC 7702 in Phoenix as an interesting example of an S0 galaxy whose inner ring is so bright that the galaxy looked like Saturn when seen through an eyepiece on the CTIO 4-m telescope (Figure 4 of Chapter 1). NGC 7702 also has a much fainter outer ring at about twice the size of the inner ring. The appearance is that of a double-ringed barred galaxy without the bar. When computer-deprojected, the inner ring zone is strongly oval which suggests that the feature is bar-like. This makes the galaxy similar to NGC 3081, but in a more advanced state of morphological evolution. The lack of star formation in the bright inner ring, and the relatively sparse environment of the galaxy, suggest that NGC 7702 is possibly a genuine example of a former spiral galaxy that "ran out of gas." There is dust in the region of the inner ring, but its effect is very weak.

The complexities of S0 galaxies literally threw a "monkey wrench" into the Hubble tuning fork. The original tuning fork described by Hubble in 1926 had only spirals and ellipticals, but then

Hubble added the S0s as transition objects between ellipticals and spirals 10 years later. Canadian astronomer Sidney van den Bergh once commented that the addition of S0s destroyed the "simple beauty" of Hubble's classification system. And indeed this is exactly what it did. Based on the Spitzer and Baade theory of environmental modification in galaxy clusters, van den Bergh in 1976[7] proposed that a likely more reliable way of viewing S0s is as a sequence of galaxy types S0a, S0b, and S0c *parallel* to the Sa, Sb, and Sc sequences (Figure 5). He suggested that the "a", "b", and "c" subtypes for spirals not be based on the appearance of the arms as well as the relative strength of the bulge, but only on the relative strength of the bulge. In this view, S0 galaxies are the end products of environmentally-modified spirals, but to allow for the possibility that modifications could be ongoing, van den Bergh placed another sequence, intermediate between spirals and S0s, called the "anemic" spiral sequence: Aa, Ab, and Ac. For example, an Ab spiral is an Sb spiral with a lower than average rate of star formation compared to an Sb spiral. The reduced rate is attributed to a deficiency of available neutral atomic hydrogen gas, which is being stripped from the system by an interaction still ongoing. Note that the classification does not depend on the presence or absence of a bar.

The best example of an anemic-looking spiral is the Ab galaxy NGC 4921, which is one of the brightest spirals in the Coma galaxy cluster about 250 million light years away; it is shown as compared to a normal Sb spiral in Figure 5. The two galaxies look very similar in their spiral structure, but the arms in NGC 4921 are smoother than those in NGC 3992 (M109), which lies only 50 million light years away in the constellation Ursa Major. Despite the factor of 5 difference in distance, the comparison is fair because the image of NGC 4921 is based on *Hubble Space Telescope* observations while that of NGC 3992 is from groundbased observations.

A problem with the van den Bergh point of view was that it was difficult initially to find genuine examples of S0c or Ac galaxies. Most S0s appeared to have a significant bulge. Nevertheless, considerable

---

[7]S. van den Bergh, *Astrophysical Journal*, **206**, 883 (1976).

(a)

(b)

(c)

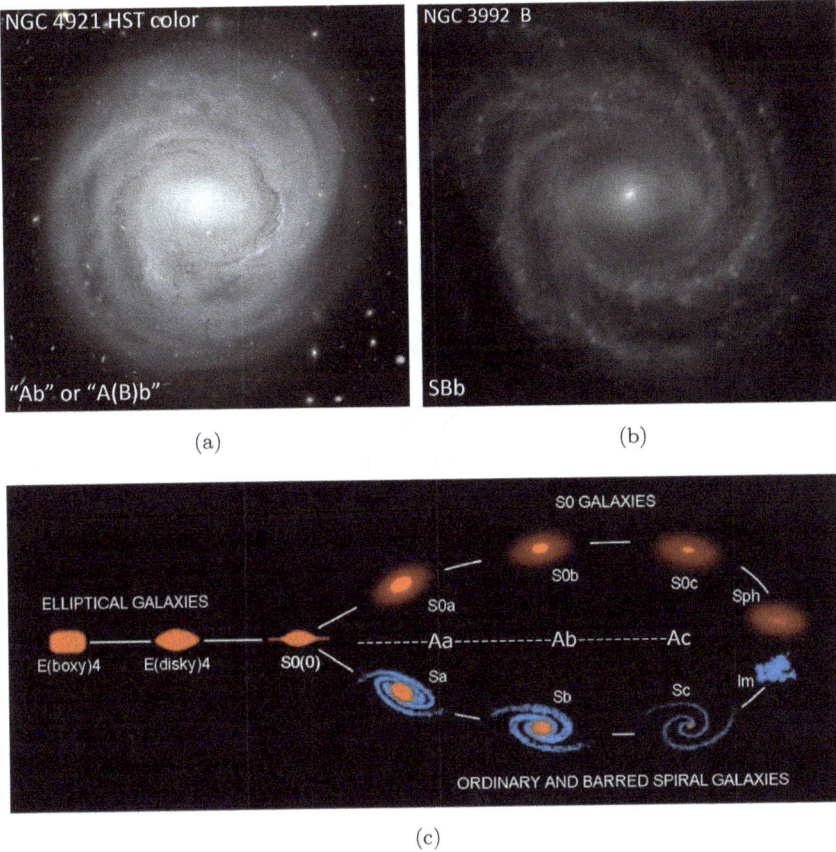

Fig. 5.   (a) NGC 4921 is a nearly-face-on spiral in the Coma galaxy cluster 250 million light years away. It is thought to be an "anemic" spiral that has been partially stripped of some of its interstellar atomic hydrogen gas (Image source: NASA, ESA, Hubble; Processing and Copyright: Kem Cook (LLNL) and Leo Shatz, with permission; (b) computer-deprojected blue light image of NGC 3992 (M109), a normal spiral that looks similar to NGC 4921 but is physically 50% smaller and not part of a dense cluster environment, and which therefore is not likely to be "anemic." NGC 4921 looks like a smoother version of M109. The M109 image has been cleaned of foreground and background objects. Image source: The de Vaucouleurs Atlas of Galaxies published by Cambridge University Press, 2007; reprinted with permission. Bottom panel: Sidney van den Bergh's revised approach to the Hubble tuning fork called "parallel sequence classification," as modified by John Kormendy and Ralf Bender. The diagram only shows nonbarred spirals; a similar diagram could be made for barred spirals. Diagram is from J. Kormendy and R. Bender, *Astrophysical Journal Supplement Series*, **198**, 2, (2012). © AAS. Reproduced with permission.

research in the past ten years has supported van den Bergh's parallel sequence view. Three groups have independently arrived at the same conclusion. In 2011, Eija Laurikainen, Heikki Salo, and coworkers published the results of a long term observing program called the "Near-Infrared S0 Survey" (NIRS0S),[8] and found several genuine S0c galaxies.[9] John Kormendy and Ralf Bender carried out several detailed studies of elliptical galaxies in the Virgo Cluster, and concluded that the abundant dwarf examples of these objects in the cluster are environmentlally-modified former late-type spiral and irregular galaxies. They called these products of environmental modification "spheroidal" galaxies, and placed them on the sequence S0a–S0b–S0c–Sph parallel to Sa-Sb-Sc-Sd-Sm-Im in the modified Hubble tuning fork (Figure 10 of Chapter 6).[10] This solved more of the missing S0c problem. In another independent study, Michele Cappellari and his coworkers examined the rotation speeds of a large number of S0 and elliptical galaxies and concluded also that parallel sequence classification is the best approach to S0 galaxies.[11]

Figure 5 also shows John Kormendy and Ralf Bender's different way of classifying elliptical galaxies from Hubble's original tuning fork. Instead of focussing on the apparent ellipticities of isophotes, these authors proposed replacing the E0-E7 classification of E galaxies with a more physical classification involving the deviations of the isophotes from perfectly elliptical shapes. Some ellipticals show "boxy" isophote shapes while others show cusped or "disky" isophote shapes. Kormendy and Bender noted that boxy elliptical galaxies have little rotation and balance their mass against gravity mainly by random motions (as in a "pressure-supported" system). Disky ellipticals on the other hand, show evidence of rotation like spiral and S0 disks. The order E (boxy) to E (disky) in the revised

---

[8]E. Laurikainen *et al.*, *Monthly Notices of the Royal Astronomical Society,* **418,** 1452 (2011).

[9]E. Laurikainen, H. Salo, *et al.*, *Astrophysical Journal,* **132,** 2634 (2006).

[10]J. Kormendy and R. Bender, *Astrophysical Journal, Supplement Series,* **198,** 2 (2012).

[11]M. Cappellari *et al.*, *Monthly Notices of the Royal Astronomical Society,* **416,** 1680, (2011).

tuning fork "handle" allows ellipticals to approach the domain of the highly-flattened, disk-shaped S0 and spiral galaxies in a more physical manner than does the apparent ellipticity of the isophotes. Note that the disks in disky ellipticals are much more subtle than those seen in S0 galaxies.

Finally, Kormendy and Bender allow for natural S0s by placing galaxies called type "S0(0)" at the juncture between ellipticals and spirals. S0(0) galaxies are usually nonbarred systems with significant bulge components and are not likely to be environmentally-modified spirals.

Chapter 22

Galactic Collisions and E Galaxies

The beautiful Hubble Space Telescope image of NGC 6050 (Figure 1) in the Hercules galaxy cluster, Abell 2151, shows two spiral galaxies that appear to be colliding. While it is possible that this is in fact happening, it is unlikely because the two spirals are practically overlapping, yet show little evidence of distortion. We might expect that a real galactic collision would do violence to a galaxy's structure, causing its shape to be distorted due to the strong tidal gravitational forces involved. NGC 6050 is more likely a case of simple superposition: two galaxies appearing nearly along the same line of sight, and not as physically close to each other as they look. But let us now imagine that the two spirals in NGC 6050 are in fact heading towards a full-fledged collision. How would the process progress? What would be the end product(s)?

As the two galaxies approach each other, they would swing around and, depending on their relative masses and speeds, fling long, spiral-shaped features known as tidal tails out into intergalactic space. An excellent example with such features is the Antennae galaxy pair NGC 4038-39, shown in Figure 2. The magnificent arcs in this system extend more than a hundred thousand light years, and appear to involve star formation. Many of the stars in these features probably will escape the system eventually.

Particularly interesting are collisions that lead to *mergers*, that is, where the end-product of the collision of two spiral galaxies is a single remnant, rather than just two distorted but still distinct systems. Such events have the greatest impact on the structure of the

Fig. 1.   Close-up of NGC 6050 in the Hercules galaxy cluster Abell 2151, as imaged with the *Hubble Space Telescope*. The two galaxies look like they could be colliding, but it is more likely they are a case of superposition and not as close together as they appear. Image credit: NASA, ESA, Hubble Heritage (STScI/AURA) — ESA/Hubble Collaboration, K. Noll (STScI).

two galaxies, and the rate at which mergers occur is an important factor in galaxy evolution studies. NGC 4038-39 is probably a merger in an advanced phase. The peculiar shape of the inner mass shows that the two galaxies have already started to blend into a single mass. However, infrared images show that the nuclei of the two objects are still distinct (Figure 3).

One of the first things a collision does is disrupt the order of the motions of the stars in the two disk-shaped galaxies. The near circular star orbits in the plane of the disk of each system can be

Fig. 2. The famous Antennae galaxy NGC 4038-39, located 66 million light years away in the constellation Corvus, the Crow. The long, arc-shaped extensions are tidal tails made of stars and gas clouds that have been flung from the outer parts of two spiral disk galaxies whose near-merged bodies are the complex mass from which the tails protrude. Image data: Subaru,NAOJ/ESA/Hubble R. W. Olsen — Processing: Federico Pelliccia and Rolf Wahl Olsen; used with permission.

changed so much that their original order is completely lost, to the point that the resulting combined system is not a disk-shaped galaxy but something more three-dimensional. The more advanced the state of a merger, the more randomized the orbits can become.

Surprisingly, even though each system may have as many as a hundred billion stars, collisions of this nature do not necessarily lead to many star-star collisions. Stars in a typical spiral galaxy are so far apart compared to their sizes that star-star collisions are very improbable. But while individual stars may not collide, interstellar gas and dust clouds, being much larger than stars, do collide, and this can lead to an intense starburst. As noted for star-forming nuclear rings in Chapter 15, a starburst occurs when a large number of new

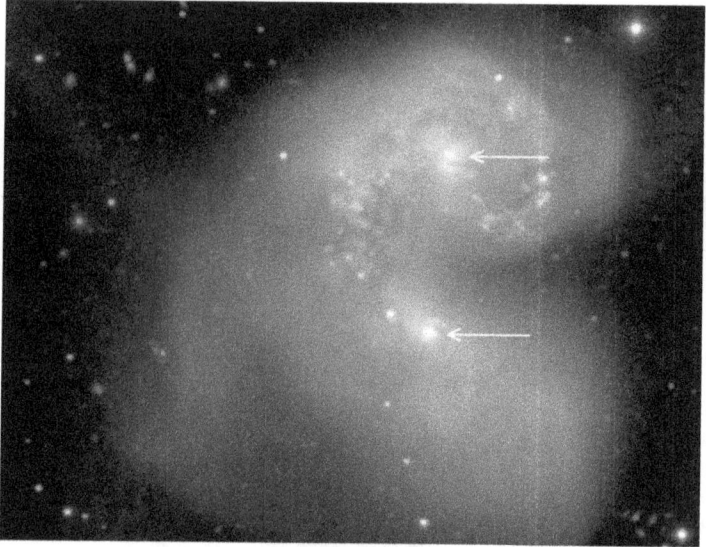

Fig. 3.   Infrared (*I*-band) image of the main body of NGC 4038-39, showing two bright infrared spots (arrows) which are likely to be the nuclei of the two former spirals that collided. This image of the central region is rotated 90° counterclockwise compared to Figure 2. Image source: The de Vaucouleurs Atlas of Galaxies published by Cambridge University Press, 2007; reprinted with permission.

stars are formed in a very short period of time. In a merger like the Antennae, the rate of star formation is ≈20 times the mass of the Sun per year,[1] which is about 10 times higher than the star formation rate in a normal Sc galaxy.[2] The rate of star formation can be so intense that the interstellar gas supply is used up in less than a billion years. After a few billion years without forming any new stars, the stars in the merger remnant age, and the orbits continue to be randomized until the system becomes fully "relaxed." The merger product would be seen as an elliptical galaxy, and indeed mergers of two comparable-sized spiral galaxies are thought to be one (and perhaps the most important) way elliptical galaxies have formed.

---

[1] Q. Zhang, S. M. Fall, and B. C. Whitmore, *Astrophysical Journal*, **561**, 727 (2001).
[2] R. C. Kennicutt, *Astrophysical Journal*, **272**, 54 (1983).

The boxy/disky dichotomy of E galaxies described in the previous chapter could have a bearing on how an elliptical galaxy formed. Boxy ellipticals could be the ones that formed from mergers of two spirals, while disky ellipticals could be cases where a pre-existing elliptical accreted material from outside into a disk plane. Perhaps as a result of this, boxy ellipticals tend to be more luminous on average than are disky ellipticals.

Today, the merger rate in the universe is much lower than it once must have been. In the remote past, galaxies were closer together than they are now, and mergers were likely more frequent. Minor mergers, where small galaxies merge with much larger galaxies, are thought to be the origin of the classical bulges seen in galaxies like the Andromeda Galaxy M31, the Sombrero Galaxy M104, and M81. It is interesting that some quasars are found in disturbed-looking galaxies that show evidence for a past violent interaction. This suggests that mergers can induce a period of nuclear activity that would have a significant impact on the star formation rate and the evolution of the merger remnant.

Has the Milky Way ever been involved in a collision? Since it formed, the Milky Way probably has been involved in many minor collisions and mergers with neighboring dwarf galaxies. An intriguing example of this involves the bright star Arcturus, the beautiful orange giant in the constellation Bootes. Arcturus is part of a nearby grouping of stars having motions and metallicity characteristics that suggest it was born in a former dwarf companion galaxy that was disrupted completely in an ancient collision with the Milky Way.[3] Thus, Arcturus is an interloper from another galaxy that collided with the Milky Way so long ago that its parent galaxy has been completely absorbed. Only subtle details of the stars in our neighborhood allude to this event. These kinds of phenomena are not just a thing of the past, however. Ongoing accetion events are happening even now. One example is the Sagittarius Dwarf Spheroidal Galaxy, recently found projected behind the star clouds

---

[3] J. Navarro *et al.*, *Astrophysical Journal*, **601**, L43 (2004).

of the Milky Way in Sagittarius.[4] Detailed analysis suggests that this dwarf is in the process of disruption.

While the Milky Way has eluded a major collision so far, in the moderately distant future this may change. It has been known since the time of Vesto Slipher (Chapter 5) that the Andromeda Galaxy is approaching the Sun at 300 km/s. If we correct the line of sight speed of Andromeda for the motion of the Sun relative to the Galactic Center, then the center of Andromeda is falling towards the center of our Galaxy at about 110 km/s. Provided that the Andromeda Galaxy has little transverse movement (or proper motion) on the sky, this would mean that Andromeda and the Milky Way are heading towards a future major collision. It has been estimated that the two galaxies will merge about 6 billion years from now.[5] Depending on the geometry of the collision and the relative speeds of the two galaxies, the Sun may or may not get flung out into intergalactic space on a tidal tail. Even if it did, there would be little for us to worry about. The probability of the Sun colliding with another star would not be significantly increased by the collision of the two galaxies. The most significant thing that will likely happen to the Sun in 6 billion years is that it will have evolved into a red giant by then.

The study of interacting and merging galaxies is one of the most important topics of extragalactic astronomy today. Computer codes have become so sophisticated that astronomers can follow the evolution of any interacting system, including equal and unequal mass pairs, as seen from any particular vantage point at any particular time after the interaction begins. These have shown that mergers not only can produce elliptical galaxies, but also disk-shaped remnants, and that there is a wide variety of possibilities of merger types.

22.1. Living in an E Galaxy

What would it be like if we lived in an elliptical galaxy? What kinds of stars would make up our constellations? In an elliptical

---

[4]R. Ibata, G. Gilmore, and M. J. Irwin, *Nature*, **370**, 194 (1994).
[5]R. P. van der Marel *et al.*, *Astrophysical Journal*, **753**, 9 (2012).

galaxy, the most common luminous stars we would see, and the ones that would dominate our constellations, would be red giants like Aldebaran and Arcturus. There would likely be no stars as massive as our own Sun, and the most luminous main sequence stars would be whitish-yellow in color (compared to yellow-white for the Sun). Stars whose luminosity is powered by core helium fusion (called "horizontal branch" stars) could add some whitish color to the mix. The general color of the system would be determined by how metal-rich or metal-poor the stars are; massive ellipticals would have more metal-rich stars and would be redder overall than lower mass ellipticals. There would also be no bright galactic nebulosities like the Orion Nebula, nor any open star clusters like the Pleiades. There would be some planetary nebulae, and globular clusters would likely be very common. One of the best examples illustrating these points is the giant Virgo Cluster elliptical galaxy M87, which has at least 300 planetary nebulae and more than ten thousand detectable globular clusters. Many of the most luminous elliptical galaxies in the Virgo Cluster have similar systems.

Another type of object that would be missing from our night sky if we lived in an elliptical galaxy would be normal Cepheid variable stars like those discovered by Henrietta Leavitt in the Small Magellanic Cloud. Normal Cepheids in our Galaxy are generally stars more massive than the Sun. The main pulsating variables we would likely see would be "cluster variables," or RR Lyrae variables, the radial pulsators having periods less than 1 day. Unless an elliptical galaxy has a companion like the SMC, the Cepheid P–L relation would not be as readily discoverable as it was from our vantage point in the Milky Way. Short-period pulsators may not be as metal-poor in an E galaxy as in a globular cluster.

Perhaps the greatest difference we would notice if we lived in an E galaxy is the absence of significant amounts of interstellar dust and gas. Although there are exceptions, many, if not most, ellipticals are relatively free of the effects of dust, such that there would be no "Zone of Avoidance" like we see from within our Galaxy. We would notice that the bright central region and nucleus of the galaxy are plainly visible to the naked eye. We would not see a Milky Way-like band going around the sky, just a smooth, regular distribution of stars with more stars in the half of the sky including the galactic center than in the opposite half. The central region would be diffuse or highly resolved depending on how far away it is from us. Corrections to apparent brightness measurements for extinction and reddening would be minimal, meaning any line of sight would take us through the entire galaxy. If we were as excentrically-situated in an E galaxy as we are in our Galaxy, then looking towards the bright center we would see out the other side of the galaxy, into the unresolved glow of billions of faint, low mass stars.

Chapter 23

Spirals and the Extragalactic Distance Scale

The final topic I wish to discuss in this book is the extragalactic distance scale, the series of steps needed to estimate the Hubble constant in the Hubble law, the relation between the line of sight speeds of galaxies and their distances. This was discussed in the context of Hubble's work in Chapter 5, but here I want to discuss the topic from my experience working with Gérard de Vaucouleurs in the late 1970s and 1980s.

Spiral galaxies figure prominently in this story because spirals tend to have the most luminous stellar populations, including the brightest Cepheid variable stars and the brightest blue stars. Spirals are replete with star-forming regions whose characteristics can serve as indicators of distance. The rotation speeds of spirals deduced from 21 cm radio emission can also be used to get distances. The nearest major galaxies are all spirals, so the extragalactic distance scale necessarily begins with them.

It wasn't long ago that the extragalactic distance scale was a topic of considerable stress among astronomers. There were different points of view as to the "right way" to derive the Hubble constant, which often led to openly expressed and (arguably) sometimes impolite disagreements in journals or in referee reports and authors' responses to those reports. People debated the best "distance indicators" to use, and there were concerns about biases or other effects that might leave an estimate of the Hubble constant to be either too large or too small. The effects of galactic extinction, both in our Galaxy and in other galaxies, were until recently difficult to contend with. There can be no doubt that the extragalactic distance scale was a formidable challenge, and that no one who was brave enough to get involved in it would necessarily come out unscathed.

When I first met Gérard de Vaucouleurs in the spring semester of 1977, he was actively working on the distance scale. It makes perfect sense that this would have been a topic that he would have focussed on, because nearly everything he had done up to that time was in support of such work. In the 25 years prior, he had actively collected any basic published data on galaxies that appeared in journal articles or in observatory publications. At the same time, he and his wife Antoinette had an active observing program at McDonald Observatory to get photoelectric photometry of as many galaxies as possible. All of this information led Gérard to prepare several series' of detailed papers in rather short periods of time, beginning in 1975–1976 with "Supergalactic Studies" I–V, in which he outlines the evidence for the peculiar way nearby galaxies are distributed within a system he called the "Local Supercluster," a gigantic, flattened galaxy of galaxies. In 1977–1979, he wrote nine papers in a series titled "Contributions to Galaxy Photometry," where he described how various sources of apparent brightness measurements were compared in order to define a standard brightness system for galaxies: the B_T or total blue-light magnitude system. All of this work was in support of the Reference Catalogues, which were described in Chapter 7.

Then in 1978 de Vaucouleurs began a series of seven papers titled "The Extragalactic Distance Scale." He began the series in reaction to a separate series of papers by Allan Sandage (1926–2010) of Mount Wilson and Palomar Observatores, and Gustav A. Tammann (1932–2019) of the Uninversity of Basel, Switzerland, titled "Steps to the Hubble Constant," begun in 1974 and carried to seven papers by 1982. In these papers, Sandage and Tammann presented their case for a value of the Hubble constant nearly a factor of 10 lower than what Hubble had derived in the 1920s. In their final paper in the series, Sandage and Tammann concluded that the Hubble constant is $50 \pm 5\,\text{km/s/Mpc}$, which corresponds to $15\,\text{km/s/million light years}$.

This means that for every increase in line of sight speed by 15 km/s, the distance is greater by 1 million light years.

In a 1982 monograph written at Mount Stromlo Observatory, de Vaucouleurs noted that "several astronomers soon began to examine critically the distance ladder built by Sandage and Tammann and found in it much that left to be desired; it relies on a long chain of assumptions, extrapolations and, in several instances, circular arguments which raised serious doubts about the reliability of their conclusions." de Vaucouleurs noted that in September 1976 he had presented evidence for a higher value of the Hubble constant, 100 km/s/Mpc (31 km/s/million light years) at a meeting of the International Astronomical Union in Grenoble, France, and then later at another meeting in Paris. He also noted that other astronomers had collected data that led similarly to a higher value of the Hubble constant. But, for de Vaucouleurs, the "straw that broke the camel's back" was Gustav Tammann's reaction to these alternate findings when he quotedly said: "I cannot imagine the Hubble constant to be above 70." de Vaucouleurs threw this comment right back at Tammann by noting that he and Sandage had acknowledged that they could not really assess the systematic error of their estimate of the Hubble constant. de Vaucouleurs found Tammann's comment to be so audacious that he put it on a blackboard in his office to remind him of what he was up against.

The extragalactic distance scale is like a ladder with several rungs. The lowest rung involves distances to the nearest galaxies, while the higher rungs involve more distant galaxies. The scale begins with reliable measurement of the average Earth-Sun distance, 92,957,000 miles (called the "astronomical unit"). This is the baseline for measurement of trigonometric parallax, which leads to a distance of 46 pc for the Hyades star cluster. The distances to other clusters are obtained by matching the brightnesses of their main sequence stars to those in the Hyades. Some of these clusters have Cepheid variable stars, thus allowing calibration of the Cepheid period–luminosity (P–L) relation. Finding Cepheids in nearby galaxies and measuring their pulsation period then gives the distances to these galaxies, the first rung on the ladder.

Because Cepheid variables can be calibrated this way, and because the Cepheid P–L relation has a small scatter, they are known as *primary distance indicators*. Sandage and Tammann based their distance scale on Cepheids alone, which de Vaucouleurs considered to be risky given difficulties people had had with calibrating the P–L relation allowing for star colors and heavy element content. de Vaucouleurs felt there were at least four other primary distance indicators that could be used: common novae, cluster or RR Lyrae variables, supergiant stars like Rigel, and eclipsing binary stars. de Vaucouleurs' philosophy was to "spread the risks" by not putting too much emphasis on a single distance indicator.

The galaxies whose distances can be estimated using primary distance indicators are not necessarily the ones that can be used to derive the Hubble constant. On the scale of nearby galaxies, the "Hubble flow" is not necessarily smooth. Galaxies in nearby groups and clusters could have motions due to their orbits and gravitational interactions with other galaxies, motions that have little to do with the Hubble flow. To access the smooth Hubble flow, it was essential to bootstrap our way to more distant galaxies. This was done with secondary and tertiary distance indicators. Secondary distance indicators are individual objects like the brightest red giants, globular clusters, blue stars, red supergiants, and the largest HII regions. These can't be calibrated accurately in the Milky Way, but can be calibrated in nearby galaxies which have primary indicator distances. Secondary distance indicators can be detected to considerably greater distances than primary indicators, and therefore provide access to the Hubble flow. But to do even better, it is necessary to use tertiary distance indicators, whose calibration is based on galaxies having distances known from primary and secondary distance indicators.

The most important tertiary distance indicator used by de Vaucouleurs and others was the maximum rotation speeds of spiral galaxies, which correlate well with the total luminosity (or absolute magnitude). The correlation is in the sense that the higher the maximum rotation speed, the more luminous and larger (and therefore the more massive) the galaxy. Thus, the maximum rotation speed acts as a distance-independent index of total luminosity, physical size, and mass.

Maximum rotation speeds can be derived in large numbers from 21 cm observations with single-dish radio telescopes. A single dish radio telescope (as opposed to a multi-dish radio interferometer) does not usually provide a detailed radio image (like that for NGC 5850 in Figure 5 of Chapter 8) because the resolution is generally too poor. Instead, it gives the spectrum of 21 cm emission basically integrated over most or all of the galaxy, depending on the object's angular size. For a face-on galaxy, the spectrum would be a single peak of brightness, like an emission line in the optical part of the spectrum. In contrast, a tilted, symmetric galaxy shows a "double-horned" 21 cm emission profile (Figure 1), that can be much broader than a face-on emission peak. The reason for the two peaks in the tilted galaxy spectrum is because half of the galaxy is coming towards us and half is receding from us; this is how rotation projects to the line of sight (for example, see Figure 5 of Chapter 14). The horizontal arrows in Figure 1 show the width, W, of the 21-cm profile. To a good approximation, this width is equal to twice the maximum rotation speed projected to the line of sight. In 1977, American astronomers, R. Brent Tully and J. Richard Fisher showed that, after correction for

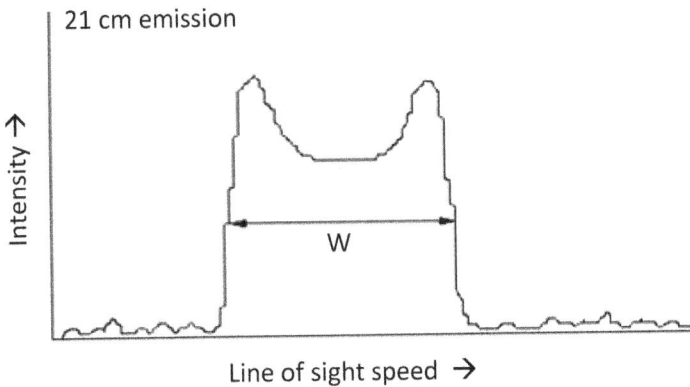

Fig. 1.   Schematic of a 21-cm emission profile for a tilted spiral galaxy.

tilt, W could be used to get distances to galaxies, through a method that has been christened as the "Tully-Fisher" relation.[1]

In 1960, Canadian astronomer Sidney van den Bergh noticed that the simple appearance of spiral galaxies correlated with total galaxy luminosity and physical size.[2] Just by looking at how well organized the structure of a spiral galaxy was, he was able to distinguish a large, high luminosity spiral from a smaller, lower luminosity one. To account for this, he added something called the "luminosity class" to the Hubble classification. The most luminous spirals were assigned to luminosity class I (supergiant spirals), while the smallest and least luminous were assigned to luminosity class V. Everything else was assigned to categories II, III, and IV in between these extremes. Some examples are shown in Figure 2. The notation van den Bergh used is the same as stellar astronomers used for the luminosity classification of stars. Two other examples of luminosity effects in spirals are shown in Figures 15 and 16 of Chapter 17.

Luminosity classifications figured prominently in both Sandage's and de Vaucouleurs's distance scales. Large-scale plates were used by

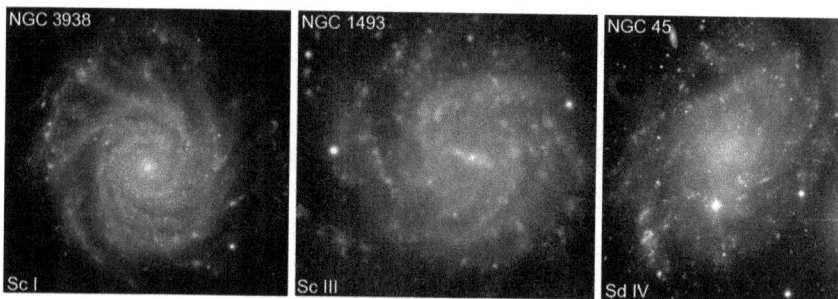

Fig. 2.   Examples of van den Bergh luminosity classifications, where class I is the most luminous and class V is the least luminous, at a given galaxy type. Classes II, III, and IV are intermediate between these. Luminosity class V spirals are not generally found because dwarf spirals are very rare. Images source: The de Vaucouleurs Atlas of Galaxies published by Cambridge University Press, 2007; reprinted with permission.

---

[1]R. B. Tully and J. R. Fisher, *Astronomy and Astrophysics*, **54**, 661 (1977).
[2]S. van den Bergh, *Astrophysical Journal*, **131**, 215, 558 (1960).

Sandage and Tammann[3] to judge luminosity classes as accurately as possible for distance scale work.

Another important distance indicator that was recognized at this time was the maximum luminosity of Type Ia supernovae. As described in Chapters 4 and 13, both common novae and Type Ia supernovae involve mass transfer onto a white dwarf in a close binary system. The remarkable aspect of Type Ia supernovae is that they all achieve nearly the same luminosity, about 4 billion blue-light solar luminosities, when they are at maximum brightness. This collosal luminosity allows them to be seen to much greater distances than most other distance indicators, well into the domain of the smooth "Hubble flow." They can also be seen in any type of galaxy: spiral, elliptical, S0, or irregular. It was the study of Type Ia supernovae that led to the realization that the universe is accelerating its rate of expansion.[4]

In the end, the factor of two difference in Hubble constants between Sandage and Tammann on one hand, and de Vaucouleurs on the other, most likely was due to their very different approaches. Sandage and Tammann used comparatively few distance indicators in their analysis, while de Vaucouleurs used many more. The discrepancy of the two camps persisted until the launch of the *Hubble Space Telescope* in 1990. At the time of his death, de Vaucouleurs had conceded that his estimate of the Hubble constant, 100 km/s/Mpc, was likely too high due in part to biases in the samples of galaxies he used for his analysis. He accepted that the constant could be as low as 85 km/s/Mpc, still much higher than Sandage and Tammann's value.

23.1. The Extragalactic Distance Scale Today

Since the battle between Sandage and de Vaucouleurs in the 1970s, considerable progress has been made on the distance scale and the

---

[3]A. Sandage and G. Tammann, *A Revised Shapley-Ames Catalogue of Bright Galaxies* (Carnegie institution of Washington Publication No. 635, 1981).
[4]A. Riess *et al.*, *Astronomical Journal*, **116**, 1009 (1998); B. Schmidt *et al.*, *Astrophysical Journal*, **507**, 46 (1998).

value of the Hubble constant. *The Hubble Space Telescope* made it possible to extend the Cepheid primary distance scale calibration to well-beyond the local galaxies like the Andromeda and Triangulum spirals, making it possible to calibrate former tertiary indicators like maximum rotation speed and Type Ia supernovae as secondary indicators. Resolving the large disgreements of the past was a key project for the Hubble telescope.[5] Using the most effective distance indicators, including some that are mainly applicable to non-spiral galaxies, the modern value of the Hubble constant has been estimated as 73 ± 5 km/s/Mpc (22 km/s/million light years). It is ironic that the final value came out intermediate between the Sandage and de Vaucouleurs values. With this value, a reasonably accurate distance can be obtained for any galaxy having a known line of sight speed. Fortunately, it is not necessary for us to derive such a distance ourselves. The NASA/IPAC Extragalactic Database (NED; ned@ipac.caltech.edu) does this for us and much more.

---

[5]W. Freedman *et al.*, *Astrophysical Journal*, **427**, 628 (1994); 28 total papers 1994–2000.

Epilogue

The discovery of spiral nebulae 175 years ago was a monumental event in the history of science. Since that time, we have learned a great deal about spiral galaxies, such as the role they play in star formation, what they tell us about star evolution and death, and how they tell us about the expanding universe we live in. We even have the pleasure of calling a spiral galaxy home. The magnificent appearance of the Milky Way, afforded by a visit to a mountaintop in Chile on a dark night in July, belies a mystique that carries clues to our own origin. Seeing the star clouds in the Milky Way is like looking at something truly ancient and mysterious. It is mind-boggling to think about all that likely happened in the Milky Way since it formed, and that somehow, we sprang out of the interstellar gas and dust clouds of the huge galactic disk. We do not know if the existence of life like ours is unique to the Milky Way, or even if it is unique within the Milky Way. We do know that our Galaxy achieved the conditions to allow advanced forms of life to develop on one out of 100 billion possible planets, and in particular a form of life that could look out into the Universe and ask: What is it? Where did it come from? How did it achieve its current state? What is its future?

An interesting topic of discussion recently has been the concept of a *galactic habitability zone*, that is, a range of distances from the galactic center that may be favorable for the development of advanced forms of life. Every star in the Milky Way has its own habitability zone, defined as the range of distances from the star where the temperature on an Earth-like planet would allow liquid water to exist on the surface. The galactic habitability zone has other requirements. One is that it contains an adequate abundance of heavy chemical elements, the main required elements being carbon, nitrogen, oxygen, sodium, magnesium, phosphorus, sulfur, chlorine, potassium, calcium, manganese, iron, cobalt, copper, zinc, selenium, molybdenum, and iodine.[1] These are the elements that seem to be essential for life on Earth. It is a well-known characteristic of spiral galaxies that the metallicity of interstellar gases (that is, how enriched these gases are with heavy chemical elements) decreases with increasing distance from the galactic center, such that at some point the abundance of required heavy elements might drop too low.

Another requirement of the galactic habitability zone is a minimum number of catastrophic astrophysical events, such as nearby supernova explosions or gamma ray bursts. Because massive stars are so rare even in a spiral galaxy, and because the type of binary system that has the most massive white dwarfs is also uncommon, the probability of the Sun being close to a catastrophic supernova event (that is, one where the explosion occurs within about 30 light years) is very low. The greatest threat would occur when the Sun passed through a major spiral arm because, as we have seen, star-forming regions generally trace spiral arms in many disk-shaped galaxies, and these regions are where the most massive and youngest stars in a galaxy are located.

The launching of the *James Webb Space Telescope* (JWST) in 2021 has the potential to add a great deal to our knowledge of galaxies and many other processes and objects in astronomy. The JWST has interesting characteristics. First, the telescope has the largest primary mirror, 6.5 m in diameter, of any space telescope. This will give it a light gathering power more than 5 times greater than the *Hubble Space Telescope*. Second, the telescope mirror will be constructed from 18 hexagonal sections made of the element berylium, each piece coated with a very thin layer of gold for maximum reflectivity in the infrared, the main wavelength domain

---

[1]B. A. Averill and P. Eldredge, *Chemistry: Principles, Patterns, and Applications* (Prentice Hall, 2006).

of operation planned for the telescope. Lastly, the telescope will be placed not in Earth orbit like the *Hubble Space Telescope*, but in a peculiar "halo orbit" around a point 9.3 million miles from Earth on the Earth–Sun line. This point is called the L_2 Lagrangian point and is one of five special points that arise in the combined gravitational field of two massive objects. (These points were discussed in the context of barred galaxies in Chapters 15 and 16.) The nature of the L_2 point is such that as the Earth orbits the Sun, the L_2 point still remains 9.3 million miles from Earth along the Earth–Sun line. The reason for putting the JWST out there is to keep the telescope as cold as possible so as to minimize the influence of the infrared radiation that the telescope will itself emit. All ordinary objects emit thermal infrared radiation, and the key to reducing the effect of this emission on infrared observations is to shield the JWST from the heat radiated by the Sun and the Earth.

The JWST will shed new light on many issues. Of particular relevance to the topic of this book will be the detection of finer detail in the images of high-redshift galaxies that can be used to study how galaxies assembled themselves after the Big Bang, and that will provide more information on the structure of the earliest spirals. The greater light-gathering power of the JWST will also allow the study of fainter stellar populations in nearby galaxies, and facilitate such studies in more distant galaxies. The telescope will open up higher resolution studies of infrared galaxy morphology, and especially will be able to image optical and ultraviolet light that has been redshifted into the infrared. Finally, the infrared domain is the best for studying star-forming regions in great detail, and for penetrating the dust that pervades our Galaxy and spirals in general.

With all this, the study of spiral galaxies will continue to provide new insights into the structure, origin, and evolution of the universe.

Appendix A

Glossary

AAS — acronym for American Astronomical Society.

absolute magnitude — The apparent magnitude a star would have if it could be viewed from a standard distance of 10 parsecs. A logarithmic representation of intrinsic brightness, or luminosity.

active galaxy — A galaxy showing strong emission lines in the spectrum of its nucleus and often evidence of violent interstellar gas motions.

apparent brightness — A quantitative measure of how bright a celestial object appears. Physically, apparent brightness has units of watts per square meter, but astronomers almost never specify it this way. The magnitude system is used instead.

apparent magnitude — A logarithmic representation of apparent brightness, rooted in ancient observations which assigned first magnitude to the brightest stars in the naked eye sky, and sixth magnitude to the dimmest. The modern magnitude system defines 5 magnitudes to be a factor of 100 in apparent brightness, where 1 magnitude corresponds to a factor of 2.512. The system is such that smaller or negative magnitudes correspond to brighter objects. Magnitudes larger than 6 would require a telescope to be seen.

Andromeda Galaxy — The nearby Local Group spiral known as M31 (NGC 224), located 2.5 million light years away in the constellation Andromeda, the chained princess.

AURA — acronym for Association of Universities for Research in Astronomy.

axially symmetric galaxy — A disk-shaped galaxy which is perfectly symmetric around its axis of rotation; a flat galaxy with no noncircular structures.

Backwards Galaxy — The spiral NGC 4622 that was found to be rotating in a reverse sense from what was expected.

Balmer Series — The series of emission or absorption lines that arises from quantum leaps to and from the first excited energy level of hydrogen.

baryonic matter — Ordinary matter consisting of protons and neutrons.

Big Bang — The term astronomers use for the singular event that led to the creation of the Universe in a burst of high energy radiation.

broad-line region — In an active galaxy, the inner part of the accretion disk of a supermassive black hole which is the source of broad permitted emission lines.

brown dwarf — A substellar object more massive than a planet, but not massive enough to be a star.

bulge — The central mass component of a disk-shaped galaxy.

Cassegrain focus — The point where light rays are brought to a focus in a telescope design where a curved secondary mirror reflects the light from the primary mirror back into a hole in the primary mirror.

cD galaxy — A supergiant amorphous galaxy often found at the center of a rich galaxy cluster. Believed to have formed by "cannibalism" of numerous smaller galaxies over a long period of time.

Cepheid — A type of variable star showing periodic brightness variations attributable to radial pulsations in size and the conse- quent changes in surface temperature. The light curve often shows a characteristic rapid rise from minimum to maximum apparent brightness, followed by a slower decline from maximum to minimum brightness.

Cepheid period–luminosity relation — The well-defined correlation between the period of the brightness variations of a Cepheid variable star and its average luminosity. The correlation is in the sense that longer period Cepheids have higher luminosities.

CGCG number — The number of a galaxy in Fritz Zwicky's *Catalogue of Galaxies and of Clusters of Galaxies.*

charge-coupled device (CCD) — A modern light-sensitive electrical device that converts photons into charge, and then reads out the charge as digital values at different positions in a two-dimensional image.

classical bulge — A bulge in a disk-shaped galaxy thought to have been formed in early mergers of smaller galaxies. Made entirely of Population II stars, and typically found in galaxies like M31, M81, and the Sombrero Galaxy M104.

color index map — A black and white map of the difference in magnitude between two filters, called a color index. The color indices $B-V$ (blue minus visual) and $B-I$ (blue minus near-infrared) are often used for such maps, which are coded (in this book) such that blue features (like star-forming regions) are dark and red features (like dust lanes) are light.

corotation resonance (CR) — The location in a disk-shaped galaxy where the angular speed of rotation of stars and gas clouds equals the pattern speed of a bar or spiral.

cosmology — The study of the structure, origin, and evolution of the universe as a distinct object. Observational cosmology involves measuring the Hubble constant, studying the morphology of high redshift galaxies, examining the distribution of galaxies, and deriving other parameters of the universe, including the average density of matter and energy.

CTIO — Acronym for Cerro Tololo Inter-American Observatory, a major collection of large, professional telescopes located in northern Chile. The telescopes range from 0.4 to 4 m in primary mirror diameter.

dark energy — A hypothetical energy that characterizes the vacuum of space and which is believed to be the reason the universe is accelerating its rate of expansion.

dark matter — Matter in a galaxy that is detectable only through its gravitational influence on the motions of visible objects. Does not reflect, emit, or absorb easily detectable electromagnetic radiation.

declination — A sky coordinate similar to latitude on Earth. It is specified as an angle north or south of the celestial equator, which is the extension of the Earth's equator onto the celestial sphere. In the 18th and 19th centuries, declination was specified relative to the direction of the north celestial pole and was at those times called the "north polar distance."

degeneracy pressure — A type of pressure that arises in the core of a star when identical particles (like electrons or neutrons) are being forced by gravity to occupy the same positions. Based on the quantum mechanical rule known as the Pauli Exclusion Principle, the particles must increase their speeds, and this can provide enough outward pressure to balance the mass of the star core against gravity, independent of the actual temperature.

density wave — A pattern in a disk-shaped galaxy that rotates with a uniform angular speed (called the "pattern speed"). The pattern could be a spiral or a bar.

diffuse nebula — A generally irregular cloud of ionized gas and dust, like the Orion Nebula.

disk — The highly-flattened sub-component of a spiral or S0 galaxy, and generally where most of the stellar mass in the system resides. In spirals, the disk is also where most of the interstellar gas and dust is found.

Doppler effect — The shifts in the wavelengths of absorption or emission lines in the spectrum of an object, due to a component of relative motion along the line of sight.

envelope (stars) — The outer layers of a star that surround the fusing core.

envelope (S0 galaxies) — The outer disk light of an S0 galaxy.

epicycle orbit — An orbit whose shape is the combination of a pure circular motion and a small retrograde epicycle that represents the first order deviation from pure circular motion.

ESA — acronym for European Space Agency.

ESO — acronym for European Southern Observatory.

event horizon — An imaginary spherical region, surrounding the singularity of a black hole, where the escape speed equals the speed of light. If an object passes through the event horizon, it will lose its ability to communicate with the outside universe.

extragalactic nebula — a nebula (usually a galaxy) lying away from the band of the Milky Way and most prevalent near the Galactic poles.

Fabry–Perot interferometer — A scientific instrument that uses an etalon, an optical element consisting of two exactly parallel, highly reflective pieces of glass, to detect Doppler wavelength shifts of spectral lines. Superior to slit spectroscopy in that it can give the velocity field of a galaxy, not just line of sight speeds along a given slit poisition.

GC — The short acronym for Sir John Herschel's *General Catalogue of Nebulae and Clusters of Stars*, a compilation of 5079 objects that served to provide names (GC numbers) and descriptions for all nebulae and star clusters discovered up to 1864. It was eventually superceded by the *New General Catalogue of Nebulae and Clusters of Stars* (see "NGC").

galactic — Related either to other galaxies or galaxies in general.

Galactic — Related specifically to the Milky Way.

Galactic extinction — The reduction in apparent brightness of a celestial object due to absorption and scattering of light by interstellar dust in our Galaxy.

galactic nucleus — The dense bright center of a massive galaxy that from a great distance can appear star-like. In a normal galaxy, the light of the nucleus is dominated by starlight and the spectrum mainly shows absorption lines. There may also be interstellar ionized gas that would add emission lines to the spectrum.

Galactic plane — The symmetry plane around which the bulk of the stellar mass in the Milky Way is found.

Galactic poles — The directions on the sky perpendicular to the Galactic plane.

galactocentric distance — distance from the center of a galaxy, in either angular or linear units.

galaxy — A massive system of stars, interstellar gas, dust, and dark matter bound by gravity into a single physical unit. Ranging from a few thousand to more than a hundred thousand light years in size, and including from a few billion to more than a hundred billion stars, galaxies are the most fundamental concentrations of matter in the universe.

Galaxy — Authoritative term for the Milky Way. There are other galaxies, but the Milky Way is the Galaxy.

Galaxy Zoo — A crowd-sourcing project that used Sloan Digital Sky Survey color images to classify hundreds of thousands of galaxies in a short period of time.

gamma rays — The most dangerously energetic electromagnetic waves, having a wavelength of less than 0.01 nm.

gamma ray burst — A phenomenon where an extremely intense, short burst of gamma rays is produced by a violent cosmic event, like the collapse of an extremely massive star into a black hole.

globular cluster — A spherically-shaped, massive star cluster consisting of as many as a million very old, generally metal-poor low mass stars. About 150 globular clusters occupy the halo of the Milky Way. The ages of these clusters range from 10 to 12 billion years.

Hanny's Voorwerp — The name given to a peculiar intergalactic cloud near the disturbed spiral galaxy IC 2497. Named after Hanny van Arkel, a Dutch school teacher who discovered the object while participating in the Galaxy Zoo project.

Hertzsprung–Russell (HR) diagram — A graph of star brightness versus star color or, more specifically, luminosity versus surface temperature. Such a graph is embedded with clues to stellar evolution, as it brings attention to main sequence stars, red giants, supergiants, horizontal branch stars, and white dwarfs.

high redshift galaxy — A galaxy having a redshift $z \geq 0.5$.

HII region — A photoionized cloud of interstellar gas and dust that is bathed in ultraviolet light from hot, embedded massive main sequence stars. Also known as a star-forming region.

horizontal branch star — A giant, yellowish-colored star in an advanced state of star evolution that in the HR diagram appears in a horizontal region above the main sequence and to the left of the red giants. Horizontal branch stars are believed to be powered by core helium fusion, having exhausted all of the core hydrogen they were born with.

Hubble constant — The proportionality constant between the line of sight recession speed of a galaxy and its distance.

Hubble-Lemaitre law — The correlation between the line of sight recession speed of a galaxy and its distance.

Hubble flow — The part of a galaxy's bodily motion due to the expansion of space itself.

Hubble tuning fork — Edwin P. Hubble's classification system for galaxies, where the "handle" of the tuning fork includes ellipticals of increasing ellipticity, connecting to spirals in two parallel sequences of nonbarred and barred types. S0 galaxies are placed at the juncture between the handle and the prongs.

HUDF — acronym for Hubble Ultra Deep Field.

hypergiant star — An extremely massive star that shows evidence of a strong stellar wind and mass loss.

IC — The short acronym for John Louis Emil Dreyer's 1895 *Index Catalogue of Nebulae and Clusters of Stars* and 1908 *Second Index Catalogue of Nebulae and Clusters of Stars*, two compilations of 5365 objects that served to provide names (IC numbers) and descriptions for all nebulae and star clusters discovered after the publication of the NGC in 1888.

infrared radiation — Low energy electromagnetic waves having wavelengths in the range 750 nm to 1 mm. A domain intermediate between visible light and microwaves.

inner 4:1 resonance (I4R) — The position in the inner parts of a disk-shaped galaxy where a star or gas cloud would undergo four radial excursions for every single orbit in the reference frame corotating with a bar or spiral pattern. In a galaxy with low central concentration, the I4R may not exist.

inner Lindblad resonance (ILR) — The position in the inner parts of a disk-shaped galaxy where a star or gas cloud would undergo two radial excursions for every single orbit in the reference frame corotating with a bar or spiral pattern. In a galaxy with low central concentration, an ILR is not likely to exist.

inner ring — A ringed-shaped pattern usually seen enveloping the bar of a barred spiral galaxy. Typically elongated with an intrinsic minor-to-major axis ratio of 0.8 ± 0.1 and aligned parallel to the bar.

interstellar extinction — The dimming of apparent star brightnesses caused by the scattering and absorption of light by grains of interstellar dust. Extinction depends on wavelength such that dust preferentially scatters and absorbs shorter wavelength light as opposed to longer wavelength light.

interstellar reddening — The reddening of starlight caused by the selective scattering and absorption of blue light as opposed to red light, due to the fact that interstellar dust grains have sizes comparable to the wavelength of blue light ($\sim$440 nm). Reddening increases the B–V or B–I color indices and makes stars look cooler than they actually are.

island universe — The 19th century term for a galaxy, a distant star system like the Milky Way.

isophote — A contour of constant surface brightness in an image.

KPNO — Acronym for Kitt Peak National Observatory, a major collection of large, professional telescopes located in southern Arizona. The telescopes range from 0.9 to 4 m in primary mirror diameter.

leading arm — A spiral arm that opens outward in the direction of rotation.

lens — 1. A piece of glass, curved parabolicly on both sides in double convex fashion, that acts as the light-gathering element of a refracting telescope.
2. A morphological feature, typically found in S0 galaxies, that has a shallow brightness gradient interior to a sharp edge.

lenticular — having the shape of a double-convex lens seen side-on.

Leviathan of Parsonstown — The name given to the 72-inch telescope built by William Parsons, the Third Earl of Rosse, on the grounds of Birr Castle in the mid-1840s. Best known as a biblical sea monster, the term "Leviathan" is meant to convey exceptional size and telescopic power. Parsonstown is an alternate name for the town of Birr in Offaly County, Ireland, and is meant to convey the influence of the Parsons family.

light curve — A graph of the apparent magnitude of an object versus time.

light-gathering power — A quantitative measure of how effectively a telescope can reveal faint stars and nebulae. The light-gathering power is proportional to the square of the diameter of the primary mirror, and is such that a 16-inch telescope gathers four times as much light as an 8-inch telescope, and a 160-inch telescope gathers 100 times as much light as a 16-inch telescope, etc.

light year — The distance, 5.878 trillion miles, a beam of light travels in one year at its speed of 300,000 kilometers per second in vacuum.

line of nodes — The line of intersection between the plane of a disk-shaped galaxy and the plane of the sky.

Local Group — The small group of galaxies to which the Milky Way, the Andromeda Galaxy, the Triangulum Galaxy, the Magellanic Clouds, and a host of dwarf galaxies belong.

luminosity — The amount of radiant energy emitted from the surface of a star every second. Units are watts. The luminosity of the Sun is 400 trillion trillion Watts.

MACHO — Acronym for "massive compact halo object," referring to massive but small objects (like planets or other very dim baryonic entities) in the Galactic halo detectable individually only through gravitational microlensing. Once considered a prime candidate for the nature of dark matter.

Maffei galaxies — Two nearby external galaxies lying in the direction of the Galactic plane that were discovered in 1968 by Italian astronomer Paolo Maffei. The apparent brightness and color of the two galaxies are heavily affected by foreground Galactic extinction and reddening.

Magellanic Clouds — Two bright, nebulous objects visible to the naked eye near the south celestial pole that were brought to the attention of northern hemisphere, western people by members of Ferdinand Magellan's crew after their voyage to the southern hemisphere in the early 1500s. The clouds are now known to be companion galaxies to the Milky Way.

main sequence star — In the Hertzsprung-Russell diagram, any star lying on a band going from the upper left (luminous, hot stars) to the lower right (dim, cool stars). The luminosity of all main sequence stars is powered by fusion of hydrogen into helium in the core region.

main sequence lifetime — The length of time a star shines as a main sequence star. The lifetime is a strong function of the mass in the sense that the higher the mass the much shorter the main sequence lifetime.

material arm — A spiral-ahaped arm in a galaxy that is made of the same stars as it evolves. Material arms are expected to be shortlived because they would quickly wind up due to the non-solid-body rotation of a typical disk-shaped galaxy. Material arms can be triggered by a gravitational encounter with another galaxy.

merger — The violent collision between two galaxies that ends in the production of a single remnant.

Messier 51 — The name of a double nebula in Canes Venatici, catalogued as number 51 in Charles Messier's famous catalogue, that is bright and easy to see with a small telescope, but which is best known as the first nebula discovered to have spiral structure. The discovery was made in the spring of 1845 with the Leviathan of Parsonstown.

metal — In astronomy, any chemical element heavier than helium.

metallicity — The relative proportion of elements heavier than helium in a star or stellar population.

microwaves — Low energy electromagnetic waves having a wavelength in the range 1 mm to about 30 cm.

Milky Way — The term given to our home galaxy based on its appearance as a bright band of mostly unresolved starlight going all around the sky. The term is also used in a general way for our galaxy when compared with other galaxies.

narrow-line region — In an active galaxy, the extended surroundings of the accretion disk of a supermassive black hole which is the source of narrow permitted and forbidden emission lines.

nebula — A cloud-like object seen in the night sky. Nebulae can be actual clouds of glowing gas, or distant star systems too far away to be resolved into their individual stars.

NGC — The short acronym for John Louis Emil Dreyer's *New General Catalogue of Nebulae and Clusters of Stars*, a compilation of 7840 objects that served to provide names (NGC numbers) and descriptions for all nebulae and star clusters discovered up to 1887. It was eventually supplemented by the first and second *Index Catalogues of Nebulae and Clusters of Stars* (see "IC").

NASA — acronym for National Aeronautics and Space Administration.

NOAO — acronym for National Optical Astronomy Observatories.

non-circular motions — movements in a galaxy which involve departures from pure circular rotation.

non-circular structures — Features in the mass distribution of a galaxy that depart from a circular shape.

nova — a "new star" that almost suddenly appears and then more slowly fades with time. Physically thought to be a white dwarf receiving mass from a companion star in a very close binary system.

nuclear ring — A small, ring-shaped pattern usually associated with a bar and about one-10th the size of the bar in a disk-shaped galaxy. Typically round or only slightly elongated with an intrinsic minor-to-major axis ratio of 0.9 ± 0.1 and, if oval, usually misaligned with the bar.

open cluster — A generally irregular grouping of less than a thousand stars, loosely bound together by gravity, that orbits mostly within the plane of the Milky Way. Ages range from a few million up to 5 billion years. Most open clusters disperse (i.e. break up) on a time scale of 200 million years.

Orion Nebula — One of the nearest examples of a photoionzed nebula, easily visible to the naked eye in the Sword of Orion. The nebula surrounds a small group of bright stars known as the Trapezium, one of which is hot enough to photoionize the surrounding interstellar gases.

orbital resonance — A location in a disk-shaped galaxy where stars and interstellar gas and dust clouds move in step with a perturbing pattern such as a spiral or a bar.

outer 4:1 resonance (O4R) — The position in the outer parts of a disk-shaped galaxy where a star or gas cloud would undergo four radial excursions for every single orbit in the reference frame corotating with a bar or spiral pattern. In this frame, material moves in a sense opposite to the pattern rotation.

outer Lindblad resonance (OLR) — The position in the outer parts of a disk-shaped galaxy where a star or gas cloud would undergo two radial excursions for every single orbit in the reference frame corotating with a bar or spiral pattern. In this frame, material moves in a sense opposite to the pattern rotation.

outer ring — A ring-shaped pattern usually associated with a bar and about twice the size of the bar in a disk-shaped galaxy. Typically elongated with an intrinsic minor-to-major axis ratio of 0.9±0.1 and aligned either parallel or perpendicular to the bar.

parallel sequence galaxy classification — A revision of the Hubble tuning fork where S0 galaxies are placed on a sequence of types parallel to spirals, rather than at the juncture of spirals and ellipticals. Unlike the tuning fork, the position along the sequence is based solely on the degree of central concentration for both spirals and S0s. Not to be confused with the parallel prongs of the Hubble tuning fork due to the subdivision of spirals into barred and nonbarred types.

parsec — The distance of a star having a trigonometric parallax of 1 arcsec, corresponding to 3.26 light years. A kiloparsec is 1,000 pc, while a megaparsec is 1,000,000 pc.

pattern speed — The angular speed of rotation of a spiral or bar pattern in a disk-shaped galaxy.

Perseus arm — The nearest of two major spiral arms in the Milky Way, passing through the northern constellations of Auriga, Perseus, and Cassiopeia.

PGC number — The number of a galaxy in the *Principal Galaxy Catalogue.*

phi-type spiral — A spiral galaxy resembling the Greek letter ϕ, first described by Heber Curtis in 1918 and later renamed barred spirals by Hubble in 1926.

photoionization — A process that occurs when an atom absorbs an energetic ultraviolet photon and loses an electron as a result. The photoionization of hydrogen requires a photon of wavelength 91.2 nm or less. A cloud where most of the hydrogen is ionized is known as an HII region.

photometer — An electronic device designed to quantitatively measure the apparent brightnesses of celestial objects. A standard photometer uses the photoelectric effect to count photons.

photon — A particle of light, characterized by an energy, a wavelength, and a speed of 300,000 km/s in vacuum.

photoelectric effect — A physical effect where photons striking a piece of metal (photocathode) cause electrons in the metal to be dislodged.

photometry — The process of measuring the apparent brightnesses of celestial objects.

planetary nebula — The glowing cloud of gas that forms when a dying low mass star ejects its envelope gases into space, exposing the hot, degenerate core. The envelope gases glow with an emission-line spectrum caused by ultraviolet light from the core. When still fairly young (less than 10,000 years old), some of these objects resemble planets, and as a result were given the name "planetary nebulae" by early visual observers.

polar ring galaxy — Generally a low luminosity disk-shaped system where a small companion has been disrupted into a polar orbit around the disk. The disrupted companion looks like a ring and can itself be disk-shaped.

Population I star — A star belonging to the disk of a typical spiral galaxy. Such stars follow nearly circular orbits around the galactic center, have a wide range of ages (from a few million to more than 5 billion years), and are generally metal-rich. From a great distance, the youngest Population I stars are the ones most easily resolved with large telescopes.

Population II star — A star belonging to the (classical) bulge or halo of a spiral galaxy. Such stars follow non-circular orbits often highly-inclined to the galactic plane, and are generally extremely old and metal-poor. Most elliptical galaxies are pure population II systems.

primary mirror — A piece of glass, curved parabolicly on one side in concave fashion and coated with highly-reflective material, that acts as the light-gathering element of a reflecting telescope.

prime focus — The point in front of a telescope's primary mirror where light rays from an object are brought to a focus.

progenitor — In the context of supernovae, the actual star that exploded.

proper motion — Angular movement of a celestial object on the sky, due to the tangential component of the object's space velocity.

pseudobulge — A bulge in a disk-shaped galaxy that likely formed by the slow inward movement of stars and interstellar gas clouds due to the presence of noncircular structures such as a bar or a spiral pattern. Such a bulge generally involves Population I material, such as interstellar gas and dust and young stars, as well as spiral structure and inner secondary bars.

pseudoring — A ring-like morphological feature of disk-shaped galaxies that appears to be made of tightly-wrapped spiral structure; usually an incomplete ring.

quasar — A remote active galaxy whose active nucleus is so bright it drowns out the light of the host galaxy.

radial pulsation — A physical process whereby a star undergoes periodic changes in size, due to difficulties of smooth energy flow from deep inside. The size changes are accompanied by surface temperature changes, which lead to periodic changes in apparent brightness.

radio waves — Extremely low energy electromagnetic waves having wavelengths in the range 30 centimeters to 100 kilometers.

ram pressure stripping — A process thought to occur in a cluster of galaxies where the interstellar medium of a disk-shaped galaxy interacts with the intra-cluster medium, leading to the stripping of all interstellar material from the disk galaxy.

radial velocity — The component of a celestial object's space velocity projected to the line of sight. Also known as line of sight speed, it is the only part of an object's motion detectable through the Doppler effect.

red giant — A low-mass star that has used up all its core hydrogen and is powered by shell hydrogen fusion around a degenerate helium core. A cool, evolved low mass star about 100 times the size of the Sun.

red supergiant — A high-mass star that has used up all its core hydrogen and is powered by multiple shell fusions; generally more than 500 times the size of the Sun.

resolving power — The ability of a telescope to resolve fine details in an object. Theoretically depends only on the diameter of the primary mirror of a reflecting telescope, such that the larger the telescope mirror, the better the resolving power. However, resolving power can be greatly reduced by atmospheric turbulence (known as "seeing").

resonance ring — A galactic ring whose appearance and location link it to a specific orbital resonance with a bar, oval, or spiral pattern in a disk-shaped galaxy.

right ascension — A sky coordinate similar to longitude on Earth. Often specified in time units, it is measured eastward from the direction of the vernal equinox, the point on the sky where the Sun is located on the first day of northern spring.

ring — A morphological feature of disk-shaped galaxies where the luminosity distribution outlines a distinct ring-shaped pattern. Normal rings are of three types: outer, inner, and nuclear, in order of decreasing relative size.

rotation curve — A graph of the orbital speed versus the distance from the center of a galaxy, assuming the movements are perfectly circular.

SDSS — acronym for Sloan Digital Sky Survey.

secular evolution — The slow evolutionary changes in the structure of a galaxy attributable to weak interactions between disk stars and gas clouds and the presence of large-scale internal perturbations like bars and spiral arms. Weak, external interactions in a cluster environment can also produce secular changes.

Seyfert galaxy — A galaxy whose nucleus shows strong emission lines and evidence of violent interstellar gas motions.

speculum metal — A special alloy of tin and copper that, when polished, has a highly-reflective surface. It was the material of choice for the large, curved mirrors used in 18th century telescopes, and was also used in 19th century telescopes until about 1880, when silver-coated glass began to be used.

spectrograph — A scientific instrument designed to disperse the light of an object into its component colors and record the spectrum on a detector, such as a photographic plate or a digital imaging device like a CCD camera.

spectroscopy — The scientific study of light dispersed into its component colors, as in a rainbow. The dispersing element can be a prism (a triangular piece of glass) or a grating (an optical element consisting of a large number of finely spaced lines or grooves). Spectroscopy is used to deduce stellar surface temperatures and chemical compositions, the rotation of galaxies, and the expansion of the universe.

spiral, flocculent — A spiral pattern where the arms lack continuity and global extent.

spiral, grand-design — A galaxy showing a strong, global two-armed spiral pattern.

spiral, multi-armed — A galaxy often having a well-defined, inner two-armed spiral and a multi-armed outer spiral pattern.

stellar stream — A group of stars that formed together but which is in an advanced state of dispersal. In our Galaxy, often detected as moving groups, or groups of stars, not necessarily recognizeable as clusters, moving through space together.

STScI — acronym for Space Telescope Science Institute.

supermassive black hole — A region of space occupied by a black hole much more massive than one produced by stellar evolution. Typical supermassive black holes range from a few million to a few billion solar masses.

supernova — Like a nova, but on a much higher scale of violence. Physically can arise from the explosion of a high mass star or a white dwarf.

supernova remnant — The expanding and irregular gases hurtling into space after a supernova explosion. The Crab Nebula M1 in Taurus is an example.

surface photometry — The process of measuring the distribution of luminosity in a galaxy,

thick disk — A sub-component of a disk-shaped galaxy that is thicker (by a factor of 2-3) and fainter than the main disk, and that is made mostly of very old, relatively metal-poor, stars. Originally discovered in studies of edge-on S0 galaxies.

thin disk — The main disk in any spiral or S0 galaxy, and the feature most easily seen from a great distance. Most of the stellar mass (as well as the interstellar gas and dust) in a typical disk-shaped galaxy is in the thin disk.

tidal arm — A spiral-shaped arm generated in a gravitational interaction between a massive disk-shaped galaxy and (often) a smaller companion galaxy.

trailing arm — A spiral arm that opens outward opposite the direction of rotation.

Triangulum Spiral — The nearby Local Group spiral known as M33 (NGC 598), located 2.8 million light years away in the constellation Triangulum, the triangle.

trigonometric parallax — The apparent shift in position of a nearby star relative to more distant background stars due to the Earth's orbital movement around the Sun. The baseline for trigonometric parallax is the average Earth–Sun distance, 92,957,000 miles.

Twenty-one (21)-centimeter radiation — Long wavelength radio waves emitted from very cold interstellar atomic hydrogen atoms in the ground state, due to an electron spin-flip transition.

Type Ia supernova — A supernova whose spectrum lacks lines due to hydrogen but includes a strong absorption line due to silicon. A phenomenon that is likely a result of excessive mass accretion onto a white dwarf in a close binary system.

Type II supernova — A supernova whose spectrum includes lines due to hydrogen. A stellar explosion resulting from violent core collapse in a massive star.

Type II-P Supernova — A Type II supernova where the light curve shows a prominent plateau during the first three months or so after maximum apparent brightness.

UGC number — The number of a galaxy in the Uppsala General Catalogue of Galaxies. UGCA refers to the number in a supplement to the Uppsala General Catalogue.

ultraviolet light — Energetic electromagnetic waves having wavelengths in the range 10–400 nm.

velocity field — A two-dimensional map of the line of sight speed across the image of a galaxy. Usually, only points having detectable Hα emission appear in such a map.

vertical motion — The motion of a star perpendicular to (or in and out of) the Galactic plane.

visible light — Electromagnetic waves having wavelengths in the range 400–750 nm, the wavelength domain of the human eye. Intermediate between ultraviolet light and infrared radiation.

white dwarf — The former core of a low-mass star that silently ejected all of its envelope gases into space. A hot sphere of degenerate matter that can either be made of helium or carbon and oxygen, among other possibilities.

white nebula — The term given, in the 1860s, to nebulae which showed a largely continuous spectrum, as opposed to those which showed an emission-line spectrum.

WIMPs — acronym for "weakly-interacting massive particles," a type of exotic particle that is currently a prime candidate for the nature of dark matter.

WISE — acronym for Wide-field Infrared Survey Explorer.

WIYN — acronym for Wisconsin-Indiana-Yale-NOAO 3 m telscope.

X-rays — Very energetic electromagnetic waves having wavelengths in the range 0.01–10 nm, intermediate between gamma rays and ultraviolet light.

Zone of Avoidance — The regions of the sky within about 10° of the Galactic plane where almost no external galaxies were seen either visually or on the blue-sensitive photographic plates of the early 20th century.

Appendix B

Seeing Spiral Galaxies: A Tribute to Lord Rosse

William Parsons, the Third Earl of Rosse, his son, Laurence, the Fourth Earl, and their various assistant observers were the first astronomers to experience what it was like to visually detect the spiral structure of galaxies. Not only this, but they were also the first to appreciate the richness of galaxy morphology as a whole, as evidenced in their beautiful sketches (Figures 9–11, Chapter 2). Although Parsons (like everyone else at the time) was not certain of what the nebulae were, he was motivated to build a large telescope to observe them because he believed they were galaxies, not just shining clouds in space. He could have focussed his large telescope exclusively on the finding of new nebulae, and had he done this he undoubtedly would have discovered many that were missed by the Herschels and others. The list of new nebulae would have gone well beyond those that were found serendipitously near previously catalogued objects. Instead, he decided to focus on the details of bright, already catalogued nebulae, which brought galaxy morphology to a level that no one previously experienced.

Naturally, one might wonder: What would it be like to observe the spirals today? Having a "Lord Rosse experience" today does not require a 72-inch telescope. A telescope with a modern mirror half the size of the Leviathan's speculum metal mirror and located at a site with a more favorable climate can effectively reveal all of the Rosse spirals as well as many more the Birr Castle observers did not see.

Exploring the spirals can be an educational adventure allowing one to connect to the great observers of the past.

The visual detection of spiral structure in galaxies is not in the domain of small telescopes (that is, telescopes with a mirror less than 10–12.5 inches in diameter). Spiral arms generally have low surface brightness, and are fine details that a small telescope may not be able to resolve. The arms' visibility depends largely on their contrast with the surrounding galaxy light as well as their brightness relative to the sky background. The slightest light pollution can render spiral structure, even as bright as that seen in M51, invisible. The critical factor is being able to observe from a very dark site having a relatively dry climate. The ideal site is a remote mountain-top location in the desert southwestern United States. The Davis Mountains in west Texas provide such a site for McDonald Observatory.

Table B.1 provides a list of 132 of the best (or most interesting) nearby spirals. The list includes 74 objects noted to have a spiral appearance, or suspected to be spiral, as seen by the Birr Castle observers with the Leviathan. The list also covers both celestial hemispheres and includes several spectacular spirals (like NGC 1300, 1365, and 1566) that could not be observed from Birr Castle. Some of the listed objects, like NGC 253, 891, 3628, 4565, 4594, 4631, and 7814, are exceptional edge-on spirals where the planar dust lane or other edge-on structure can be seen. Several others, such as NGC 3081, 3351, 4725, 4736, and 4900, involve rings or pseudorings. The spirals observed with the CTIO 4-m telescope are also included.[1]

For each galaxy, the table provides the following information: Column 1: the name of the galaxy in the New General Catalogue that was described in Chapter 3; Columns 2 and 3: the right ascension in time units and the declination in angular units; Column 4: the Hubble type in the system of Figure 10 of Chapter 6; Column 5: the total visual magnitude (from RC3), where smaller numbers

---

[1]For visual observations of many of the galaxies in Table B.1, made at McDonald and Siding Spring Observatories from 1977 to 1984, see the webpage https://rbuta.people.ua.edu or R. Buta, *The Deep-Sky Observer, Quarterly Journal of the Webb Society*, **131**, 1 (2003); **137**, 19 (2005); **142**, 12 (2006).

Table B.1. The best nearby spirals.

NGC	Right ascension (2020)	Declination (2020)	Hubble type	Visual magnitude	Surface brightness (relative)	Constellation	Notes
7814	00^h 04^m.3	16° 15′.4	Sab	10.6	2.0	Pegasus	
23	00 10.9	26 02.0	SBa	12.0	4.9	Pegasus	
55	00 16.1	−39 06.5	SBm	7.9	—	Sculptor	
224	00 43.8	41 22.7	Sb	3.4	0.7	Andromeda	
253	00 48.5	−25 10.8	Sc	7.2	0.9	Sculptor	
278	00 53.2	47 39.5	Sb	10.8	8.8	Cassiopeia	Birr Castle
300	00 55.8	−37 34.5	Sd	8.1	—	Sculptor	
337	01 00.8	−07 28.2	SBd	11.6	1.5	Cetus	Birr Castle
598	01 35.0	30 45.8	Scd	5.7	0.3	Triangulum	Birr Castle
628	01 37.8	15 53.1	Sc	9.4	0.7	Pisces	Birr Castle
772	02 00.4	19 06.2	Sb	10.3	0.9	Aries	Birr Castle
891	02 23.8	42 26.2	Sb	9.9	0.4	Andromeda	
908	02 24.0	−21 08.6	Sc	10.2	1.3	Cetus	
925	02 28.5	33 40.0	Sd	10.1	0.3	Triangulum	
1068	02 43.7	00 04.3	Sb	8.9	12.2	Cetus	Birr Castle
1073	02 44.7	01 27.6	SBc	11.0	0.3	Cetus	
1084	02 47.0	−07 29.7	Sc	10.7	4.6	Eridanus	Birr Castle
1097	02 47.2	−30 11.3	SBb	9.5	2.8	Fornax	
1087	02 47.4	−00 24.8	Sc	10.9	1.4	Cetus	Birr Castle
1232	03 10.7	−20 30.4	Sc	9.9	0.7	Eridanus	
1313	03 18.5	−66 25.5	SBd	8.7	1.2	Reticulum	
1300	03 20.6	−19 20.4	SBbc	10.4	0.5	Eridanus	
1365	03 34.4	−36 04.3	SBb	9.6	1.5	Fornax	
1433	03 42.7	−47 09.5	SBab	9.9	1.2	Horologium	
IC 342	03 48.8	68 09.4	Scd	9.1	—	Camelopardalis	

(*Continued*)

Table B.1.   (*Continued*)

NGC	Right ascension (2020)	Declination (2020)	Hubble type	Visual magnitude	Surface brightness (relative)	Constellation	Notes
1512	04 04.6	−43 17.8	SBa	10.3	1.0	Horologium	
1532	04 12.9	−32 49.4	SBb	9.8	0.9	Eridanus	
1566	04 20.5	−54 53.5	Sbc	9.7	2.2	Dorado	
1637	04 42.5	−02 49.2	Sc	10.8	0.9	Eridanus	
2403	07 38.8	65 33.2	Scd	8.5	1.3	Camelopardalis	Birr Castle
2500	08 03.4	50 40.9	SBd	11.6	—	Lynx	
2525	08 06.6	−11 29.2	SBc	11.6	0.9	Puppis	Birr Castle
2537	08 14.7	45 55.8	SBm	11.7	2.2	Lynx	Birr Castle
2776	09 13.6	44 52.4	Sc	11.6	1.3	Lynx	Birr Castle
2903	09 33.3	21 24.7	Sbc	9.0	2.3	Leo	Birr Castle
2997	09 46.5	−31 17.0	Sc	10.1	—	Antlia	Birr Castle
3021	09 52.1	33 27.6	Sbc	12.5	—	Leo Minor	
3031	09 57.2	68 58.3	Sab	6.9	3.0	Ursa Major	Birr Castle
3081	10 00.4	−22 55.3	S	12.0	1.8	Hydra	
3162	10 14.6	22 38.4	Sbc	11.6	1.2	Leo	Birr Castle
3184	10 19.5	41 19.4	Scd	9.8	0.6	Ursa Major	Birr Castle
3198	10 21.1	45 27.1	SBc	10.3	0.6	Ursa Major	Birr Castle
3310	10 40.0	53 23.9	Sbc	10.8	16.0	Ursa Major	Birr Castle
3338	10 43.2	13 38.6	Sc	11.1	0.8	Leo	Birr Castle
3344	10 44.6	24 49.1	Sbc	9.9	1.0	Leo Minor	Birr Castle
3351	10 45.0	11 35.9	SBb	9.7	2.0	Leo	Birr Castle
3359	10 47.9	63 07.0	SBc	10.6	0.6	Ursa Major	Birr Castle
3368	10 47.8	11 42.9	Sab	9.2	2.6	Leo	Birr Castle
3395	10 50.9	32 52.5	Scd	12.1	—	Leo Minor	Birr Castle
3486	11 01.5	28 52.1	Sc	10.5	1.0	Leo Minor	Birr Castle

(*Continued*)

Table B.1. (*Continued*)

NGC	Right ascension (2020)	Declination (2020)	Hubble type	Visual magnitude	Surface brightness (relative)	Constellation	Notes
3556	11 12.7	55 33.7	SBcd	10.0	0.7	Ursa Major	Birr Castle
3623	11 20.0	12 59.0	Sa	9.3	2.1	Leo	Birr Castle
3627	11 21.3	12 52.9	Sb	8.9	3.8	Leo	Birr Castle
3628	11 21.3	13 28.8	Sb	9.5	—	Leo	Birr Castle
3631	11 22.2	53 03.7	Sc	10.4	—	Ursa Major	Birr Castle
3646	11 22.8	20 03.7	Sbc	11.1	0.8	Leo	Birr Castle
3642	11 23.5	58 58.0	Sbc	11.2	0.6	Ursa Major	Birr Castle
3664	11 25.4	03 13.1	SBm	12.8	0.4	Leo	
3718	11 33.7	52 57.4	SBa	10.8	0.9	Ursa Major	Birr Castle
3726	11 34.4	46 55.0	Sc	10.4	—	Ursa Major	
3729	11 34.9	53 01.0	SBa	11.4	—	Ursa Major	
3808	11 41.8	22 19.0	Sc	13.4	—	Leo	
3810	11 42.0	11 21.6	Sc	10.8	2.4	Leo	Birr Castle
3893	11 49.7	48 36.0	Sc	11.2	—	Ursa Major	Birr Castle
3938	11 53.9	44 00.8	Sc	10.4	1.0	Ursa Major	Birr Castle
3953	11 54.9	52 13.0	SBbc	10.1	1.0	Ursa Major	Birr Castle
4017	11 59.8	27 20.6	Sbc	13.0	—	Coma Berenices	
4027	12 00.5	−19 22.8	SBdm	11.1	1.9	Corvus	Birr Castle
4038	12 02.9	−18 58.6	Sc	10.3	—	Corvus	
4051	12 04.2	44 25.2	Sbc	10.2	0.7	Ursa Major	Birr Castle
4088	12 06.6	50 25.9	Sbc	10.6	1.4	Ursa Major	Birr Castle
4102	12 07.4	52 36.0	Sb	12.0	—	Ursa Major	Birr Castle
4123	12 09.2	02 46.0	SBc	11.4	0.3	Virgo	Birr Castle
4151	12 11.6	39 17.8	Sab	10.8	14.3	Canes Venatici	Birr Castle

(*Continued*)

Table B.1.   (*Continued*)

NGC	Right ascension (2020)	Declination (2020)	Hubble type	Visual magnitude	Surface brightness (relative)	Constellation	Notes
4214	12 16.7	36 13.0	Sm	9.8	1.4	Canes Venatici	Birr Castle
4254	12 19.8	14 18.4	Sc	9.9	2.5	Coma Berenices	Birr Castle
4258	12 20.0	47 11.6	Sbc	8.4	2.0	Canes Venatici	
4303	12 22.9	04 21.7	Sbc	9.7	1.5	Virgo	Birr Castle
4321	12 23.9	15 42.7	Sbc	9.4	1.0	Coma Berenices	Birr Castle
4395	12 26.8	33 26.1	Sm	10.2	0.2	Canes Venatici	
4490	12 31.6	41 31.8	SBd	9.8	2.6	Canes Venatici	Birr Castle
4501	12 33.0	14 18.7	Sb	9.6	2.1	Coma Berenices	Birr Castle
4535	12 35.4	08 05.3	Sc	10.0	0.7	Virgo	
4536	12 35.5	02 04.7	Sbc	10.6	0.7	Virgo	Birr Castle
4565	12 37.3	25 52.5	Sb	9.6	1.5	Coma Berenices	
4569	12 37.8	13 03.2	Sab	9.5	1.2	Virgo	
4594	12 41.0	−11 44.0	Sa	8.0	6.5	Virgo	
4618	12 42.5	41 02.5	SBm	10.8	1.1	Canes Venatici	Birr Castle
4625	12 42.8	41 09.9	Sm	12.3	—	Canes Venatici	Birr Castle
4631	12 43.1	32 25.9	SBd	9.2	0.9	Canes Venatici	Birr Castle
4654	12 44.9	13 01.0	Scd	10.5	1.3	Virgo	
4656	12 44.9	32 03.7	SBm	10.5	0.4	Canes Venatici	Birr Castle
4725	12 51.4	25 23.5	Sab	9.4	1.3	Coma Berenices	Birr Castle
4736	12 51.8	41 00.7	Sab	8.2	22.9	Canes Venatici	Birr Castle
4826	12 57.7	21 34.6	Sab	8.5	4.7	Coma Berenices	
4861	12 60.0	34 45.3	SBm	12.3	—	Canes Venatici	
4900	13 01.7	02 23.6	SBc	11.4	1.6	Virgo	Birr Castle
5033	13 14.4	36 29.3	Sc	10.2	0.5	Canes Venatici	Birr Castle

Table B.1.   (*Continued*)

NGC	Right ascension (2020)	Declination (2020)	Hubble type	Visual magnitude	Surface brightness (relative)	Constellation	Notes
5055	13 16.7	41 55.8	Sbc	8.6	2.3	Canes Venatici	Birr Castle
5112	13 22.8	38 38.0	SBcd	12.1	—	Canes Venatici	Birr Castle
5204	13 30.4	58 18.9	Sm	11.3	0.8	Ursa Major	Birr Castle
5194	13 30.7	47 05.6	Sbc	8.4	2.1	Canes Venatici	Birr Castle
5236	13 38.1	−29 58.1	Sc	7.5	—	Hydra	
5248	13 38.5	08 47.0	Sbc	10.3	1.8	Bootes	Birr Castle
5247	13 39.1	−17 59.0	Sbc	10.0	0.5	Virgo	
5371	13 56.5	40 21.9	Sbc	10.6	1.1	Canes Venatici	Birr Castle
5364	13 57.2	04 55.1	Sbc	10.5	0.7	Virgo	
5378	13 57.7	37 42.2	SBa	12.6	—	Canes Venatici	Birr Castle
5383	13 57.9	41 44.9	SBb	11.4	1.1	Canes Venatici	
5394	13 59.4	37 21.5	SBb	13.0	—	Canes Venatici	
5395	13 59.5	37 19.7	Sb	11.4	0.8	Canes Venatici	
5457	14 03.9	54 15.2	Scd	7.9	0.5	Ursa Major	Birr Castle
5427	14 04.5	−06 07.6	Sc	11.4	1.7	Virgo	Birr Castle
5713	14 41.2	−00 22.6	Sbc	11.2	2.0	Virgo	Birr Castle
5746	14 46.0	01 52.2	Sb	10.3	0.9	Virgo	
5921	15 22.9	04 59.9	SBbc	10.8	0.9	Serpens	Birr Castle
5985	15 40.0	59 16.1	Sb	11.1	0.8	Draco	Birr Castle
5996	15 47.9	17 49.5	S	12.8	—	Serpens	
6215	16 52.8	−59 01.6	Sc	11.5	3.5	Ara	
6221	16 54.5	−59 14.9	SBc	9.9	1.6	Ara	
6300	17 18.9	−62 50.4	SBb	10.2	1.5	Ara	
6764	19 08.8	50 57.9	SBbc	11.9	0.6	Cygnus	

(*Continued*)

Table B.1. (*Continued*)

NGC	Right ascension (2020)	Declination (2020)	Hubble type	Visual magnitude	Surface brightness (relative)	Constellation	Notes
6744	19 11.6	−63 49.4	Sbc	9.1	—	Pavo	
6872	20 19.0	−70 42.3	SBb	11.8	1.9	Pavo	
6946	20 35.3	60 13.4	Scd	8.8	0.6	Cygnus	Birr Castle
7479	23 06.0	12 25.8	SBc	10.8	0.9	Pegasus	Birr Castle
7531	23 15.9	−43 29.4	Sbc	11.3	2.6	Grus	
7640	23 23.1	40 57.3	SBc	11.3	0.5	Andromeda	
7678	23 29.5	22 32.0	Sc	11.8	1.1	Pegasus	Birr Castle
7714	23 37.3	02 15.9	SBb	12.5	4.9	Pisces	
7741	23 44.9	26 11.2	SBcd	11.3	0.5	Pegasus	Birr Castle
7757	23 49.8	04 17.1	Sc	12.7	0.5	Pisces	Birr Castle

mean brighter cases; those brighter than 11th magnitude can be seen with a telescope as small as a 4-inch; Column 6: the average surface brightness of the inner parts of the galaxy relative to the surface brightness of the dark night sky (based on RC3 data); the larger this number, the higher the surface brightness[2]; Column 7: the constellation in which the object is located; and Column 8: a note indicating if the object was discovered to be spiral at Birr Castle.

The right ascension and declination listed in the table are strictly valid only for the year 2020. However, the effects of precession are small and these coordinates will work for a couple of decades past 2020 at least.

---

[2]For many of the galaxies in the list, there is insufficient information in RC3 to derive a mean inner surface brightness. This is because large, bright galaxies often could not have their total brightnesses reliably estimated from photoelectric multi-aperture photometry, owing to their large angular sizes.

Index